Python编程快速上手 2
趣味小项目轻松学

THE BIG BOOK OF SMALL PYTHON PROJECTS

【美】阿尔·斯维加特（Al Sweigart）◎著　荣耀　张嘉豪◎译

人民邮电出版社

北京

图书在版编目（CIP）数据

Python编程快速上手：趣味小项目轻松学. 2 / (美) 阿尔·斯维加特 (Al Sweigart) 著；荣耀，张嘉豪译. -- 北京：人民邮电出版社，2022.9
ISBN 978-7-115-59350-4

Ⅰ. ①P… Ⅱ. ①阿… ②荣… ③张… Ⅲ. ①软件工具—程序设计 Ⅳ. ①TP311.561

中国版本图书馆CIP数据核字(2022)第089063号

版权声明

Copyright © 2021 by Al Sweigart. Title of English-language original: *The Big Book of Small Python Projects: 81 Easy Practice Programs*, ISBN 9781718501249, published by No Starch Press Inc. 245 8th Street, San Francisco, California United States 94103. The Simplified Chinese-edition Copyright ©2022 by Posts and Telecom Press Co., Ltd under license by No Starch Press Inc. All rights reserved.

本书简体中文字版由美国 No Starch 出版社授权人民邮电出版社有限公司出版。未经出版者书面许可，不得复制或抄袭本书任何部分的内容。

版权所有，侵权必究。

◆ 著　　[美] 阿尔·斯维加特 (Al Sweigart)
　　译　　荣　耀　张嘉豪
　　责任编辑　吴晋瑜
　　责任印制　王　郁　焦志炜

◆ 人民邮电出版社出版发行　北京市丰台区成寿寺路 11 号
　　邮编　100164　电子邮件　315@ptpress.com.cn
　　网址　https://www.ptpress.com.cn
　　大厂回族自治县聚鑫印刷有限责任公司印刷

◆ 开本：800×1000　1/16
　　印张：21　　　　　　　　　2022 年 9 月第 1 版
　　字数：493 千字　　　　　　2022 年 9 月河北第 1 次印刷
　　著作权合同登记号　图字：01-2021-4698 号

定价：99.80 元

读者服务热线：(010)81055410　印装质量热线：(010)81055316
反盗版热线：(010)81055315
广告经营许可证：京东市监广登字 20170147 号

内容提要

本书旨在以"最短小精悍的代码+最丰富的创造力"的方式向读者展示 81 个简单、有趣的实践项目。如果你已经掌握了基本的 Python 语法,并且准备开始编写程序,那么阅读本书会让你觉得既有趣又很受启发。

本书给出的 81 个 Python 项目,可以助你快速上手 Python 编程完成数字艺术、游戏、动画、计数程序等方面的任务。一旦了解了代码是如何工作的,你就可以动手重新编写程序,并通过添加自定义的内容来不断实践。需要说明的是,书中这些项目涉及的程序大多是用少于 256 行的代码实现的,如蜗牛赛跑、诱饵标题生成器、DNA 可视化,而且每个项目都被设计成独立的 Python 源文件,可供读者轻松地在网上分享。

本书适合所有想通过 Python 学习编程的读者使用,尤其适合有 Python 基础、需要快速实现编程的读者。

作者简介

阿尔·斯维加特（Al Sweigart）是软件开发人员、知名作者和 Python 软件基金会的研究员。他曾在美国加利福尼亚州奥克兰电子游戏博物馆、艺术和数字娱乐博物馆担任教育主管。他撰写了多本编程图书，如畅销书《Python 编程快速上手——让繁琐工作自动化》《Python 游戏编程快速上手》等。

技术审稿人简介

Sarah Kuchinsky 女士是一名企业培训师和顾问。她用 Python 开发各种应用，包括卫生系统建模、游戏开发以及任务自动化。Sarah 是 North Bay Python 会议的联合创始人、PyCon 开发者大会的主题演讲的负责人以及 PyLadies Silicon Valley 的主要组织者。她拥有管理科学与工程和数学双学位。

译者序

Python 是一门独特的编程语言，对于初学者和它的爱好者来说，它多少带有些玩乐的色彩——学习门槛低，容易上手，几十行代码就能实现了不起的效果。但是，当用于专业软件开发和数据分析、机器学习建模时，Python 又立刻展现出作为专业编程语言严肃的一面——语法简洁富有表达力，开发效率高，运行性能好。Python 具有较好的跨平台性，其生态中不乏丰富的开源工具。

不少写给初学者的编程指导书都是以枯燥乏味的小程序片段讲述编程的原理和概念，往往难以激发初学者学习编程的兴趣（兴趣是最好的老师）。这本书跟作者写的其他几本畅销书类似，均为项目导向式风格。每个小项目自成一章，先交代项目背景，接着给出程序的执行效果，然后讲解其工作原理，中间穿插一些语言语法或模块用法的解释，最后展示完整的项目代码，并提供一些思考和练习题。显然，这并不是一本系统讲解 Python 语言特性的入门书，不过如果你已经有一些语言基础，但在编写程序时不知从何下手，这本书就是很好的启发式读物了。

编程是一种创造性的活动，可以锻炼人的逻辑思维能力，而 Python 能让初学者和爱好者更容易发挥自己的创造性。无论对于有志于从事软件开发的学生，还是对于其他非职业程序员但又需要进行编程的人员（比如从事数据分析、AI 模型训练的专业人员），Python 都是当前进入编程世界的最佳起点。纵观编程语言发展史，Python 可以算是一门"古老"的编程语言了，对于熟悉它的程序员来说，Python 在大数据和人工智能时代的大放异彩让人始料未及。不过凡事有因有果。Python 没有想把自己发展成为无所不能的语言，对新语言特性的加入保持克制，始终保持语言核心特性简洁，能通过库完成的全都交给库（这些库通常使用高性能 C 或 C++ 编写）。在更大的场景看，Python 恰如其分地发挥了"黏合剂"的作用。

阿尔·斯维加特（Al Sweigart）曾是一位软件工程师，现在全职从事 Python 编程布道，他聪明、有才华，在写书方面很有经验，且深谙畅销书的写作"套路"。本书中的代码写得很漂亮，充分展现了 Python 语言简洁、强大的特点，也初步展示了对一些数据结构和算法简单、精巧的使用。他崇尚开源（也乐于接受捐赠），开发了多个 Python 开源模块，并将自己的全部著作慷慨分享在个人官网上供读者免费阅读，帮助了很多 Python 初学者和爱好者。

本书由我和张嘉豪翻译，嘉豪做事认真踏实，富有责任心和进取心。感谢洪伊晓晨、余志昊以及晏金陈在翻译后期奉献的辅助工作。感谢浣石公司的朱艳女士在翻译过程中提供的协助。

你们的奉献得以让本书以更佳的面貌呈现在读者面前。

请开始学习 Python 编程吧，体验编程的乐趣，自由展示你的才华和创意。祝编程之旅愉快！

荣耀 博士
于南京大行宫

前　　言

如果只是跟随 print('Hello, world!')这样的教程学习，你会觉得编程异常简单。你可能看过一些面向初学者的优秀图书或者相关的在线课程，也做过一些练习，而且已对大部分技术术语烂熟于心。然而，你很可能会发现，一旦离开"安乐窝"去编写自己的程序，就会寸步难行。面对空白的编辑器窗口，你可能会茫然、发呆，对如何动手编写 Python 程序一筹莫展。

问题的关键在于，遵循"教程非常适合学习"的主旨，与"学习从头开始编写原创程序"是两码事。通常，对于此阶段的建议是学习开源软件的源代码，或者创建你自己的项目。但有些开源项目并未提供良好的文档，也不见得特别适合新手学习。虽然开源项目会激励你创建自己的项目，但你不能从中得到系统的指导。

本书为你提供如何应用编程概念的实践示例，其中包含 80 多个游戏、模拟和艺术等方面的程序。重要的是，它们并非代码片段，而是完整的、可运行的 Python 程序。你可以复制并运行这些代码，以理解其工作方式，也可以尝试自己进行更改。作为练习，你还可以按照自己的意愿重新编写这些代码。一段时间后，你会逐渐获得自己的编程灵感，更重要的是，你将知道如何着手创建程序。

如何设计小程序

编程已被证明是一项强大的技能，它"创造"了价值数十亿美元的科技公司和惊人的技术进步。我们创建自己的软件时很容易好高骛远，"贪多嚼不烂"的后果只能是程序烂尾，而开发者也会倍感沮丧。其实，你无须成为计算机天才也能编写有趣且富有创意的程序。

本书中的 Python 程序遵循以下几个设计原则，有助于编程新手理解源代码。

❏ **小型**：本书中大多数项目的程序不超过 256 行代码，这样做是为了让读者更容易理解。选择 256 这个行数限制没什么特别的含义，不过是因为 256 是 2 的 8 次方，而 2 的幂通常被认为是程序员的幸运数字。

❏ **基于文本**：文字比图形简单。源代码和程序输出都是文本，这样便于你搞清楚代码中的 print('Thanks for playing!')和屏幕上输出的 Thanks for playing!之间的关系。

❏ **免安装**：每个程序都保存为单独的 Python 源文件，其扩展名为.py，如 tictactoe.py，你不需要额外运行安装程序。

❏ **丰富**：本书共有 81 个程序，包括棋盘类、卡牌类、艺术类、模拟类、谜题类、迷宫类和诙谐类程序等。

❑ 简单：这些程序的编写方式易于初学者理解。每当不得不在使用复杂的高性能算法编写代码和编写简单、直接的代码之间做出选择时，我总是选择后者。

基于文本的程序可能看起来很老套，但采用这种编程风格可以避开下载图形、安装附加库以及管理项目文件夹等带来的干扰和陷阱，只需关注代码本身。

读者对象

本书是为两类读者写的。第一类是已经学习了 Python 和基础编程知识但仍然不知道如何自己编写程序的人。他们可能会觉得自己不适合编程。他们也许能够完成教程中的练习，但仍然难以想象一个完整的程序是什么样子的。通过先"照抄"然后重新创建本书中的游戏，他们就能明白自己所学的编程概念是如何构成各种真实程序的。

第二类是虽然刚接触编程但兴趣浓厚并勇于探索的人。他们通常希望能立即着手编写游戏、模拟和数字运算类的程序。这类读者会在实践过程中快速成长，或者他们已经会用其他语言编程，只是不熟悉 Python 而已。虽然本书不能替代完整的 Python 入门课程，但涵盖了对 Python 基础知识的简要介绍，以及如何在程序运行时使用调试器检查程序的内部运作机制。

有经验的程序员也可能喜欢本书中介绍的程序，不过请别忘了，本书是为初学者写的。

关于本书

本书的大部分内容专注于特色程序，同时提供了包括通用编程和 Python 知识在内的额外学习资源。本书包括如下内容。
❑ 项目：共有 81 个项目。每个项目独立介绍，包括项目名称、描述、程序运行输出示例以及程序源代码。某些项目还提供了一些对代码进行实验性修改以定制程序的建议。
❑ 附录 A：标签索引，列出按项目标签分类的所有项目。
❑ 附录 B：字符映射表，列出程序可以输出的符号（例如心形、线条、箭头和方块）的字符代码。

如何学习书中的程序

本书不像传统教程那样教授 Python 或编程概念，而是提供一种"做中学"的方法，鼓励你亲自动手输入程序，运行程序（像玩游戏一样），并在调试器中了解其内部工作原理。

本书的重点不在于详细解释编程语言的语法，而是展示执行实际可靠的程序示例，无论是纸牌游戏、动画，还是数学谜题的探索，均是如此。因此，我建议遵循以下步骤来学习。
1. 下载程序并运行，查看程序的效果。
2. 从一个空白文件开始，自己手动输入（不要复制和粘贴！）本书中的游戏代码。
3. 再次运行程序，然后返回并修改你可能引入的任何拼写错误或弥补缺陷。
4. 在调试器下运行程序，以便仔细地逐行运行代码，从而理解其作用。
5. 找到标有!的注释，查看你可以修改的代码，然后在下次运行程序时观察自己的修改对程序有何影响。

6. 尝试自己从头开始创建程序。不必完全复制程序代码，你可以在程序中融入自己的想法。

复制本书中的代码时，不必输入注释（#符号后面的文本），因为这些是给程序员看的，Python 会忽略这些注释。不过，请尽量保持你的程序代码与本书中的程序代码行号相同，以便更容易进行对比。

本书中的每个程序都有一组描述标签，例如"棋盘类""模拟类""艺术类""双人游戏类"等。对这些标签的解释以及标签和项目之间的交叉索引参见附录 A。不过，项目是按原书英文字母顺序列出的。

下载并安装 Python

Python 既是编程语言的名称，也是运行 Python 代码的解释器软件的名称。Python 软件可以免费下载和使用。你可以通过终端或命令提示符窗口检查是否已经安装了 Python。在 Windows 上，打开命令提示符窗口，然后运行 py --version。如果看到如下输出，则说明 Python 已安装：

```
C:\Users\Al>py --version
Python 3.9.1
```

在 macOS 和 Linux 上，打开终端程序，然后运行 python3 --version。如果看到如下输出，则说明 Python 已安装：

```
$ python3 --version
Python 3.9.1
```

本书使用 Python 3。Python 2 和 Python 3 不完全兼容，本书中的程序至少需要 Python 3.1.1（2009 年发布）才能运行。

下载并安装 Mu Editor

在使用 Python 软件运行程序之前，你需要在文本编辑器或集成开发环境（IDE）应用程序中输入 Python 代码。如果你是初学者，建议使用 Mu Editor（简称 Mu）作为 IDE，它简单易用，也没有那么多"让人困扰"的高级选项。

请下载适合你的操作系统的安装程序，然后双击安装文件运行它。如果你使用的是 macOS，运行安装程序会打开一个窗口，必须在其中将 Mu 图标拖到 Applications 文件夹图标上才能继续安装。如果你使用的是 Ubuntu，则需要将 Mu 作为 Python 包安装。在这种情况下，打开一个新的终端窗口并运行 pip3 install mu-editor 进行安装，然后使用 mu-editor 命令运行它。单击下载页面的 Python 包部分的 Instructions 按钮，可以获取翔实的说明信息。

运行 Mu Editor

安装完成后，可以按照如下步骤运行 Mu。
- 在 Windows 7 或更高版本上，单击屏幕左下角的开始图标，在搜索框中输入 mu，并在输出的结果中选择 Mu。
- 在 macOS 上，打开 Finder 窗口，单击应用程序，然后单击 mu-editor。

- 在 Ubuntu 上，按 Ctrl-Alt-T 打开终端窗口，然后运行 python3 -m mu。

Mu 首次运行时，会出现一个选择模式窗口，其中包含以下选项：Adafruit CircuitPython、BBC micro:bit、Pygame Zero 和 Python 3。在本书中，我们选择 Python 3，当然，也可以随时通过单击编辑器窗口顶部的模式按钮更改模式。

现在我们可以在 Mu 的主窗口中输入代码，然后通过顶部的按钮保存、打开和运行程序文件。

运行 IDLE 和其他编辑器

我们可以任选编辑器编写 Python 代码。集成开发和学习环境（IDLE）软件与 Python 一起安装，如果由于某种原因无法安装或运行 Mu，它可以作为备用编辑器。IDLE 的用法如下。

- 在 Windows 7 或更高版本上，单击屏幕左下角的开始图标，在搜索框中输入 idle，然后选择 IDLE (Python GUI)。
- 在 macOS 上，打开 Finder 窗口，单击应用程序→Python 3.9→IDLE。
- 在 Ubuntu 上，选择 Applications→Accessories→Terminal，然后输入 idle3。也可以单击屏幕顶部的应用程序，选择 Programming，然后单击 IDLE 3。
- 在树莓派上，单击左上角的树莓派菜单按钮，然后单击 Programming 和 Python 3 (IDLE)。我们还可以从 Programming 菜单下选择 Thonny Python IDE。

我们还可以使用其他几种免费编辑器输入和运行 Python 代码。

- Thonny，一个面向初学者的 Python IDE。
- PyCharm 社区版，这是专业开发人员使用的 Python IDE。

安装 Python 第三方模块

本书中的大多数程序只需要使用 Python 标准库，它是随 Python 自动安装的。然而某些程序需要使用第三方模块，例如 pyperclip、bext、playsound 以及 pyttsx3。这些都可以通过安装 bigbookpython 模块一次性安装。

对于 Mu Editor，我们必须安装 1.1.0-alpha 版本或更高版本。截至 2020 年，我们可以在 Code with Mu 官方网站下载页面顶部的 "Try the Alpha of the Next Version of Mu" 部分找到此版本。安装完成后，单击窗口左下角的齿轮图标，可以调出 Mu Administration 窗口。选择 Third Party Packages 选项卡，在文本框中输入 bigbookpython，然后单击 OK 按钮，即可安装本书程序中用到的所有第三方模块。

对于 Visual Studio Code 或 IDLE 编辑器，打开编辑器并通过交互式 Shell 运行以下 Python 代码：

```
>>> import os, sys
>>> os.system(sys.executable + ' -m pip install --user bigbookpython')
0
```

如果一切正常，第二条指令执行完成后，会输出一个 0。如果你看到一条错误消息或其他数字，请尝试运行以下代码，其中没有--user：

```
>>> import os, sys
>>> os.system(sys.executable + ' -m pip install bigbookpython')
0
```

无论使用哪种编辑器，我们都可以尝试运行 import pyperclip 或 import bext，以检查安装是否有效。如果这些导入代码的运行没有产生错误消息，则说明这些模块安装正确，你将能够运行本书中使用这些模块的程序。

复制本书中的代码

编程是一种技能，这种技能可以通过编程活动来提高。请勿仅阅读本书中的代码或仅将其简单地复制、粘贴到你的计算机上，你应该花些时间自己将代码输入编辑器——在编辑器中打开一个新文件并输入代码。为了能更好地理解这些程序，请尝试在调试器下运行它们。

输入源代码并运行成功后，请尝试对代码进行更改。标有!的注释给你提供了更改建议，每个项目还给出了较大改动的建议。

接下来，请尝试从头开始创建程序（不要查看本书中的源代码）。你的代码不必与书中的程序代码完全相同，你可以发明自己的版本！

一旦完全理解了本书中的程序，你可以着手创建自己的程序。大多数电子游戏和应用程序都很复杂，需要由程序员、艺术家和设计师组成的团队联合开发。不过，许多生活中的棋盘、纸牌和纸笔游戏通常很简单，你不妨尝试编写相应的程序。

从终端运行程序

本书用到 bext 模块的程序，其文本输出会以彩色显示。然而，从 Mu、IDLE 或其他编辑器中运行它们时，这些颜色不会显示出来。这些程序应该通过终端（也称为命令行）窗口运行。在 Windows 上，可以通过"开始"菜单打开命令提示符窗口。在 macOS 上，可以通过 Spotlight 运行终端。在 Ubuntu 上，可以通过 Ubuntu Dash 或按 Ctrl-Alt-T 快捷键运行终端。

如果终端窗口出现，则应该使用 cd（更改目录）命令将当前目录更改为包含.py 文件的文件夹（"目录"是"文件夹"的另一种说法）。例如，如果使用的是 Windows，并将 Python 程序保存在 C:\Users\Al 文件夹内，则执行以下命令：

```
C:\>cd C:\Users\Al

C:\Users\Al>
```

要运行 Python 程序，应在 Windows 命令提示符窗口中运行 python yourProgram.py，在 macOS 或 Linux 终端窗口中则应运行 python3 yourProgram.py。请将 yourProgram.py 替换为真实的 Python 程序名称，示例如下：

```
C:\Users\Al>python guess.py
Guess the Number, by Al Sweigart al@inventwithpython.com

I am thinking of a number between 1 and 100.
You have 10 guesses left. Take a guess.
--snip--
```

你可以通过按 Ctrl-C 键（而不是关闭终端窗口）来终止从终端运行的程序。

在手机或平板电脑上运行程序

理想情况下，你可以使用一台配备全键盘的笔记本电脑或台式计算机编写代码，因为从某种程度上讲，使用手机甚至平板电脑默认的键盘输入代码挺麻烦的。尽管没有适用于 Android 或 iOS 的成熟 Python 解释器，但有些网站提供了可以在 Web 浏览器中使用的在线 Python 交互式 Shell。

这些网站支持本书中的大多数项目，但是不支持使用第三方模块（例如 bext、pyperclip、pyttsx3 以及 playsound）的程序，也不支持需要使用 open()函数读写文件的程序。如果程序代码中有这些内容，则这个程序无法使用在线 Python 解释器运行。不过，本书中的大多数程序可以正常在线运行。

帮助

除非请私人导师或程序员朋友来回答你的编程问题，否则你只能靠自己找到问题的答案。幸运的是，你的问题大多是有人问过的。"自己找到答案"是程序员要学习的一项重要技能。

如果你发现自己经常需要到互联网上查找编程问题的答案，请不要为此沮丧。你可能会觉得到互联网上找答案而不是自己记住关于编程知识的一切属于"作弊"行为，但其实只要你在不断学习，就不是"作弊"。因为即使是专业的软件开发人员，也会经常在互联网上搜索答案。

当程序尝试执行无效指令时，它会显示一条称为回溯的错误消息。回溯会告诉你发生了哪种错误以及错误发生在哪一行代码。下面是一个程序示例，它在计算每个人应该得到多少块比萨时出错：

```
Traceback (most recent call last):
  File "pizza.py", line 5, in <module>
    print('Each person gets', (slices / people), ' slices of pizza.')
ZeroDivisionError: division by zero
```

对于这个回溯提示的错误，你也许没有意识到问题在于 people 变量被设置为 0 导致表达式 slices/people 出现除零错误。错误消息通常很短，甚至不是完整的句子，其用意在于提醒而非完整解释。如果你是第一次遇到错误消息，可将其复制并粘贴到搜索引擎中进行搜索，通常会返回有关错误的含义及错误出现可能的原因的详细说明。

如果直接搜索无法找到问题的解决方案，你可以将问题发布到在线论坛或向某人发送电子邮件，以寻求帮助。为了使这个过程尽可能高效，请具体、清晰地描述问题，这意味着提供完整的源代码和详细的错误消息，解释你已经尝试过的做法，并确切说明你所使用的操作系统和 Python 版本。最终发布的答案不仅可以解决你的问题，还有助于将来遇到相同问题的程序员找到你的帖子。

输入代码

打字打得飞快并不是成为一名程序员必备的技能，但这无疑是大有裨益的。有些人不会盲打，打起字来就像小鸡啄米。打字快可以让编程变得不那么麻烦。在学习使用本书中的程序时，你肯定希望自己的眼睛看着代码而不是盯着键盘。

你可以在某些网站上学习如何打字，比如 TypingClub。好的打字程序会在计算机屏幕上显示键盘和一双手，让你在练习过程中避免养成低头看键盘的坏习惯。与其他技能一样，打字需要练习，你大可借编写代码来提升自己的打字技能。

善用键盘快捷键，可以免于将鼠标指针移动到菜单上再执行操作。快捷键通常写成"Ctrl-C"

的形式，意味着按下键盘任一侧的 Ctrl 键的同时按下 C 键，而不是先按一次 Ctrl 键，再按一次 C 键。

使用鼠标打开应用程序顶部（在 Windows 和 Linux 中）或屏幕顶部（在 macOS 中）的菜单栏，你可以找到常用的快捷方式，例如 Ctrl-S（保存）和 Ctrl-C（复制）。磨刀不误砍柴工，花时间学习使用这些快捷键是值得的。

其他快捷键就没有那么重要了。例如，Windows 和 Linux 上的 Alt-Tab，macOS 上的 command-TAB，允许将焦点切换到另一个应用程序的窗口中。你可以按住 Alt 或 command 键并重复按 TAB 选择要切换成的特定窗口。

复制和粘贴

剪贴板是操作系统中的一项功能，可以临时存储用于粘贴的数据。这些数据可以是文本、图像、文件或其他类型的信息，这里只描述如何处理文本数据。复制文本会将当前选定文本的副本放置到剪贴板上。粘贴文本会将剪贴板上的文本输入文本光标当前所在的位置，就像你自己即刻输入的那样。复制和粘贴文本可让你免于重新输入计算机上已有的文本，无论文本是一行还是数百页。

要想复制和粘贴文本，首先要选择（或高亮显示）要复制的文本。你可以通过按住鼠标主（右）键（设置为"右撇子"鼠标的即左键）并在文本上拖动来完成文本选择。不过，按住 Shift 键并使用键盘快捷键移动光标通常更快、更精确。许多应用程序允许双击选定一个词。通常还可以三击选择整行或整段文本。

下一步是按 Ctrl-C（在 Windows 上）或 command-C（在 macOS 上）将所选文本复制到剪贴板。剪贴板只能保存一份复制的文本，复制文本操作会替换掉之前剪贴板上的任何内容。

最后，将光标定位到你希望文本出现的位置，然后按 Ctrl-V（在 Windows 上）或 command-V（在 macOS 上）以粘贴文本。你可以根据需要粘贴多次。复制的文本会一直保留在剪贴板上，直到复制新文本将其替换掉。

查找和替换文本

谷歌公司的搜索人类学家 Dan Russell 曾在其发表的一篇文章中谈到，他研究人们的计算机使用习惯时发现，约 90% 的人不知道可以按 Ctrl-F（在 Windows 和 Linux 上）或 command-F（在 macOS 上）在应用程序中搜索单词。这是一个非常有用的功能，不仅存在于代码编辑器中，还存在于文字处理器、Web 浏览器、电子表格应用程序以及几乎所有能够显示文本的程序中。你可以按 Ctrl-F 调出查找文本框，输入希望在程序中查找的单词。通常，按 F3 键会重复该搜索并高亮显示下一个出现的该单词。与手动滚动页面查找单词相比，该功能可以为我们节省大量时间。

编辑器还具有查找和替换功能，该功能通常对应 Ctrl-H 或 command-H 快捷键，可用于定位一段文本并将其替换为另一段文本。如果希望重命名变量或函数，这个功能很有用。不过，我们要小心使用查找和替换功能，因为可能会错误替换与查找条件匹配的别的文本。

调试器

调试器是一种工具，它可以逐行运行程序，让我们能够检查程序变量的当前状态。它是跟踪程序缺陷的宝贵工具。这里解释 Mu Editor 调试器的功能。不用担心，基本每款调试器都提供了相同的功能，即使它们的用户界面看上去不一样。

在调试器中启动程序要使用 IDE 中的 Debug 而不是 Run 菜单项。调试器将被启动并暂停在程序的第一行。调试器通常都有以下按钮：Continue、Step In、Step Over、Step Out 和 Stop。

单击 Continue 按钮会使程序正常运行，直到终止或遇到断点（稍后会介绍断点）。如果你已完成调试并希望程序正常继续运行，可以单击 Continue 按钮。

单击 Step In 按钮会使调试器运行下一行代码，然后再次暂停。如果下一行代码是函数调用代码，调试器将"步入"该函数并跳转到该函数的第一行代码。

单击 Step Over 按钮会使调试器运行下一行代码，类似于 Step In 按钮。但是，如果下一行代码是函数调用代码，单击 Step Over 按钮将"跳过"函数中的代码。函数的代码全速运行，一旦函数调用返回，调试器就会暂停。Step Over 按钮比 Step In 按钮更常用。

单击 Step Out 按钮会使调试器全速运行代码，直到从当前函数返回。如果你已经使用 Step In 按钮进入了一个函数调用，现在只想继续执行指令直到退出，单击 Step Out 按钮即可。

如果你希望完全停止调试且不想继续运行程序的其余部分，可单击 Stop 按钮。Stop 按钮可以用于立即终止程序。

可以在特定行上设置断点，让程序以正常速度运行，直到到达断点所在行，此时，调试器会暂停，让我们检查变量并继续逐步运行各行代码。在大多数 IDE 中，可以通过双击窗口左侧的行号设置断点。

当前存储在程序变量中的值显示在调试器的调试窗口中的某处。然而，调试程序的一种常用方法是输出调试，即添加 print() 调用以显示变量的值，然后重新运行程序。尽管这种调试方法简单、方便，但其通常比使用调试器低效。使用输出调试方法，我们必须添加 print() 调用，重新运行程序，然后删除 print() 调用。然而，重新运行程序后，你通常会发现需要添加更多的 print() 调用以便查看其他变量的值。这意味着需要重新运行程序，而这次运行可能会表明需要添加另一轮的 print() 调用，以此类推。此外，忘记自己添加的某些 print() 调用并不罕见，这需要额外的删除 print() 调用的操作。对于简单的程序缺陷，输出调试很方便，但从长远来看，使用真正的调试器可以节省你的时间。

结语

编程是一项有趣的、创造性的开发技能。无论你已经掌握 Python 语法的基础知识，还是想一头扎入真正的 Python 程序，本书中的项目都能够给你新的启发，让你明白用几页代码能实现些什么。

学习这些程序的最佳方法绝非仅仅阅读代码或将其复制、粘贴到计算机上。花些时间将本书中的代码手动输入编辑器，培养编写代码的肌肉记忆。这么做会减慢速度，因此你可以在输入时思考每一行代码，而不仅仅是用眼睛浏览。使用搜索引擎查找你不明白的任何指令，或在交互式 Shell 中对其进行实验。

最后，从头开始创建程序，然后进行修改以实现你自己想要的功能。这些练习会为你应用编程概念创建可运行的实际程序打下坚实的基础。最重要的是，希望你乐在其中！

资源与支持

本书由异步社区出品，社区（https://www.epubit.com）为你提供相关资源和后续服务。

配套资源

本书为读者提供源代码。读者可在异步社区本书页面中单击 配套资源 ，跳转到下载界面，按提示进行操作即可。注意：为保证购书读者的权益，该操作会给出相关提示，要求输入提取码进行验证。

勘误

作者和编辑尽最大努力来确保书中内容的准确性，但难免会存在疏漏。欢迎读者将发现的问题反馈给我们，帮助我们提升图书的质量。

如果读者发现错误，请登录异步社区，搜索到本书页面，输入勘误信息，单击"提交"按钮即可。本书的作者和编辑会对读者提交的勘误进行审核，确认并接受后，将赠予读者异步社区的 100 积分（积分可用于在异步社区兑换优惠券、样书或奖品）。

扫码关注本书

扫描下方二维码，读者将会在异步社区微信服务号中看到本书信息及相关的服务提示。

与我们联系

我们的联系邮箱是 contact@epubit.com.cn。

如果读者对本书有任何疑问或建议，请发送邮件给我们，并请在邮件标题中注明本书书名，以便我们更高效地做出反馈。

如果读者有兴趣出版图书、录制教学视频，或者参与图书翻译、技术审校等工作，可以发邮件给我们；有意出版图书的作者也可以到异步社区投稿（直接访问 www.epubit.com/ contribute 即可）。

如果读者所在学校、培训机构或企业想批量购买本书或异步社区出版的其他图书，也可以发邮件给我们。

如果读者在网上发现有针对异步社区出品图书的各种形式的盗版行为，包括对图书全部或部分内容的非授权传播，请将怀疑有侵权行为的链接通过邮件发给我们。这一举动是对作者权益的保护，也是我们持续为广大读者提供有价值的内容的动力之源。

关于异步社区和异步图书

"异步社区"是人民邮电出版社旗下 IT 专业图书社区，致力于出版精品 IT 图书和相关学习产品，为作译者提供优质出版服务。异步社区创办于 2015 年 8 月，提供大量精品 IT 图书和电子书，以及高品质技术文章和视频课程。更多详情请访问异步社区官网 https://www.epubit.com。

"异步图书"是由异步社区编辑团队策划出版的精品 IT 专业图书的品牌，依托于人民邮电出版社近 40 年的计算机图书出版积累和专业编辑团队，相关图书在封面上印有异步图书的 LOGO。异步图书的出版领域包括软件开发、大数据、人工智能、测试、前端、网络技术等。

异步社区

微信服务号

目　　录

项目 1　Pico Fermi Bagels 猜数字游戏 …… 1
项目 2　生日悖论 ………………………… 5
项目 3　位图消息 ………………………… 8
项目 4　21 点纸牌游戏 …………………… 11
项目 5　弹跳 DVD 标志 ………………… 17
项目 6　恺撒密码 ………………………… 22
项目 7　恺撒黑客 ………………………… 25
项目 8　日历生成器 ……………………… 27
项目 9　盒子里的胡萝卜 ………………… 31
项目 10　cho-han 骰子游戏 ……………… 36
项目 11　诱饵标题生成器 ………………… 39
项目 12　Collatz 序列 …………………… 43
项目 13　康威生命游戏 …………………… 45
项目 14　倒计时 …………………………… 48
项目 15　地穴冒险 ………………………… 50
项目 16　钻石 ……………………………… 52
项目 17　骰子数学 ………………………… 55
项目 18　掷骰子 …………………………… 60
项目 19　数字时钟 ………………………… 63
项目 20　数字流 …………………………… 65
项目 21　DNA 可视化 …………………… 68
项目 22　小鸭子 …………………………… 71

项目 23　蚀刻绘图器 ……………………… 76
项目 24　因数查找器 ……………………… 81
项目 25　"快速拔枪" ……………………… 84
项目 26　斐波那契数列 …………………… 86
项目 27　虚拟水族箱 ……………………… 89
项目 28　Flooder 游戏 …………………… 96
项目 29　森林火灾模拟 …………………… 102
项目 30　四子棋 …………………………… 106
项目 31　猜数字 …………………………… 111
项目 32　"上当受骗" ……………………… 114
项目 33　黑客小游戏 ……………………… 116
项目 34　"绞刑架"与"断头台" ………… 121
项目 35　六边形网格 ……………………… 126
项目 36　沙漏 ……………………………… 128
项目 37　饥饿的机器人 …………………… 133
项目 38　"我指证" ………………………… 139
项目 39　朗顿蚂蚁 ………………………… 146
项目 40　火星文 …………………………… 151
项目 41　幸运星 …………………………… 154
项目 42　魔法幸运球 ……………………… 160
项目 43　播棋 ……………………………… 163
项目 44　二维版移动迷宫 ………………… 169

项目 45	三维版移动迷宫 …………… 174		项目 64	7 段显示模块 …………… 243
项目 46	掷 100 万次骰子结果统计模拟器 …………… 181		项目 65	"闪灵地毯" …………… 246
项目 47	蒙德里安艺术品生成器 …… 184		项目 66	简单替换密码 …………… 249
项目 48	3 扇门问题 …………… 189		项目 67	正弦消息 …………… 253
项目 49	乘法表 …………… 194		项目 68	滑动拼图 …………… 256
项目 50	99 瓶牛奶 1 …………… 196		项目 69	蜗牛赛跑 …………… 261
项目 51	99 瓶牛奶 2 …………… 198		项目 70	虚拟算盘 …………… 264
项目 52	数字系统计数器 …………… 201		项目 71	声音模拟 …………… 269
项目 53	元素周期表 …………… 204		项目 72	"海绵宝宝的嘲弄" …… 272
项目 54	儿童隐语 …………… 207		项目 73	数独 …………… 274
项目 55	强力球彩票 …………… 210		项目 74	语音合成 …………… 279
项目 56	素数 …………… 214		项目 75	3 张牌蒙特 …………… 281
项目 57	进度条 …………… 217		项目 76	井字棋 …………… 285
项目 58	彩虹 …………… 220		项目 77	汉诺塔 …………… 288
项目 59	石头剪刀布 …………… 223		项目 78	脑筋急转弯 …………… 292
项目 60	石头剪刀布(无敌版) …… 226		项目 79	2048 …………… 298
项目 61	ROT13 密码 …………… 229		项目 80	弗吉尼亚密码 …………… 304
项目 62	旋转立方体 …………… 231		项目 81	水桶谜题 …………… 308
项目 63	乌尔皇室游戏 …………… 236		附录 A	标签索引 …………… 312
			附录 B	字符映射表 …………… 314

项目 1

Pico Fermi Bagels
猜数字游戏

在 Pico Fermi Bagels 这个逻辑推理游戏中,你要根据线索猜出一个三位数。游戏会根据你的猜测给出以下提示之一:如果你猜对一位数字但数字位置不对,则会提示"Pico";如果你同时猜对了一位数字及其位置,则会提示"Fermi";如果你猜测的数字及其位置都不对,则会提示"Bagels"。你有 10 次猜数字机会。

运行程序

运行 bagels.py,输出如下所示:

```
Bagels, a deductive logic game.
By Al Sweigart al@inventwithpython.com

I am thinking of a 3-digit number. Try to guess what it is.
Here are some clues:
When I say:    That means:
  Pico         One digit is correct but in the wrong position.
  Fermi        One digit is correct and in the right position.
  Bagels       No digit is correct.
I have thought up a number.
 You have 10 guesses to get it.
Guess #1:
> 123
Pico
Guess #2:
> 456
Bagels
Guess #3:
> 178
Pico Pico
--snip--
Guess #7:
> 791
Fermi Fermi
Guess #8:
> 701
You got it!
Do you want to play again? (yes or no)
> no
Thanks for playing!
```

工作原理

切记，这个程序使用的不是整数值，而是包含数字的字符串值。例如，'426'与426是不同的值。之所以这么做，是因为我们需要与秘密数字进行字符串比较，而不是进行数学运算。记住，0可以作为前导数字，在这种情况下，字符串'026'与'26'不同，但整数026与26相同。

```python
1. """
2. Pico Fermi Bagels 猜数字游戏, 作者：Al Sweigart al@inventwithpython.com
3. 一个逻辑推理游戏，你必须根据线索猜数字
4. 本游戏的一个版本在《Python游戏编程快速上手》中有相应介绍
5. 标签：简短，游戏，谜题
6. """
7.
8. import random
9.
10. NUM_DIGITS = 3  # (!) 请试着将 3 设置为 1 或 10
11. MAX_GUESSES = 10  # (!) 请试着将 10 设置为 1 或 100
12.
13.
14. def main():
15.     print('''Bagels, a deductive logic game.
16. By Al Sweigart al@inventwithpython.com
17.
18. I am thinking of a {}-digit number with no repeated digits.
19. Try to guess what it is. Here are some clues:
20. When I say:    That means:
21.   Pico         One digit is correct but in the wrong position.
22.   Fermi        One digit is correct and in the right position.
23.   Bagels       No digit is correct.
24.
25. For example, if the secret number was 248 and your guess was 843, the
26. clues would be Fermi Pico.'''.format(NUM_DIGITS))
27.
28.     while True:  # 主循环
29.         # secretNum 存储了玩家所要猜测的秘密数字
30.         secretNum = getSecretNum()
31.         print('I have thought up a number.')
32.         print(' You have {} guesses to get it.'.format(MAX_GUESSES))
33.
34.         numGuesses = 1
35.         while numGuesses <= MAX_GUESSES:
36.             guess = ''
37.             # 保持循环，直到玩家输入正确的猜测数字
38.             while len(guess) != NUM_DIGITS or not guess.isdecimal():
39.                 print('Guess #{}: '.format(numGuesses))
40.                 guess = input('> ')
41.
42.             clues = getClues(guess, secretNum)
43.             print(clues)
44.             numGuesses += 1
45.
46.             if guess == secretNum:
47.                 break  # 玩家猜对了数字，结束当前循环
48.             if numGuesses > MAX_GUESSES:
49.                 print('You ran out of guesses.')
50.                 print('The answer was {}.'.format(secretNum))
51.
52.         # 询问玩家是否想再玩一次
53.         print('Do you want to play again? (yes or no)')
```

```
54.          if not input('> ').lower().startswith('y'):
55.              break
56.      print('Thanks for playing!')
57.
58.
59. def getSecretNum():
60.     """返回唯一一个长度为 NUM_DIGITS 且由随机数字组成的字符串"""
61.     numbers = list('0123456789')  # 创建数字 0~9 的列表
62.     random.shuffle(numbers)  # 将它们随机排列
63.
64.     # 获取秘密数字列表中的前 NUM_DIGITS 位数字
65.     secretNum = ''
66.     for i in range(NUM_DIGITS):
67.         secretNum += str(numbers[i])
68.     return secretNum
69.
70.
71. def getClues(guess, secretNum):
72.     """返回一个由 Pico、Fermi、Bagels 组成的字符串，用于猜测一个三位数"""
73.     if guess == secretNum:
74.         return 'You got it!'
75.
76.     clues = []
77.
78.     for i in range(len(guess)):
79.         if guess[i] == secretNum[i]:
80.             # 正确的数字位于正确的位置
81.             clues.append('Fermi')
82.         elif guess[i] in secretNum:
83.             # 正确的数字不在正确的位置
84.             clues.append('Pico')
85.     if len(clues) == 0:
86.         return 'Bagels'  # 没有正确的数字
87.     else:
88.         # 将 clues 列表按字母顺序排序，使其不会泄露数字的信息
89.         clues.sort()
90.         # 返回一个由 clues 列表中所有元素组成的字符串
91.         return ' '.join(clues)
92.
93.
94. # 程序运行入口（如果不是作为模块导入的话）
95. if __name__ == '__main__':
96.     main()
```

输入源代码并运行几次后，请试着对其进行更改。标有!的注释给你提供了更改建议。你也可以自己尝试执行以下操作。

- 通过更改 NUM_DIGITS 修改秘密数字的位数。
- 通过更改 MAX_GUESSES 修改玩家的最大猜测次数。
- 尝试创建秘密数字中包含字母和数字的程序版本。

探索程序

请尝试找出以下问题的答案。可尝试对代码进行一些修改，再次运行程序，并查看修改后的效果。

1. 如果更改 NUM_DIGITS 常量，则会发生什么？

2. 如果更改 MAX_GUESSES 常量，则会发生什么？
3. 如果将 NUM_DIGITS 设置为大于 10 的数字会发生什么？
4. 如果将第 30 行的 secretNum = getSecretNum() 替换为 secretNum = '123'，会发生什么？
5. 如果删除或注释掉第 34 行的 numGuesses = 1，会得到什么错误消息？
6. 如果删除或注释掉第 62 行的 random.shuffle(numbers)，会发生什么？
7. 如果删除或注释掉第 73 行的 if guess == secretNum:和第 74 行的 return'You got it!'，会发生什么？
8. 如果注释掉第 44 行的 numGuesses += 1，会发生什么？

项目 2

生日悖论

"生日悖论"又称为"生日问题",是指即使在很少的一群人中两人生日相同的概率也是惊人的。在一个70人的组中,两人生日相同的概率高达99.9%。即使在一个只有23人的小组中,这个概率也高达50%。以下程序用于进行多个概率实验,以确定不同规模小组中生日相同的概率值。这类实验称为蒙特卡罗实验。在这类实验中,我们需要进行多次随机试验,以了解可能的结果。

运行程序

运行 birthdayparadox.py,输出如下所示:

```
Birthday Paradox, by Al Sweigart al@inventwithpython.com
--snip--
How many birthdays shall I generate? (Max 100)
> 23

Here are 23 birthdays:
Oct 9, Sep 1, May 28, Jul 29, Feb 17, Jan 8, Aug 18, Feb 19, Dec 1, Jan 22,
May 16, Sep 25, Oct 6, May 6, May 26, Oct 11, Dec 19, Jun 28, Jul 29, Dec 6,
Nov 26, Aug 18, Mar 18

In this simulation, multiple people have a birthday on Jul 29

Generating 23 random birthdays 100,000 times...
Press Enter to begin...
Let's run another 100,000 simulations.
0 simulations run...
10000 simulations run...
--snip--
90000 simulations run...
100000 simulations run.
Out of 100,000 simulations of 23 people, there was a
matching birthday in that group 50955 times. This means
that 23 people have a 50.95 % chance of
having a matching birthday in their group.
That's probably more than you would think!
```

工作原理

运行100000次模拟可能需要耗费一些时间,这就是后文的完整程序中第91行和第92行报告另外 10000 次模拟已经完成的原因。该反馈可以让用户确信程序没有"僵死"。注意，某些整数（例如第 91 行的 10_000 以及第 89 行和第 99 行的 100_000）带有下划线。这些下划线没有特殊含义，但 Python 允许这么做，这样程序员可以让整数更易于阅读。换句话说，将 100_000（而不是 100000）读成"十万"更容易一些。

```python
1.  """生日悖论，作者：Al Sweigart al@inventwithpython.com
2.  探索生日悖论的惊人概率
3.  标签：简短，数学，模拟"""
4.
5.  import datetime, random
6.
7.
8.  def getBirthdays(numberOfBirthdays):
9.      """返回一个随机生日日期对象的数字列表"""
10.     birthdays = []
11.     for i in range(numberOfBirthdays):
12.         # 年份对于模拟并不重要，假设所有生日在同一年即可
13.         startOfYear = datetime.date(2001, 1, 1)
14.
15.         # 随机获取一年中的一天
16.         randomNumberOfDays = datetime.timedelta(random.randint(0, 364))
17.         birthday = startOfYear + randomNumberOfDays
18.         birthdays.append(birthday)
19.     return birthdays
20.
21.
22. def getMatch(birthdays):
23.     """返回在生日列表中多次出现的日期对象"""
24.     if len(birthdays) == len(set(birthdays)):
25.         return None  # 若所有生日都不同，则返回 None
26.
27.     # 将这个生日与其他生日进行比较
28.     for a, birthdayA in enumerate(birthdays):
29.         for b, birthdayB in enumerate(birthdays[a + 1 :]):
30.             if birthdayA == birthdayB:
31.                 return birthdayA  # 返回相同的生日
32.
33.
34. # 显示介绍信息
35. print('''Birthday Paradox, by Al Sweigart al@inventwithpython.com
36.
37. The birthday paradox shows us that in a group of N people, the odds
38. that two of them have matching birthdays is surprisingly large.
39. This program does a Monte Carlo simulation (that is, repeated random
40. simulations) to explore this concept.
41.
42. (It's not actually a paradox, it's just a surprising result.)
43. ''')
44.
45. # 创建一个按照月份排序的元组
46. MONTHS = ('Jan', 'Feb', 'Mar', 'Apr', 'May', 'Jun',
47.           'Jul', 'Aug', 'Sep', 'Oct', 'Nov', 'Dec')
48.
49. while True:  # 要求玩家输入有效的总数
50.     print('How many birthdays shall I generate? (Max 100)')
51.     response = input('> ')
52.     if response.isdecimal() and (0 < int(response) <= 100):
53.         numBDays = int(response)
54.         break  # 玩家输入了有效的总数，结束循环
55. print()
56.
```

```
57. # 显示生成的生日
58. print('Here are', numBDays, 'birthdays:')
59. birthdays = getBirthdays(numBDays)
60. for i, birthday in enumerate(birthdays):
61.     if i != 0:
62.         # 每个生日之间用逗号隔开
63.         print(', ', end='')
64.     monthName = MONTHS[birthday.month - 1]
65.     dateText = '{} {}'.format(monthName, birthday.day)
66.     print(dateText, end='')
67. print()
68. print()
69.
70. # 确定是否存在两个相同的生日
71. match = getMatch(birthdays)
72.
73. # 显示结果
74. print('In this simulation, ', end='')
75. if match != None:
76.     monthName = MONTHS[match.month - 1]
77.     dateText = '{} {}'.format(monthName, match.day)
78.     print('multiple people have a birthday on', dateText)
79. else:
80.     print('there are no matching birthdays.')
81. print()
82.
83. # 运行100000次模拟
84. print('Generating', numBDays, 'random birthdays 100000 times...')
85. input('Press Enter to begin...')
86.
87. print('Let\'s run another 100000 simulations.')
88. simMatch = 0  #模拟中有多少相同的生日
89. for i in range(100_000):
90.     # 每10000次模拟后输出当前进度
91.     if i % 10_000 == 0:
92.         print(i, 'simulations run...')
93.     birthdays = getBirthdays(numBDays)
94.     if getMatch(birthdays) != None:
95.         simMatch = simMatch + 1
96. print('100000 simulations run.')
97.
98. # 显示模拟结果
99. probability = round(simMatch / 100_000 * 100, 2)
100. print('Out of 100,000 simulations of', numBDays, 'people, there was a')
101. print('matching birthday in that group', simMatch, 'times. This means')
102. print('that', numBDays, 'people have a', probability, '% chance of')
103. print('having a matching birthday in their group.')
104. print('That\'s probably more than you would think!')
```

探索程序

尝试找出以下问题的答案。尝试对代码进行一些修改并重新运行程序以查看修改代码后的效果。

1. 程序是如何表示生日的？（提示：请查看第13行代码。）
2. 如何取消程序最多生成100个生日的限制？
3. 如果删除或注释掉第53行的 numBDays = int(response) 会得到什么错误消息？
4. 如何让程序显示完整的月份名称，例如显示 January 而不是 Jan？
5. 如何让"× simulations run..."每执行1000次模拟（而不是每执行10000次模拟）出现1次？

项目 3

位图消息

本项目中的程序用多行字符串表示位图，即"每个像素"只有两种可能颜色的二维图像，以确定如何显示来自用户的消息。在位图中，空格字符代表一个空格，其他字符都被用户消息中的字符所替换。输出的位图有点像地图的样子，你也可以将其修改为自己喜欢的任何图案。因为使用的是简单的二元"空格或消息"字符系统，所以本项目很适合初学者。你可以尝试不同的消息，看看效果如何！

运行程序

运行 bitmapmessage.py，输出如下所示：

```
Bitmap Message, by Al Sweigart al@inventwithpython.com
Enter the message to display with the bitmap.
> Hello!

Hello!Hello!Hello!Hello!Hello!Hello!Hello!Hello!Hello!Hello!He
  lo!Hello!Hello    l !He lo e      llo!Hello!Hello!Hello!He
 llo!Hello!Hello   He lo H   l !Hello!Hello!Hello!Hello!H
 el    lo!Hello!Hello!He       lo!Hello!Hello!Hello!Hel
         o!Hello!Hello           lo  e lo!H ll !Hello!Hello!H  l
           !Hello!He             llo!Hel    Hello!Hello!Hell ! e
          Hello!He              ello!Hello!Hello!Hell H
  l       H llo! ell            ello!Hello!Hello el o
             lo!H  l            ello!Hello!Hell   ell!He    o
              !Hello            llo!Hello!Hel     el He    o
              !Hello!H          lo!Hello!Hell    l !H ll o
              ello!Hel          Hello!He        H ll o Hell
              ello!Hell         ello!H  l       Hell !H l o!
              ello!Hell         ello!H l o        o!H l     H
               lo!Hel           ello! el          o!Hel    H
               lo!He             llo! e           llo!Hell
               llo!H             llo!             llo!Hello
               llo!               ll              lo!Hell   e
               llo                                    l   e
               ll   l                    H
Hello!Hello!Hello!Hello!Hello!Hello!Hello!Hello!Hello!Hello!He
```

工作原理

图案顶部和底部两行各有的 68 个字符作为标尺，用于帮助我们正确对齐。不过，即使你在输入图案时打错了字符，该程序仍然可以正确运行。

调用第 40 行的 `bitmap.splitlines()` 方法会返回一个字符串列表，每个字符串都是多行 bitmap 字符串中的一行。使用多行字符串易于将位图修改为你喜欢的任何图案。程序会填充图案中的所有非空格字符。这也是使用星号、句点或其他字符运行效果都相同的原因。

第 48 行代码 `message[i % len(message)]` 会使 message 中的文本重复出现。当 i 从 0 增加到大于 `len(message)` 的数值时，表达式 `i % len(message)` 再次为 0。这会导致 `message[i % len(message)]` 随着 i 的增加而重复 message 中的字符。

```python
1.  """位图消息，作者: Al Sweigart al@inventwithpython.com
2.  根据提供的位图图像显示文本消息
3.  标签: 小，初学者，艺术"""
4.
5.  import sys
6.
7.  # (!) 尝试将这个多行字符串更改为你喜欢的任何图案
8.
9.  # 此字符串的顶部和底部各有 68 个字符
10. bitmap = """
11. ....................................................................
12.    **************   *  ***  **  *    *******************************
13.   ************************ **  ** *  *  ******************************* *
14.  **     *******************       ******************************
15.           ***************      **  * ****  ** ***************** *
16.             *********           *******    ***************** * *
17.             ********            **********************************
18.     *    * **** ***              *******************  ******  **  *
19.            ****   *                ****************   ***  ***  *
20.            ******                   *************     **   **  *
21.            ********                  *************       *  ** ***
22.             ********                   ********            * *** ****
23.             *********                   ******  *           **** ** * **
24.             *********                   ****** * *            ***  *   *
25.              ******                     *****  **              ****    *
26.              *****                       ****                 ********
27.              *****                        ****                *********
28.              ****                          **                  *******   *
29.              ***                                                *     *
30.              **    *                        *
31. ...................................................................."""
32.
33. print('Bitmap Message, by Al Sweigart al@inventwithpython.com')
34. print('Enter the message to display with the bitmap.')
35. message = input('> ')
36. if message == '':
37.     sys.exit()
38.
39. # 循环遍历 bitmap 字符串中的每一行
40. for line in bitmap.splitlines():
41.     # 循环遍历一行中的每一个字符
42.     for i, bit in enumerate(line):
43.         if bit == ' ':
44.             # 输出空格，以表示位图中的空格
45.             print(' ', end='')
46.         else:
47.             # 输出 message 中的一个字符
48.             print(message[i % len(message)], end='')
49.     print()   # 输出一行空行
```

输入源代码并运行几次后，请尝试对其进行更改。你可以通过修改 bitmap 变量中的字符串来创建全新的图案。

探索程序

请尝试找出以下问题的答案。你需要尝试对代码进行一些修改，再次运行程序，查看修改后的效果。

1. 如果玩家输入空字符串，会发生什么？
2. bitmap 字符串中的非空格字符是什么，重要吗？
3. 第 42 行的 i 变量代表什么？
4. 如果删除或注释掉第 49 行的 print()，会出现什么错误？

项目 4

21点纸牌游戏

21 点纸牌游戏又称为"黑杰克"。玩家努力让手中的牌的点数之和不超过 21 点且尽可能大。在本项目中,我们将使用文本字符绘制的图像——称为 ASCII 艺术画。美国信息交换标准代码(ASCII)定义了文本字符到数字代码的映射,在 Unicode 被取代之前,计算机是用 ASCII 编码的。本项目程序中的扑克牌是 ASCII 艺术画的一个例子:

```
 ___   ___
|A  | |10 |
|♣  | | ♦ |
|__A| |_10|
```

如果你对 21 点纸牌游戏的其他规则和历史感兴趣,请阅读维基百科的相关词条。

运行程序

运行 blackjack.py,输出如下所示:

```
Blackjack, by Al Sweigart al@inventwithpython.com

    Rules:
      Try to get as close to 21 without going over.
      Kings, Queens, and Jacks are worth 10 points.
      Aces are worth 1 or 11 points.
      Cards 2 through 10 are worth their face value.
      (H)it to take another card.
      (S)tand to stop taking cards.
      On your first play, you can (D)ouble down to increase your bet
      but must hit exactly one more time before standing.
      In case of a tie, the bet is returned to the player.
      The dealer stops hitting at 17.
Money: 5000
How much do you bet? (1-5000, or QUIT)
> 400
Bet: 400

DEALER: ???
 ___   ___
|## | |2  |
|###| | ♥ |
|_##| |__2|
```

```
PLAYER: 17
 ___   ___
|K  | |7  |
|♠  | |♦  |
|__K| |__7|


(H)it, (S)tand, (D)ouble down
> h
You drew a 4 of ♦.
--snip--
DEALER: 18
 ___   ___   ___
|K  | |2  | |6  |
|♦  | |♥  | |♠  |
|__K| |__2| |__6|

PLAYER: 21
 ___   ___   ___
|K  | |7  | |4  |
|♠  | |♦  | |♦  |
|__K| |__7| |__4|

You won $400!
--snip--
```

工作原理

键盘上不存在纸牌套装符号,这就是我们调用 chr()函数来创建这些符号的原因。传递给 chr() 的整数称为 Unicode 代码点,这是根据 Unicode 标准标识字符的唯一编号。Unicode 常常会被误解。Ned Batchelder 在 2012 PyCon US 上的演讲 "Pragmatic Unicode, or How Do I Stop the Pain?"对 Unicode 进行了详尽介绍,附录 B 提供了可以在 Python 程序中使用的 Unicode 字符的完整列表。

```
 1. """21 点纸牌游戏,作者: Al Sweigart al@inventwithpython.com
 2. 经典的纸牌游戏,也被称为"黑杰克"(这个版本没有分牌或保险)
 3. 标签: 大,游戏,卡牌"""
 4.
 5. import random, sys
 6.
 7. # 创建常量
 8. HEARTS   = chr(9829) # 字符 9829 表示 ♥
 9. DIAMONDS = chr(9830) # 字符 9830 表示 ♦
10. SPADES   = chr(9824) # 字符 9824 表示 ♠
11. CLUBS    = chr(9827) # 字符 9827 表示 ♣
12. BACKSIDE = 'backside'
13.
14.
15. def main():
16.     print('''Blackjack, by Al Sweigart al@inventwithpython.com
17.
18. Rules:
19.     Try to get as close to 21 without going over.
20.     Kings, Queens, and Jacks are worth 10 points.
21.     Aces are worth 1 or 11 points.
```

```
22.        Cards 2 through 10 are worth their face value.
23.        (H)it to take another card.
24.        (S)tand to stop taking cards.
25.        On your first play, you can (D)ouble down to increase your bet
26.        but must hit exactly one more time before standing.
27.        In case of a tie, the bet is returned to the player.
28.        The dealer stops hitting at 17.''')
29.
30.    money = 5000
31.    while True:    # 主循环
32.        # 判断玩家的点数是否用完了
33.        if money <= 0:
34.            print("You're broke!")
35.            print("Good thing you weren't playing with real money.")
36.            print('Thanks for playing!')
37.            sys.exit()
38.
39.        # 让玩家输入这一轮的点数
40.        print('Money:', money)
41.        bet = getBet(money)
42.
43.        # 给庄家和玩家各发两张牌
44.        deck = getDeck()
45.        dealerHand = [deck.pop(), deck.pop()]
46.        playerHand = [deck.pop(), deck.pop()]
47.
48.        # 处理玩家的动作
49.        print('Bet:', bet)
50.        while True:    # 保持循环，直到玩家停牌或爆掉
51.            displayHands(playerHand, dealerHand, False)
52.            print()
53.
54.            # 检查玩家牌的点数是否超过 21 点
55.            if getHandValue(playerHand) > 21:
56.                break
57.
58.            # 获取玩家的下一步操作，即拿牌（H）、停牌（S）或加倍（D）
59.            move = getMove(playerHand, money - bet)
60.
61.            # 处理玩家的动作
62.            if move == 'D':
63.                # 玩家将点数加倍
64.                additionalBet = getBet(min(bet, (money - bet)))
65.                bet += additionalBet
66.                print('Bet increased to {}.'.format(bet))
67.                print('Bet:', bet)
68.
69.            if move in ('H', 'D'):
70.                # 选择另一张牌或将点数加倍
71.                newCard = deck.pop()
72.                rank, suit = newCard
73.                print('You drew a {} of {}.'.format(rank, suit))
74.                playerHand.append(newCard)
75.
76.                if getHandValue(playerHand) > 21:
77.                    # 玩家爆掉
78.                    continue
79.
80.            if move in ('S', 'D'):
81.                # 玩家在回合内停牌或将点数加倍
82.                break
83.
```

```
84.          # 处理庄家的动作
85.          if getHandValue(playerHand) <= 21:
86.              while getHandValue(dealerHand) < 17:
87.                  # 庄家拿牌
88.                  print('Dealer hits...')
89.                  dealerHand.append(deck.pop())
90.                  displayHands(playerHand, dealerHand, False)
91.
92.                  if getHandValue(dealerHand) > 21:
93.                      break  # 庄家爆掉
94.                  input('Press Enter to continue...')
95.                  print('\n\n')
96.
97.          # 显示最后所有人的点数
98.          displayHands(playerHand, dealerHand, True)
99.
100.         playerValue = getHandValue(playerHand)
101.         dealerValue = getHandValue(dealerHand)
102.         # 判断玩家是赢、输还是平局
103.         if dealerValue > 21:
104.             print('Dealer busts! You win ${}!'.format(bet))
105.             money += bet
106.         elif (playerValue > 21) or (playerValue < dealerValue):
107.             print('You lost!')
108.             money -= bet
109.         elif playerValue > dealerValue:
110.             print('You won ${}!'.format(bet))
111.             money += bet
112.         elif playerValue == dealerValue:
113.             print('It\'s a tie, the bet is returned to you.')
114.
115.         input('Press Enter to continue...')
116.         print('\n\n')
117.
118.
119. def getBet(maxBet):
120.     """询问玩家这一轮选择的点数"""
121.     while True:    # 继续询问，直到玩家输入有效的金额数据
122.         print('How much do you bet? (1-{}, or QUIT'.format(maxBet))
123.         bet = input('> ').upper().strip()
124.         if bet == 'QUIT':
125.             print('Thanks for playing!')
126.             sys.exit()
127.
128.         if not bet.isdecimal():
129.             continue  # 如果玩家没有输入点数，请再次询问
130.
131.         bet = int(bet)
132.         if 1 <= bet <= maxBet:
133.             return bet  # 玩家输入了有效的点数
134.
135.
136. def getDeck():
137.     """返回52张牌的 (rank,suit) 元组"""
138.     deck = []
139.     for suit in (HEARTS, DIAMONDS, SPADES, CLUBS):
140.         for rank in range(2, 11):
141.             deck.append((str(rank), suit))  # 添加编号牌
142.         for rank in ('J', 'Q', 'K', 'A'):
143.             deck.append((rank, suit))  # 添加脸谱牌和王牌
144.     random.shuffle(deck)
145.     return deck
```

```
146.
147.
148. def displayHands(playerHand, dealerHand, showDealerHand):
149.     """展示玩家和庄家的牌。如果 showDealerHand 为 False, 则隐藏庄家的第一张牌"""
150.     print()
151.     if showDealerHand:
152.         print('DEALER:', getHandValue(dealerHand))
153.         displayCards(dealerHand)
154.     else:
155.         print('DEALER: ???')
156.         # 隐藏庄家的第一张牌
157.         displayCards([BACKSIDE] + dealerHand[1:])
158.
159.     # 显示玩家的牌
160.     print('PLAYER:', getHandValue(playerHand))
161.     displayCards(playerHand)
162.
163.
164. def getHandValue(cards):
165.     """返回所有牌的点数。脸谱牌的点数为 10, 王牌的点数为值 11 或 1 (王牌的点数选择最合适的)"""
166.     value = 0
167.     numberOfAces = 0
168.
169.     # 计算非王牌的点数
170.     for card in cards:
171.         rank = card[0]  # card 是一个包含大小和花色的元组
172.         if rank == 'A':
173.             numberOfAces += 1
174.         elif rank in ('K', 'Q', 'J'):  # 脸谱牌的点数为 10
175.             value += 10
176.         else:
177.             value += int(rank)  # 编号牌上的数字就是其对应的点数
178.
179.     # 计算王牌的点数
180.     value += numberOfAces  # 每张王牌加 1 点
181.     for i in range(numberOfAces):
182.         # 如果可以在不爆掉的情况下再增加 10 点, 请执行如下代码
183.         if value + 10 <= 21:
184.             value += 10
185.
186.     return value
187.
188.
189. def displayCards(cards):
190.     """显示 cards 列表中的所有纸牌"""
191.     rows = ['', '', '', '', '']  # 每行显示的文本格式
192.
193.     for i, card in enumerate(cards):
194.         rows[0] += ' ___  '  # 输出纸牌的最上面一行
195.         if card == BACKSIDE:
196.             # 输出纸牌背面
197.             rows[1] += '|## | '
198.             rows[2] += '|###| '
199.             rows[3] += '|_##| '
200.         else:
201.             # 输出纸牌正面
202.             rank, suit = card  # card 是一个元组
203.             rows[1] += '|{} | '.format(rank.ljust(2))
204.             rows[2] += '| {} | '.format(suit)
205.             rows[3] += '|_{}| '.format(rank.rjust(2, '_'))
206.
207.     # 在屏幕上输出每一行
```

```
208.    for row in rows:
209.        print(row)
210.
211.
212. def getMove(playerHand, money):
213.     """询问玩家的下一步操作,并返回 H(表示拿牌)、S(表示停牌)、D(表示将点数加倍)"""
214.     while True:  # 继续循环,直到玩家输入正确的下一步操作
215.         # 确定玩家可以进行哪些操作
216.         moves = ['(H)it', '(S)tand']
217.
218.         # 如果玩家恰好有两张牌,提醒他们在第一步将点数加倍
219.         if len(playerHand) == 2 and money > 0:
220.             moves.append('(D)ouble down')
221.
222.         # 获取玩家的下一步操作
223.         movePrompt = ', '.join(moves) + '> '
224.         move = input(movePrompt).upper()
225.         if move in ('H', 'S'):
226.             return move  # 玩家输入了有效的下一步操作
227.         if move == 'D' and '(D)ouble down' in moves:
228.             return move  # 玩家输入了有效的下一步操作
229.
230.
231. # 程序运行入口(如果不是被作为模块导入的话)
232. if __name__ == '__main__':
233.     main()
```

输入源代码并运行几次后,请尝试对其进行更改。21 点纸牌游戏有几条自定义规则。例如,如果前两张牌具有相同的点数,玩家可以将它们分成两手牌并分别选择点数。此外,如果玩家的前两张牌(黑桃王牌和梅花杰克牌)得到"21 点",则玩家将赢得 10 倍的奖金。

探索程序

请尝试找出以下问题的答案。你需要尝试对代码进行一些修改,再次运行程序,查看修改后的效果。

1. 怎样才能让玩家以不同的筹码来玩游戏?
2. 程序如何防止玩家输入的点数大于他们拥有的筹码?
3. 程序如何表示单张牌?
4. 程序如何表示一手牌?
5. `rows` 列表(在第 191 行创建)中的字符串代表什么?
6. 如果删除或注释掉第 144 行的 `random.shuffle(deck)`,会发生什么?
7. 如果将第 108 行的 `money -= bet` 改为 `money += bet`,会发生什么?
8. 如果 `displayHands()` 函数中的 `showDealerHand` 设置为 `True`,会发生什么?如果它为 `False`,又会发生什么?

| 项目 5 | 弹跳DVD标志

如果你是 20 世纪出生的人,应该会对那些"古董"DVD 播放器记忆犹新。DVD 暂停播放时,DVD 标志会斜着从屏幕边缘弹跳出来。在本项目中,我们将编写程序,通过在 DVD 标志每次碰到屏幕边缘时改变其方向来模拟彩色弹跳 DVD 标志。此外,程序将"跟踪"标志击中屏幕一角的次数。最终实现的效果是一个有趣的视觉动画,特别有意思的是当标志与角落完美对齐的神奇时刻。

因为程序用到了 bext 模块,我们无法在集成开发环境(IDE)或编辑器中运行这个程序,必须在命令提示符窗口或终端窗口运行才能正确地输出。

运行程序

运行 bouncingdvd.py,输出如下所示:

18　项目 5　弹跳 DVD 标志

工作原理

你也许还记得数学课上学到的笛卡儿坐标系。与此类似，在编程中，x 坐标代表一个物体的水平位置，y 坐标代表它的垂直位置。与数学中不同的是，编程中的原点(0, 0)位于屏幕的左上角，并且 y 坐标向下表示增加。x 坐标则随着对象向右移动而增加，和数学中的含义相同，如下所示：

bext 模块的 goto()函数有着相似的工作逻辑：调用 bext.goto(0, 0)会将文本光标置于终端窗口的左上角。我们使用带有 color、direction、x 和 y 键值的 Python 字典来表示每个跳动的 DVD 标志。x 和 y 的值是整数，代表标志在窗口中的位置。由于这些值被传递给 bext.goto()，因此增大这些值会导致标志向右、向下移动；减小这些值则会导致标志向左、向上移动。

```
 1. """弹跳 DVD 标志, 作者：Al Sweigart al@inventwithpython.com
 2. 一个弹跳的 DVD 标志动画
 3. 按 Ctrl-C, 程序结束
 4.
 5. 注意：不要在程序运行时调整终端窗口的大小
 6. 标签：大, 艺术, bext 模块"""
 7.
 8. import sys, random, time
 9.
10. try:
11.     import bext
12. except ImportError:
13.     print('This program requires the bext module, which you')
14.     print('can install by following the instructions at')
15.     print('https://pypi.org/project/Bext/')
16.     sys.exit()
17.
```

```
18. # 创建常量
19. WIDTH, HEIGHT = bext.size()
20. # 如果不添加自动换行符，我们无法在 Windows 上输出到最后一列，所以宽度减 1
21.
22. WIDTH -= 1
23.
24. NUMBER_OF_LOGOS = 5    # (!) 尝试将 5 更改为 1 或 100
25. PAUSE_AMOUNT = 0.2    # (!) 尝试将 0.2 更改为 1.0 或 0.0
26. # (!) 尝试减少此列表中的颜色
27. COLORS = ['red', 'green', 'yellow', 'blue', 'magenta', 'cyan', 'white']
28.
29. UP_RIGHT   = 'ur'
30. UP_LEFT    = 'ul'
31. DOWN_RIGHT = 'dr'
32. DOWN_LEFT  = 'dl'
33. DIRECTIONS = (UP_RIGHT, UP_LEFT, DOWN_RIGHT, DOWN_LEFT)
34.
35. # 标志字典的关键值
36. COLOR = 'color'
37. X = 'x'
38. Y = 'y'
39. DIR = 'direction'
40.
41.
42. def main():
43.     bext.clear()
44.
45.     # 生成一些标志
46.     logos = []
47.     for i in range(NUMBER_OF_LOGOS):
48.         logos.append({COLOR: random.choice(COLORS),
49.                       X: random.randint(1, WIDTH - 4),
50.                       Y: random.randint(1, HEIGHT - 4),
51.                       DIR: random.choice(DIRECTIONS)})
52.         if logos[-1][X] % 2 == 1:
53.             # 确保 X 是偶数，以便标志可以弹跳至角落
54.             logos[-1][X] -= 1
55.
56.     cornerBounces = 0  # 统计一个标志有多少次弹跳至角落
57.     while True:  # 主循环
58.         for logo in logos:  # 处理 logos 列表中的每个标志
59.             # 消除标志的当前位置
60.             bext.goto(logo[X], logo[Y])
61.             print(' ', end='')  # (!) 你可以试着注释掉这一行
62.
63.             originalDirection = logo[DIR]
64.
65.             # 判断标志是否从角落反弹
66.             if logo[X] == 0 and logo[Y] == 0:
67.                 logo[DIR] = DOWN_RIGHT
68.                 cornerBounces += 1
69.             elif logo[X] == 0 and logo[Y] == HEIGHT - 1:
70.                 logo[DIR] = UP_RIGHT
71.                 cornerBounces += 1
72.             elif logo[X] == WIDTH - 3 and logo[Y] == 0:
73.                 logo[DIR] = DOWN_LEFT
74.                 cornerBounces += 1
75.             elif logo[X] == WIDTH - 3 and logo[Y] == HEIGHT - 1:
76.                 logo[DIR] = UP_LEFT
77.                 cornerBounces += 1
78.
79.             # 判断标志是否从左边缘反弹
```

```
 80.            elif logo[X] == 0 and logo[DIR] == UP_LEFT:
 81.                logo[DIR] = UP_RIGHT
 82.            elif logo[X] == 0 and logo[DIR] == DOWN_LEFT:
 83.                logo[DIR] = DOWN_RIGHT
 84.
 85.            # 判断标志是否从右边缘反弹
 86.            # （设置为 WIDTH - 3 是因为 "DVD" 有 3 个字母）
 87.            elif logo[X] == WIDTH - 3 and logo[DIR] == UP_RIGHT:
 88.                logo[DIR] = UP_LEFT
 89.            elif logo[X] == WIDTH - 3 and logo[DIR] == DOWN_RIGHT:
 90.                logo[DIR] = DOWN_LEFT
 91.
 92.            # 判断标志是否从顶部边缘反弹
 93.            elif logo[Y] == 0 and logo[DIR] == UP_LEFT:
 94.                logo[DIR] = DOWN_LEFT
 95.            elif logo[Y] == 0 and logo[DIR] == UP_RIGHT:
 96.                logo[DIR] = DOWN_RIGHT
 97.
 98.            # 判断标志是否从底部边缘反弹
 99.            elif logo[Y] == HEIGHT - 1 and logo[DIR] == DOWN_LEFT:
100.                logo[DIR] = UP_LEFT
101.            elif logo[Y] == HEIGHT - 1 and logo[DIR] == DOWN_RIGHT:
102.                logo[DIR] = UP_RIGHT
103.
104.            if logo[DIR] != originalDirection:
105.                # 如果标志反弹，则更改颜色
106.                logo[COLOR] = random.choice(COLORS)
107.
108.            # 移动标志。（X 移动 2 是因为终端字符的高度是宽度的两倍）
109.            if logo[DIR] == UP_RIGHT:
110.                logo[X] += 2
111.                logo[Y] -= 1
112.            elif logo[DIR] == UP_LEFT:
113.                logo[X] -= 2
114.                logo[Y] -= 1
115.            elif logo[DIR] == DOWN_RIGHT:
116.                logo[X] += 2
117.                logo[Y] += 1
118.            elif logo[DIR] == DOWN_LEFT:
119.                logo[X] -= 2
120.                logo[Y] += 1
121.
122.        # 显示标志从角落反弹的次数
123.        bext.goto(5, 0)
124.        bext.fg('white')
125.        print('Corner bounces:', cornerBounces, end='')
126.
127.        for logo in logos:
128.            # 在新位置绘制标志
129.            bext.goto(logo[X], logo[Y])
130.            bext.fg(logo[COLOR])
131.            print('DVD', end='')
132.
133.        bext.goto(0, 0)
134.
135.        sys.stdout.flush()  # 使用 bext 模块需要刷新缓存
136.        time.sleep(PAUSE_AMOUNT)
137.
138.
139. # 程序运行入口（如果不是作为模块导入的话）
140. if __name__ == '__main__':
141.     try:
```

```
142.        main()
143.    except KeyboardInterrupt:
144.        print()
145.        print('Bouncing DVD Logo, by Al Sweigart')
146.        sys.exit()   # 按 Ctrl-C，结束程序
```

输入源代码并运行几次后，请尝试对其进行修改。你可以根据标有!的注释对程序进行修改，也可以尝试以下操作。

- 修改 NUMBER_OF_LOGOS 的值，增加屏幕上弹跳标志的数量。
- 修改 PAUSE_AMOUNT 的值，加快或减慢标志的移动速度。

探索程序

请尝试找出以下问题的答案。你需要尝试对代码进行一些修改，再次运行程序，查看修改后的效果。

1. 如果将第 19 行的 WIDTH, HEIGHT = bext.size() 改为 WIDTH, HEIGHT = 10,5，会发生什么？
2. 如果将第 51 行的 DIR: random.choice(DIRECTIONS) 替换为 DIR: DOWN_RIGHT，则会发生什么？
3. 如何使 "Corner bounces:" 文本不显示在屏幕上？
4. 如果删除或注释掉第 56 行的 cornerBounces = 0，会得到什么错误消息？

项目 6　恺撒密码

恺撒密码是一种古老的加密技术，它通过在字母表中将字母移动一定数量的位置来加密字母。我们将移位的长度称为密钥。例如，如果密钥是 3，则 A 变为 D，B 变为 E，C 变为 F，以此类推。要解密消息，必须反向移动加密的字母。本项目中的程序允许用户根据该算法加密和解密消息。

恺撒密码并不是很复杂，这使它成为初学者的理想选择。项目 7 中的程序可以通过"蛮力"破解所有 26 个可能的密钥来解密消息，即使我们不知道原始密钥。此外，如果你使用密钥 13 加密消息，则恺撒密码变得与项目 61 的相同。

有关恺撒密码的更多信息请参见维基百科的相应词条。如果希望了解一般的密码和密码破解内容，推荐阅读拙作《Python 密码学编程》（人民邮电出版社）。

运行程序

运行 caesarcipher.py，输出如下所示：

```
Caesar Cipher, by Al Sweigart al@inventwithpython.com
Do you want to (e)ncrypt or (d)ecrypt?
> e
Please enter the key (0 to 25) to use.
> 4
Enter the message to encrypt.
> Meet me by the rose bushes tonight.
QIIX QI FC XLI VSWI FYWLIW XSRMKLX.
Full encrypted text copied to clipboard.

Caesar Cipher, by Al Sweigart al@inventwithpython.com
Do you want to (e)ncrypt or (d)ecrypt?
> d
Please enter the key (0 to 26) to use.
> 4
Enter the message to decrypt.
> QIIX QI FC XLI VSWI FYWLIW XSRMKLX.
MEET ME BY THE ROSE BUSHES TONIGHT.
Full decrypted text copied to clipboard.
```

工作原理

与大多数密码程序一样，恺撒密码程序的工作原理是将字符转换为数字，对这些数字执行一些数学运算，然后将数字转换回文本字符。在密码语境下，我们称这些文本字符为符号。符号可以包括字母、数字和标点符号，每个符号将被分配唯一的整数。在恺撒密码程序中，符号都是字母，对应的整数是它们在 SYMBOLS 字符串 'ABCDEFGHIJKLMNOPQRSTUVWXYZ' 中的位置。

```
 1. """恺撒密码，作者: Al Sweigart al@inventwithpython.com
 2. 恺撒密码是一种移位密码，通过加法和减法对字母进行加密和解密
 3. 标签: 简短，初学者，密码学，数学"""
 4.
 5. try:
 6.     import pyperclip  # pyperclip 模块能够将文本复制到剪贴板
 7. except ImportError:
 8.     pass  # pyperclip 不是必需的，不安装也是可以的
 9.
10. # SYMBOLS 是所有可能被加密或解密的符号
11. # (!) 你可以添加数字和标点符号来加密它们
12. SYMBOLS = 'ABCDEFGHIJKLMNOPQRSTUVWXYZ'
13.
14. print('Caesar Cipher, by Al Sweigart al@inventwithpython.com')
15. print('The Caesar cipher encrypts letters by shifting them over by a')
16. print('key number. For example, a key of 2 means the letter A is')
17. print('encrypted into C, the letter B encrypted into D, and so on.')
18. print()
19.
20. # 让玩家选择加密或解密
21. while True:  # 继续询问，直到玩家输入 e 或 d
22.     print('Do you want to (e)ncrypt or (d)ecrypt?')
23.     response = input('> ').lower()
24.     if response.startswith('e'):
25.         mode = 'encrypt'
26.         break
27.     elif response.startswith('d'):
28.         mode = 'decrypt'
29.         break
30.     print('Please enter the letter e or d.')
31.
32. # 让玩家输入要使用的密钥
33. while True:  # 继续询问，直到玩家输入有效的密钥
34.     maxKey = len(SYMBOLS) - 1
35.     print('Please enter the key (0 to {}) to use.'.format(maxKey))
36.     response = input('> ').upper()
37.     if not response.isdecimal():
38.         continue
39.
40.     if 0 <= int(response) < len(SYMBOLS):
41.         key = int(response)
42.         break
43.
44. # 让玩家输入要加密/解密的消息
45. print('Enter the message to {}.'.format(mode))
46. message = input('> ')
47.
48. # 恺撒密码仅适用于大写字母
49. message = message.upper()
50.
```

```
51. # 存储消息的加密/解密形式
52. translated = ''
53.
54. # 加密/解密消息中的每个符号
55. for symbol in message:
56.     if symbol in SYMBOLS:
57.         # 获取此符号的加密/解密的索引 num
58.         num = SYMBOLS.find(symbol)    # 获取符号的索引 num
59.         if mode == 'encrypt':
60.             num = num + key
61.         elif mode == 'decrypt':
62.             num = num - key
63.
64.         # 如果 num 大于 SYMBLOS 的长度或小于 0，则进行环绕处理
65.         if num >= len(SYMBOLS):
66.             num = num - len(SYMBOLS)
67.         elif num < 0:
68.             num = num + len(SYMBOLS)
69.
70.         # 将加密/解密数字的符号添加到 translated 中
71.         translated = translated + SYMBOLS[num]
72.     else:
73.         # 只需添加符号而不加密/解密
74.         translated = translated + symbol
75.
76. # 在屏幕上显示加密/解密的字符串
77. print(translated)
78.
79. try:
80.     pyperclip.copy(translated)
81.     print('Full {}ed text copied to clipboard.'.format(mode))
82. except:
83.     pass    # 如果未安装 pyperclip，则不执行任何操作
```

输入源代码并运行几次后，请尝试对其进行修改。你可以根据标有!的注释对程序进行修改，也可以通过向 SYMBOLS 字符串添加字符来扩展可加密符号。

探索程序

请尝试找出以下问题的答案。你需要尝试对代码进行一些修改，再次运行程序，查看修改后的效果。

1. 如果将第 12 行的 SYMBOLS = 'ABCDEFGHIJKLMNOPQRSTUVWXYZ' 修改为 SYMBOLS = 'ABC'，会发生什么？
2. 如果使用密钥 0 加密消息，会发生什么？
3. 如果删除或注释掉第 52 行的 translated = ''，会得到什么错误消息？
4. 如果删除或注释掉第 41 行的 key = int(response)，会得到什么错误消息？
5. 如果将第 71 行的 translated = translated + SYMBOLS[num] 改为 translated = translated + symbol，会发生什么？

项目 7

恺撒黑客

在本项目中，我们不知道密钥是什么，也可以破解使用项目 6 中恺撒密码加密的消息。恺撒密码只有 26 个可能的密钥，计算机可以轻松尝试所有可能的解密方式并将结果显示给用户。密码学称这种破解技术为"蛮力攻击"。如果希望了解有关密码和密码破解的更多信息，推荐阅读拙作《Python 密码学编程》（人民邮电出版社）。

运行程序

运行 caesarhacker.py，输出如下所示：

```
Caesar Cipher Hacker, by Al Sweigart al@inventwithpython.com
Enter the encrypted Caesar cipher message to hack.
> QIIX QI FC XLI VSWI FYWLIW XSRMKLX.
Key #0: QIIX QI FC XLI VSWI FYWLIW XSRMKLX.
Key #1: PHHW PH EB WKH URVH EXVKHV WRQLJKW.
Key #2: OGGV OG DA VJG TQUG DWUJGU VQPKIJV.
Key #3: NFFU NF CZ UIF SPTF CVTIFT UPOJHIU.
Key #4: MEET ME BY THE ROSE BUSHES TONIGHT.
Key #5: LDDS LD AX SGD QNRD ATRGDR SNMHFGS.
Key #6: KCCR KC ZW RFC PMQC ZSQFCQ RMLGEFR.
--snip--
```

工作原理

注意，本项目中的这个程序，第 20～36 行与恺撒密码程序中第 57～79 行大致相同。黑客程序实现了相同的解密代码，只是这些代码运行在 `for` 循环中，目的是尝试每一个可能的密钥。

遗憾的是，黑客程序还不够高明，无法识别何时找到了正确的密钥，需要依赖人来读取输出并识别哪一种解密方式生成了原始英文（或加密的任何书面语言）。《Python 密码学编程》（人民邮电出版社）一书详细介绍了如何编写 Python 代码来检测英文消息。

```
1. """恺撒黑客, 作者: Al Sweigart al@inventwithpython.com
2. 该程序通过对每个可能的密钥进行蛮力攻击来破解用恺撒密码加密的消息
3. 标签: 小, 初学者, 密码学, 数学"""
4.
5. print('Caesar Cipher Hacker, by Al Sweigart al@inventwithpython.com')
```

```
 6.
 7. # 让玩家输入要解密的消息
 8. print('Enter the encrypted Caesar cipher message to hack.')
 9. message = input('> ')
10.
11. # SYMBOLS 包含所有可能被加密/解密的符号
12. # （这必须与加密消息时使用的 SYMBOLS 匹配）
13. SYMBOLS = 'ABCDEFGHIJKLMNOPQRSTUVWXYZ'
14.
15. for key in range(len(SYMBOLS)):    # 循环遍历每个可能的密钥
16.     translated = ''
17.
18.     # 解密消息中的每个符号
19.     for symbol in message:
20.         if symbol in SYMBOLS:
21.             num = SYMBOLS.find(symbol)    # 获取符号的索引 num
22.             num = num - key    # 解密索引 num
23.
24.             # 如果 num 小于 0，则进行环绕处理
25.             if num < 0:
26.                 num = num + len(SYMBOLS)
27.
28.             # 将解密索引对应的符号添加到 translated 中
29.             translated = translated + SYMBOLS[num]
30.         else:
31.             # 只需添加符号而不解密
32.             translated = translated + symbol
33.
34.     # 显示正在测试的密钥及解密后的文本
35.     print('Key #{}: {}'.format(key, translated))
```

输入源代码并运行几次后，请尝试对其进行更改。记住，存储在 **SYMBOLS** 变量中的字符串必须与生成解密文本的恺撒密码程序中的 **SYMBOLS** 变量相匹配。

探索程序

请尝试找出以下问题的答案。你需要尝试对代码进行一些修改，再次运行程序，查看修改后的效果。

1. 如果删除或注释掉第 16 行的 `translated = ''`，会得到什么错误消息？
2. 如果将第 29 行的 `translated = translated + SYMBOLS[num]` 改为 `translated = translated + symbol`，会发生什么？
3. 如果将未加密的消息输入恺撒黑客程序，会发生什么？

项目 8　日历生成器

在本项目中，我们可以根据输入的年份和月份生成日历文本文件。日期和日历在编程中是一个棘手的问题，因为需要运用很多不同的规则来确定一个月的天数，哪些年份是闰年，以及特定日期属于一周中的哪一天。幸运的是，Python 的 datetime 模块可以帮助我们处理这些内容。本项目中的程序专注于为日历页生成多行字符串。

运行程序

运行 calendarmaker.py，输出如下所示：

```
Calendar Maker, by Al Sweigart al@inventwithpython.com
Enter the year for the calendar:
> 2029
Enter the month for the calendar, 1-12:
> 12
                    December 2029
...Sunday.....Monday....Tuesday...Wednesday...Thursday....Friday....Saturday..
+----------+----------+----------+----------+----------+----------+----------+
|25        |26        |27        |28        |29        |30        |1         |
|          |          |          |          |          |          |          |
|          |          |          |          |          |          |          |
|          |          |          |          |          |          |          |
|          |          |          |          |          |          |          |
+----------+----------+----------+----------+----------+----------+----------+
|2         |3         |4         |5         |6         |7         |8         |
|          |          |          |          |          |          |          |
|          |          |          |          |          |          |          |
|          |          |          |          |          |          |          |
|          |          |          |          |          |          |          |
+----------+----------+----------+----------+----------+----------+----------+
|9         |10        |11        |12        |13        |14        |15        |
|          |          |          |          |          |          |          |
|          |          |          |          |          |          |          |
|          |          |          |          |          |          |          |
|          |          |          |          |          |          |          |
+----------+----------+----------+----------+----------+----------+----------+
|16        |17        |18        |19        |20        |21        |22        |
|          |          |          |          |          |          |          |
|          |          |          |          |          |          |          |
|          |          |          |          |          |          |          |
```

```
+----------+----------+----------+----------+----------+----------+----------+
|23        |24        |25        |26        |27        |28        |29        |
|          |          |          |          |          |          |          |
|          |          |          |          |          |          |          |
|          |          |          |          |          |          |          |
+----------+----------+----------+----------+----------+----------+----------+
|30        |31        | 1        | 2        | 3        | 4        | 5        |
|          |          |          |          |          |          |          |
|          |          |          |          |          |          |          |
|          |          |          |          |          |          |          |
+----------+----------+----------+----------+----------+----------+----------+
Saved to calendar_2029_12.txt
```

工作原理

注意，`getCalendarFor()`函数用于接收年份和月份参数，返回给定月份和年份的庞大、多行月历字符串。该函数使用 `calText` 变量存储这个字符串，其中包含行、空格和日期。为了跟踪日期，设置 `currentDate` 变量包含一个 `datetime.date()`对象，通过为该对象加上或减去 `datetime.timedelta()`对象可以将 `currentDate` 设置为下一个或上一个日期。

请阅读《Python 编程快速上手——让繁琐工作自动化》，以了解 Python 的日期和时间模块等知识。

```python
1. """日历生成器，作者：Al Sweigart al@inventwithpython.com
2. 创建月历，将其保存到文本文件中且使其适合输出
3. 标签：简短"""
4.
5. import datetime
6.
7. # 创建常量
8. DAYS = ('Sunday', 'Monday', 'Tuesday', 'Wednesday', 'Thursday',
9.         'Friday', 'Saturday')
10. MONTHS = ('January', 'February', 'March', 'April', 'May', 'June', 'July',
11.           'August', 'September', 'October', 'November', 'December')
12.
13. print('Calendar Maker, by Al Sweigart al@inventwithpython.com')
14.
15. while True:  # 以循环的方式要求玩家输入一个合适的年份
16.     print('Enter the year for the calendar:')
17.     response = input('> ')
18.
19.     if response.isdecimal() and int(response) > 0:
20.         year = int(response)
21.         break
22.
23.     print('Please enter a numeric year, like 2023.')
24.     continue
25.
26. while True:  # 以循环的方式要求玩家输入一个合适的月份
27.     print('Enter the month for the calendar, 1-12:')
28.     response = input('> ')
29.
30.     if not response.isdecimal():
31.         print('Please enter a numeric month, like 3 for March.')
32.         continue
33.
```

```
 34.         month = int(response)
 35.         if 1 <= month <= 12:
 36.             break
 37.
 38.         print('Please enter a number from 1 to 12.')
 39.
 40.
 41. def getCalendarFor(year, month):
 42.     calText = ''    # calText 变量用于存储日历字符串
 43.
 44.     # 将月份和年份放在日历的顶部
 45.     calText += (' ' * 34) + MONTHS[month - 1] + ' ' + str(year) + '\n'
 46.
 47.     # 将星期标签添加到日历中
 48.     # (!) 尝试将其更改为缩写形式，如 SUN、MON、TUE 等
 49.     calText += '...Sunday.....Monday....Tuesday...Wednesday...Thursday....Friday....Saturday..\n'
 50.
 51.     # 用于分隔"周"的水平线
 52.     weekSeparator = ('+----------' * 7) + '+\n'
 53.
 54.     # 空白行之间有 10 个分隔符
 55.     blankRow = ('|          ' * 7) + '|\n'
 56.
 57.     # 获取当月的第一个日期  (datetime 模块处理
 58.     # 复杂的日历内容)
 59.     currentDate = datetime.date(year, month, 1)
 60.
 61.     # 回滚 currentDate 直到它是星期日。 weekday() 返回 6（而不是 0）
 62.     # 表示是星期日
 63.     while currentDate.weekday() != 6:
 64.         currentDate -= datetime.timedelta(days=1)
 65.
 66.     while True:    # 循环遍历一个月里的每一周
 67.         calText += weekSeparator
 68.
 69.         # dayNumberRow 是带有日期标签的行
 70.         dayNumberRow = ''
 71.         for i in range(7):
 72.             dayNumberLabel = str(currentDate.day).rjust(2)
 73.             dayNumberRow += '|' + dayNumberLabel + (' ' * 8)
 74.             currentDate += datetime.timedelta(days=1) # 循环至下一天
 75.         dayNumberRow += '|\n'    # 在星期六之后添加竖线
 76.
 77.         # 将带有日期标签的行和 3 个空白行添加到日历文本中
 78.         calText += dayNumberRow
 79.         for i in range(3):    # (!) 请试着将 4 更改为 5 或 10
 80.             calText += blankRow
 81.
 82.         # 判断是否完成了一个月的循环
 83.         if currentDate.month != month:
 84.             break
 85.
 86.     # 在日历的最底部添加水平线
 87.     calText += weekSeparator
 88.     return calText
 89.
 90.
 91. calText = getCalendarFor(year, month)
 92. print(calText)  # 显示日历
 93.
 94. # 将日历保存到文本文件中
 95. calendarFilename = 'calendar_{}_{}.txt'.format(year, month)
```

```
96. with open(calendarFilename, 'w') as fileObj:
97.     fileObj.write(calText)
98.
99. print('Saved to ' + calendarFilename)
```

输入代码并运行几次后,请尝试从头开始创建程序,在此过程中不要查看本书中的源代码。你所编写的程序不必与书中的程序完全相同,完全可以开发自己的版本!你也可以尝试弄清楚如何执行以下操作。

❑ 在是节假日的日期框内添加一些文本。
❑ 在一些日期框内添加文本,以标明是特别的日期。
❑ 输出一个没有日期框的"迷你"日历。

探索程序

请尝试找出以下问题的答案。你需要尝试对代码进行一些修改,再次运行程序,查看修改后的效果。

1. 如何让日历显示月份单词的缩写?例如,显示 Jan 而不是 January?
2. 如果删除或注释掉第 20 行的 `year = int(response)`,会得到什么错误消息?
3. 如何让日历不在顶部显示星期几信息?
4. 如何让程序不将日历保存到文本文件中?
5. 如果删除或注释掉第 92 行的 `print(calText)`,会发生什么?

项目9

盒子里的胡萝卜

本项目介绍的是一个简单而滑稽的双人"虚张声势"游戏（biuffing game[1]）。每个玩家各有一个盒子，只有一个盒子里有一根胡萝卜，每个玩家都想得到胡萝卜。第一个玩家先检查盒子，然后告诉第二个玩家自己的盒子里是否有胡萝卜。第二个玩家根据第一个玩家给出的信息，通过推理来决定是否交换盒子。

因为代码中会用字符串表示图像，即ASCII艺术画（ASCII Art[2]），所以输入程序会占用一段时间（复制和粘贴可以加快完成任务的速度），不过这个项目很简单，非常适合初学者，且用到的循环最少，也没有定义函数。

运行程序

运行 carrotinabox.py，输出如下所示：

```
Carrot in a Box, by Al Sweigart al@inventwithpython.com
--snip--
Human player 1, enter your name: Alice
Human player 2, enter your name: Bob
HERE ARE TWO BOXES:

   /        /|   /        /|
  +--------+ |  +--------+ |
  |  RED   | |  |  GOLD  | |
  |  BOX   | /  |  BOX   | /
  +--------+/   +--------+/

Alice Bob
Alice, you have a RED box in front of you.
Bob, you have a GOLD box in front of you.
Press Enter to continue...
--snip--
When Bob has closed their eyes, press Enter...
Alice here is the inside of your box:
```

---

[1] 所谓 biuffing game，即通过欺骗和推理解决虚拟冲突的游戏，如 Mafia、狼人杀等。——编辑注

[2] ASCII Art 是一种用连续排列的 ASCII 字符绘制图形的技术。我们在聊天软件中用到的字符画，字符表情就是 ASCII Art 的表现形式。——编辑注

```
        ___VV____
       |    VV   |
       |    VV   |
       |___||___|       _____
      /    ||   /|     /        /|
     +--------+ |     +--------+ |
     |  RED   | |     |  GOLD  | |
     |  BOX   | /     |  BOX   | /
     +--------+/      +--------+/
     (carrot!)
       Alice             Bob
Press Enter to continue...
```
--snip--

工作原理

以下程序要"遮住"第二个玩家的眼睛，让其看不到第一个玩家盒子里的东西。为了让第二个玩家在第一个玩家看完盒子后看不到盒子里的东西，我们需要找到一种清空屏幕的方法，这就用到了第 82 行的 `print('\n' * 100)`——该代码输出 100 个换行符，让先前输出的内容向上滚动，离开玩家视野，防止第二个玩家意外看到只应该是第一个玩家看到的内容。虽然第二个玩家可以向上滚动内容以查看之前输出的内容，但对于坐在他身旁的第一个玩家来说，这种作弊行为很容易被发现并制止。

在第 112 行、128 行和 140 行代码中，虽然竖线的间距看上去不太准确，但程序会将花括号替换为字符串 RED（末尾有空格）或 GOLD。这些字符串中的 4 个字符会使盒子的其余竖线与 ASCII 艺术画的其余部分对齐。

```
 1. """盒子里的胡萝卜，作者：Al Sweigart al@inventwithpython.com
 2. 双人"虚张声势"游戏，根据英国喜剧智力节目 8 out of 10 Cats 改编
 3. 标签：大、初学者、游戏、双人游戏"""
 4.
 5. import random
 6.
 7. print('''Carrot in a Box, by Al Sweigart al@inventwithpython.com
 8.
 9. This is a bluffing game for two human players. Each player has a box.
10. One box has a carrot in it. To win, you must have the box with the
11. carrot in it.
12.
13. This is a very simple and silly game.
14.
15. The first player looks into their box (the second player must close
16. their eyes during this.) The first player then says "There is a carrot
17. in my box" or "There is not a carrot in my box". The second player then
18. gets to decide if they want to swap boxes or not.
19. ''')
20. input('Press Enter to begin...')
21.
22. p1Name = input('Human player 1, enter your name: ')
23. p2Name = input('Human player 2, enter your name: ')
24. playerNames = p1Name[:11].center(11) + '    ' + p2Name[:11].center(11)
25.
26. print('''HERE ARE TWO BOXES:
```

```
27.    _____       _____
28.   /       /|     /       /|
29.  +--------+ |   +--------+ |
30.  |  RED   | |   |  GOLD  | |
31.  |  BOX   | /   |  BOX   | /
32.  +--------+/    +--------+/''')
33.
34. print()
35. print(playerNames)
36. print()
37. print(p1Name + ', you have a RED box in front of you.')
38. print(p2Name + ', you have a GOLD box in front of you.')
39. print()
40. print(p1Name + ', you will get to look into your box.')
41. print(p2Name.upper() + ', close your eyes and don\'t look!!!')
42. input('When ' + p2Name + ' has closed their eyes, press Enter...')
43. print()
44.
45. print(p1Name + ' here is the inside of your box:')
46.
47. if random.randint(1, 2) == 1:
48.     carrotInFirstBox = True
49. else:
50.     carrotInFirstBox = False
51.
52. if carrotInFirstBox:
53.     print('''
54.    ___VV___
55.   |   VV   |
56.   |   VV   |
57.   |___||___|       _____
58.  /    ||  /|      /       /|
59. +--------+ |    +--------+ |
60. |  RED   | |    |  GOLD  | |
61. |  BOX   | /    |  BOX   | /
62. +--------+/     +--------+/
63.  (carrot!)''')
64.     print(playerNames)
65. else:
66.     print('''
67.    _____
68.   |        |
69.   |        |
70.   |_____|       _____
71.  /        /|      /       /|
72. +--------+ |    +--------+ |
73. |  RED   | |    |  GOLD  | |
74. |  BOX   | /    |  BOX   | /
75. +--------+/     +--------+/
76.  (no carrot!)''')
77.     print(playerNames)
78.
79. input('Press Enter to continue...')
80.
81. print('\n' * 100)   # 通过输出换行符来清空屏幕
82. print(p1Name + ', tell ' + p2Name + ' to open their eyes.')
83. input('Press Enter to continue...')
84.
85. print()
86. print(p1Name + ', say one of the following sentences to ' + p2Name + '.')
87. print('  1) There is a carrot in my box.')
88. print('  2) There is not a carrot in my box.')
89. print()
```

```
 90. input('Then press Enter to continue...')
 91.
 92. print()
 93. print(p2Name + ', do you want to swap boxes with ' + p1Name + '? YES/NO')
 94. while True:
 95.     response = input('> ').upper()
 96.     if not (response.startswith('Y') or response.startswith('N')):
 97.         print(p2Name + ', please enter "YES" or "NO".')
 98.     else:
 99.         break
100.
101. firstBox = 'RED '    # 注意 D 后面的空格
102. secondBox = 'GOLD'
103.
104. if response.startswith('Y'):
105.     carrotInFirstBox = not carrotInFirstBox
106.     firstBox, secondBox = secondBox, firstBox
107.
108. print('''HERE ARE THE TWO BOXES:
109.   _____      _____
110.  /        /|   /        /|
111. +--------+ |  +--------+ |
112. |   {}   | |  |   {}   | |
113. |  BOX   | /  |  BOX   | /
114. +--------+/   +--------+/'''.format(firstBox, secondBox))
115. print(playerNames)
116.
117. input('Press Enter to reveal the winner...')
118. print()
119.
120. if carrotInFirstBox:
121.     print('''
122.     ___VV___      _____
123.    |   VV   |    |        |
124.    |   VV   |    |        |
125.    |___||___|    |_____|
126.   /    ||   /|  /        /|
127. +--------+ |  +--------+ |
128. |   {}   | |  |   {}   | |
129. |  BOX   | /  |  BOX   | /
130. +--------+/   +--------+/'''.format(firstBox, secondBox))
131.
132. else:
133.     print('''
134.     _____      ___VV___
135.    |        |    |   VV   |
136.    |        |    |   VV   |
137.    |_____|    |___||___|
138.   /        /|   /    ||   /|
139. +--------+ |  +--------+ |
140. |   {}   | |  |   {}   | |
141. |  BOX   | /  |  BOX   | /
142. +--------+/   +--------+/'''.format(firstBox, secondBox))
143.
144. print(playerNames)
145.
146. # 最终的输赢由 carrotInFirstBox 变量决定
147. if carrotInFirstBox:
148.     print(p1Name + ' is the winner!')
149. else:
150.     print(p2Name + ' is the winner!')
151.
152. print('Thanks for playing!')
```

输入源代码并运行几次后，请尝试对其进行更改。你也可以尝试以下操作。
- ❏ 将盒子和胡萝卜的 ASCII 艺术画修改得更华丽一些。
- ❏ 添加"再玩一遍？"功能，让玩家在保存得分的同时能够再次玩游戏。
- ❏ 添加第三个玩家，让第二个玩家可以对其"虚张声势"。

探索程序

请尝试找出以下问题的答案。你需要尝试对代码进行一些修改，再次运行程序，查看修改后的效果。

1. 注意，第 24 行中的代码为 p1Name[:11] 和 p2Name[:11]。请输入长度超过 11 个字母的名字，查看程序是如何显示这个名字的。
2. 如果省略第 101 行 firstBox = 'RED ' 中 D 后面的空格会发生什么？
3. 如果删除或注释掉第 81 行的 print('\n' * 100)，会发生什么？
4. 如果删除或注释掉第 98 行的 else:以及第 99 行的 break，会发生什么？

项目 10

cho-han骰子游戏

cho-han是源自日本的一种骰子游戏。一个杯子里有两枚6面骰子，摇动该杯子，玩家必须猜测骰子点数总和是偶数（cho）还是奇数（han）。获胜的玩家需从奖金中抽出一部分，作为"房间费"支付给平台。本项目仅用到了简单的随机数生成方式，以及用于确定点数总和为奇数还是偶数的基本数学知识，难度不大，非常适合初学者。

运行程序

运行 chohan.py，输出如下所示：

```
cho-han, by Al Sweigart al@inventwithpython.com

In this traditional Japanese dice game, two dice are rolled in a bamboo
cup by the dealer sitting on the floor. The player must guess if the
dice total to an even (cho) or odd (han) number.

You have 5000 mon. How much do you bet? (or QUIT)
> 400
The dealer swirls the cup and you hear the rattle of dice.
The dealer slams the cup on the floor, still covering the
dice and asks for your bet.

    cho (even) or han (odd)?
> cho
The dealer lifts the cup to reveal:
  GO - GO
   5 - 5
You won! You take 800 mon.
The house collects a 40 mon fee.
--snip--
```

工作原理

random.randint(1, 6)用于返回一个1～6的随机整数，故非常适合用于表示6面骰子。因为我们还需要显示数字1～6的日语单词，所以使用一个存储在JAPANESE_NUMBERS中的字典，将整数1~6映射到日语单词的字符串，而不是使用一个`if`语句后跟5个`elif`语句的写法。这就是第56行JAPANESE_NUMBERS[dice1]和JAPANESE_NUMBERS[dice2]能直接显示骰子结果对应的日语单词的原因。

```
 1. """cho-han, 作者: Al Sweigart al@inventwithpython.com
 2. 源自日本的一种骰子游戏
 3. 标签: 简短, 初学者, 游戏"""
 4.
 5. import random, sys
 6.
 7. JAPANESE_NUMBERS = {1: 'ICHI', 2: 'NI', 3: 'SAN',
 8.                     4: 'SHI', 5: 'GO', 6: 'ROKU'}
 9.
10. print('''Cho-Han, by Al Sweigart al@inventwithpython.com
11.
12. In this traditional Japanese dice game, two dice are rolled in a bamboo
13. cup by the dealer sitting on the floor. The player must guess if the
14. dice total to an even (cho) or odd (han) number.
15. ''')
16.
17. purse = 5000
18. while True:  # 主循环
19.     # 下注
20.     print('You have', purse, 'mon. How much do you bet? (or QUIT)')
21.     while True:
22.         pot = input('> ')
23.         if pot.upper() == 'QUIT':
24.             print('Thanks for playing!')
25.             sys.exit()
26.         elif not pot.isdecimal():
27.             print('Please enter a number.')
28.         elif int(pot) > purse:
29.             print('You do not have enough to make that bet.')
30.         else:
31.             # 判断为有效的下注
32.             pot = int(pot)  # 将pot转换为整数
33.             break  # 若获得有效的下注，则结束循环
34.
35.     # 掷骰子
36.     dice1 = random.randint(1, 6)
37.     dice2 = random.randint(1, 6)
38.
39.     print('The dealer swirls the cup and you hear the rattle of dice.')
40.     print('The dealer slams the cup on the floor, still covering the')
41.     print('dice and asks for your bet.')
42.     print()
43.     print('    CHO (even) or HAN (odd)?')
44.
45.     # 让玩家下注偶数点或奇数点
46.     while True:
47.         bet = input('> ').upper()
48.         if bet != 'cho' and bet != 'han':
49.             print('Please enter either "CHO" or "HAN".')
50.             continue
51.         else:
52.             break
53.
54.     # 显示掷骰子的结果
55.     print('The dealer lifts the cup to reveal:')
56.     print('  ', JAPANESE_NUMBERS[dice1], '-', JAPANESE_NUMBERS[dice2])
```

```
57.         print('    ', dice1, '-', dice2)
58.
59.         # 判断玩家是否获胜
60.         rollIsEven = (dice1 + dice2) % 2 == 0
61.         if rollIsEven:
62.             correctBet = 'cho'
63.         else:
64.             correctBet = 'han'
65.
66.         playerWon = bet == correctBet
67.
68.         # 显示下注结果
69.         if playerWon:
70.             print('You won! You take', pot, 'mon.')
71.             purse = purse + pot    # 判玩家赢,将投注加入玩家的钱包
72.             print('The house collects a', pot // 10, 'mon fee.')
73.             purse = purse - (pot // 10)    # 平台收取下注的10%("房间费")
74.         else:
75.             purse = purse - pot    # 判玩家输,从其钱包中减去下注的钱
76.             print('You lost!')
77.
78.         # 检查玩家是否花光了钱
79.         if purse == 0:
80.             print('You have run out of money!')
81.             print('Thanks for playing!')
82.             sys.exit()
```

输入源代码并运行几次后,请尝试对其进行更改。你也可以尝试以下操作。

- 实现维基百科文章所述的此游戏的其他版本之一,让多个玩家相互下注。实现添加由计算机控制的玩家参与游戏。
- 为某些点数(例如 7 或 2)添加额外奖金。
- 允许玩家下注特定点数并获得奖金。

探索程序

请尝试找出以下问题的答案。你需要尝试对代码进行一些修改,再次运行程序,查看修改后的效果。

1. 怎样才能让玩家以不同数量的下注开始玩?
2. 程序如何防止玩家下注的数额超过其拥有的钱?
3. 程序如何知道两个骰子点数之和是偶数还是奇数?
4. 如果将第 36 行的 random.randint(1, 6) 改为 random.randint(1, 1),会发生什么?
5. 如果将第 72 行(不是第 73 行)的 pot // 10 改为 0,平台是否还会抽取 10% 的"房间费"?
6. 如果删除或注释掉第 79 行、80 行、81 行和 82 行,会发生什么?

项目 11

诱饵标题生成器

网站一般需要吸引用户浏览广告！但是想出有创意的内容实在是太难了。幸运的是，有了诱饵标题生成器，我们可以让计算机"想出"数百万个博人眼球的标题。这些标题含金量不高且大同小异，但用户似乎对此并不介意。本项目中的程序可以根据Mad Libs风格的模板生成你想要的任意多的标题。

虽然这个程序中有很多标题模板文本，但是代码本身很简单，适合初学者学习。

运行程序

运行 clickbait.py，输出如下所示：

```
Clickbait Headline Generator
By Al Sweigart al@inventwithpython.com

Our website needs to trick people into looking at ads!
Enter the number of clickbait headlines to generate:
> 1000
Big Companies Hate Him! See How This New York Cat Invented a Cheaper Robot
What Telephone Psychics Don't Want You To Know About Avocados
You Won't Believe What This North Carolina Shovel Found in Her Workplace
--snip--
14 Reasons Why Parents Are More Interesting Than You Think (Number 1 Will Surprise You!)
What Robots Don't Want You To Know About Cats
This Florida Telephone Psychic Didn't Think Robots Would Take Her Job. She Was Wrong.
```

工作原理

以下程序使用几个函数来生成不同类型的诱饵标题。每个函数都会从 STATES、NOUNS、PLACES、WHEN 和其他列表中随机获取单词。这些函数在返回字符串之前用 `format()` 方法将获取的单词插入模板字符串。这就像一本"Mad Libs"活动手册，不同的是让计算机来填空，让程序在几秒内生成数千个诱饵标题。

```
1. """诱饵标题生成器，作者：Al Sweigart al@inventwithpython.com
2. 诱饵标题生成器
3. 标签：大，初学者，诙谐，文字"""
4.
5. import random
```

```
 6.
 7. # 创建常量
 8. OBJECT_PRONOUNS = ['Her', 'Him', 'Them']
 9. POSSESIVE_PRONOUNS = ['Her', 'His', 'Their']
10. PERSONAL_PRONOUNS = ['She', 'He', 'They']
11. STATES = ['California', 'Texas', 'Florida', 'New York', 'Pennsylvania',
12.           'Illinois', 'Ohio', 'Georgia', 'North Carolina', 'Michigan']
13. NOUNS = ['Athlete', 'Clown', 'Shovel', 'Paleo Diet', 'Doctor', 'Parent',
14.          'Cat', 'Dog', 'Chicken', 'Robot', 'Video Game', 'Avocado',
15.          'Plastic Straw','Serial Killer', 'Telephone Psychic']
16. PLACES = ['House', 'Attic', 'Bank Deposit Box', 'School', 'Basement',
17.           'Workplace', 'Donut Shop', 'Apocalypse Bunker']
18. WHEN = ['Soon', 'This Year', 'Later Today', 'RIGHT NOW', 'Next Week']
19.
20.
21. def main():
22.     print('Clickbait Headline Generator')
23.     print('By Al Sweigart al@inventwithpython.com')
24.     print()
25.
26.     print('Our website needs to trick people into looking at ads!')
27.     while True:
28.         print('Enter the number of clickbait headlines to generate:')
29.         response = input('> ')
30.         if not response.isdecimal():
31.             print('Please enter a number.')
32.         else:
33.             numberOfHeadlines = int(response)
34.             break   # 输入有效数字后结束循环
35.
36.     for i in range(numberOfHeadlines):
37.         clickbaitType = random.randint(1, 8)
38.
39.         if clickbaitType == 1:
40.             headline = generateAreMillenialsKillingHeadline()
41.         elif clickbaitType == 2:
42.             headline = generateWhatYouDontKnowHeadline()
43.         elif clickbaitType == 3:
44.             headline = generateBigCompaniesHateHerHeadline()
45.         elif clickbaitType == 4:
46.             headline = generateYouWontBelieveHeadline()
47.         elif clickbaitType == 5:
48.             headline = generateDontWantYouToKnowHeadline()
49.         elif clickbaitType == 6:
50.             headline = generateGiftIdeaHeadline()
51.         elif clickbaitType == 7:
52.             headline = generateReasonsWhyHeadline()
53.         elif clickbaitType == 8:
54.             headline = generateJobAutomatedHeadline()
55.
56.         print(headline)
57.     print()
58.
59.     website = random.choice(['wobsite', 'blag', 'Facebuuk', 'Googles',
60.                              'Facesbook', 'Tweedie', 'Pastagram'])
61.     when = random.choice(WHEN).lower()
62.     print('Post these to our', website, when, 'or you\'re fired!')
63.
64.
65. # 每个函数都会返回不同类型的标题
66. def generateAreMillenialsKillingHeadline():
67.     noun = random.choice(NOUNS)
```

```
68.        return 'Are Millenials Killing the {} Industry?'.format(noun)
69.
70.
71.    def generateWhatYouDontKnowHeadline():
72.        noun = random.choice(NOUNS)
73.        pluralNoun = random.choice(NOUNS) + 's'
74.        when = random.choice(WHEN)
75.        return 'Without This {}, {} Could Kill You {}'.format(noun, pluralNoun, when)
76.
77.
78.    def generateBigCompaniesHateHerHeadline():
79.        pronoun = random.choice(OBJECT_PRONOUNS)
80.        state = random.choice(STATES)
81.        noun1 = random.choice(NOUNS)
82.        noun2 = random.choice(NOUNS)
83.        return 'Big Companies Hate {}! See How This {} {} Invented a Cheaper {}'.
84.            format(pronoun, state, noun1, noun2)
85.
86.
87.    def generateYouWontBelieveHeadline():
88.        state = random.choice(STATES)
89.        noun = random.choice(NOUNS)
90.        pronoun = random.choice(POSSESIVE_PRONOUNS)
91.        place = random.choice(PLACES)
92.        return 'You Won\'t Believe What This {} {} Found in {} {}'.format(state, noun,
93.            pronoun, place)
94.
95.    def generateDontWantYouToKnowHeadline():
96.        pluralNoun1 = random.choice(NOUNS) + 's'
97.        pluralNoun2 = random.choice(NOUNS) + 's'
98.        return 'What {} Don\'t Want You To Know About {}'.format(pluralNoun1, pluralNoun2)
99.
100.
101.   def generateGiftIdeaHeadline():
102.        number = random.randint(7, 15)
103.        noun = random.choice(NOUNS)
104.        state = random.choice(STATES)
105.        return '{} Gift Ideas to Give Your {} From {}'.format(number, noun, state)
106.
107.
108.   def generateReasonsWhyHeadline():
109.        number1 = random.randint(3, 19)
110.        pluralNoun = random.choice(NOUNS) + 's'
111.        # number2 不应大于 number1
112.        number2 = random.randint(1, number1)
113.        return '{} Reasons Why {} Are More Interesting Than You Think (Number {} Will Surprise You!)'.
114.            format(number1, pluralNoun, number2)
115.
116.
117.   def generateJobAutomatedHeadline():
118.        state = random.choice(STATES)
119.        noun = random.choice(NOUNS)
120.
121.        i = random.randint(0, 2)
122.        pronoun1 = POSSESIVE_PRONOUNS[i]
123.        pronoun2 = PERSONAL_PRONOUNS[i]
124.        if pronoun1 == 'Their':
125.            return 'This {} {} Didn\'t Think Robots Would Take {} Job. {} Were
126.                Wrong.'.format(state, noun, pronoun1, pronoun2)
127.        else:
128.            return 'This {} {} Didn\'t Think Robots Would Take {} Job. {} Was
129.                Wrong.'.format(state, noun, pronoun1, pronoun2)
```

```
130.
131.
132. # 程序运行入口（如果不是被作为模块导入的话）
133. if __name__ == '__main__':
134.     main()
```

输入源代码并运行几次后，请尝试对其进行修改。你还可以尝试执行以下操作。

❑ 添加其他类型的诱饵标题。

❑ 添加除 NOUNS、STATES 等之外的新单词类别。

探索程序

请尝试找出以下问题的答案。你需要尝试对代码进行一些修改，再次运行程序，查看修改后的效果。

1. 如果删除或注释掉第 33 行的 numberOfHeadlines = int(response)，会得到什么错误消息？
2. 如果将第 33 行的 int(response) 改为 response，会得到什么错误消息？
3. 如果将第 18 行改为 WHEN = []，会得到什么错误消息？

项目 12

Collatz 序列

Collatz 序列，也称为 3n + 1 问题，是很简单的数学题。相应的程序对于初学者来说已经够简单。从起始数字 n 开始，请遵循以下 3 个规则来获得序列中的下一个数。

- 如果 n 是偶数，则下一个数 n = n / 2。
- 如果 n 是奇数，则下一个数 n = 3n + 1。
- 如果 n 为 1，则停止计算；否则，重复计算。

人们普遍认为（只是至今还没有相应的数学证明），从任意一个正整数开始，按照以上规则计算最终都能得到 1。

运行程序

运行 collatz.py，输出如下所示：

```
Collatz Sequence, or, the 3n + 1 Problem
By Al Sweigart al@inventwithpython.com

The Collatz sequence is a sequence of numbers produced from a starting
number n, following three rules:
--snip--
Enter a starting number (greater than 0) or QUIT:
> 26
26, 13, 40, 20, 10, 5, 16, 8, 4, 2, 1

Collatz Sequence, or, the 3n + 1 Problem
By Al Sweigart al@inventwithpython.com
--snip--
Enter a starting number (greater than 0) or QUIT:
> 27
27, 82, 41, 124, 62, 31, 94, 47, 142, 71, 214, 107, 322, 161, 484, 242, 121,
364, 182, 91, 274, 137, 412, 206, 103, 310, 155, 466, 233, 700, 350, 175, 526,
263, 790, 395, 1186, 593, 1780, 890, 445, 1336, 668, 334, 167, 502, 251, 754,
377, 1132, 566, 283, 850, 425, 1276, 638, 319, 958, 479, 1438, 719, 2158,
1079, 3238, 1619, 4858, 2429, 7288, 3644, 1822, 911, 2734, 1367, 4102, 2051,
6154, 3077, 9232, 4616, 2308, 1154, 577, 1732, 866, 433, 1300, 650, 325, 976,
488, 244, 122, 61, 184, 92, 46, 23, 70, 35, 106, 53, 160, 80, 40, 20, 10, 5,
16, 8, 4, 2, 1
```

工作原理

取模运算符 % 可以用于确定一个数是偶数还是奇数。记住，此运算符是一种"求余数"运算符。比如，23 除以 7 商为 3 余数为 2，23 对 7 取模则结果为 2。偶数除以 2 没有余数，奇数除以 2 的余数为 1。当 n 为偶数时，第 31 行 `if n % 2 == 0:` 的计算结果为 `True`；当 n 为奇数时，其计算结果为 `False`。

```
1.  """Collatz 序列, 作者: Al Sweigart al@inventwithpython.com
2.  给定起始编号, 为 Collatz 序列生成编号
3.  标签: 小, 初学者, 数学"""
4.
5.  import sys, time
6.
7.  print('''Collatz Sequence, or, the 3n + 1 Problem
8.  By Al Sweigart al@inventwithpython.com
9.
10. The Collatz sequence is a sequence of numbers produced from a starting
11. number n, following three rules:
12.
13. 1) If n is even, the next number n is n / 2.
14. 2) If n is odd, the next number n is n * 3 + 1.
15. 3) If n is 1, stop. Otherwise, repeat.
16.
17. It is generally thought, but so far not mathematically proven, that
18. every starting number eventually terminates at 1.
19. ''')
20.
21. print('Enter a starting number (greater than 0) or QUIT:')
22. response = input('> ')
23.
24. if not response.isdecimal() or response == '0':
25.     print('You must enter an integer greater than 0.')
26.     sys.exit()
27.
28. n = int(response)
29. print(n, end='', flush=True)
30. while n != 1:
31.     if n % 2 == 0:   # 如果n是偶数……
32.         n = n // 2
33.     else:  # 如果n是奇数……
34.         n = 3 * n + 1
35.
36.     print(', ' + str(n), end='', flush=True)
37.     time.sleep(0.1)
38. print()
```

探索程序

请尝试找出以下问题的答案。你需要尝试对代码进行一些修改，再次运行程序，查看修改后的效果。

1. 起始数字为 32 的 Collatz 序列中有多少个数字？
2. 起始数字为 33 的 Collatz 序列中有多少个数字？
3. 起始数字是 2 的幂（2、4、8、16、32、64、128 等）的 Collatz 序列是否总是只由偶数（除了最后的 1）组成？
4. 如果输入 0 并将其作为起始整数，会发生什么？

项目 13

康威生命游戏

康威生命游戏是一种元胞自动机模拟，会按照简单的规则来创建有趣的模式。它由数学家 John Conway 于 1970 年提出，并由 Martin Gardner 在《科学美国人》(*Scientific American*)的"数学游戏"专栏中推广。时至今日，它仍深受程序员和计算机科学家的喜爱，尽管它更像一种有趣的可视化效果而非真正的"游戏"。二维面板有"单元格"，每个单元格都遵循以下 3 条简单的规则。

❏ 具有 2 个或 3 个邻居细胞的活细胞在下一步模拟中保持活动状态。
❏ 在下一步模拟中，恰好有 3 个邻居细胞的死细胞变成活细胞。
❏ 任何其他细胞在下一步模拟中死亡或保持死亡状态。

细胞在下一步模拟中的存活或死亡状态完全取决于其当前状态。细胞不会"记住"其他之前的状态。关于这些简单规则产生的模式的研究已有很多。不幸的是，Conway 教授于 2020 年 4 月因 COVID-19 并发症去世。

运行程序

运行 conwaysgameoflife.py，输出如下所示：

```
                        0                       0              00          0 0
0          0          0 0                       0                       0 0000        0 0
00         0          0 0                       0                       0             00
00                      0   0                     00                                  00
00                    00                         0  0                                 00
                                                 00                                   0 00
                        000                     00   0 0                              0
                                                  000
                                 0                   0                         0 0
                              00          00  00        00   0
                              0000        0   00        0000     0 0
                                 0  00        0   00        00    0   00
                                 0   0        0    0       0   00    0 000
                                     0              0000        00   00000 0
00                               0                  0  000   0 000        0000      0
```

工作原理

单元格的状态存储在 cells 和 nextCells 字典变量中。两个字典都以(x, y)元组表示键，其中 x 和 y 是整数。'O'表示活细胞，''表示死细胞。第 38 行～第 42 行代码用于将这些字典的表示输出到屏幕上。cells 变量的字典表示单元格的当前状态，而 nextCells 存储下一步模拟中单元格的字典。

```
 1. """康威生命游戏，作者：Al Sweigart al@inventwithpython.com
 2. 经典的元胞自动机模拟。 按 Ctrl-C 停止
 3. 标签：简短，艺术，仿真"""
 4.
 5. import copy, random, sys, time
 6.
 7. # 创建常量
 8. WIDTH = 79   # 单元格的宽度
 9. HEIGHT = 20  # 单元格的高度
10.
11. # (!) 尝试将 ALIVE 更改为 # 或其他字符
12. ALIVE = 'O'  # 代表活细胞的字符
13. # (!) 尝试将 DEAD 更改为 . 或其他字符
14. DEAD = ' '   # 代表死细胞
15.
16. # (!) 尝试将 ALIVE 更改为 "|"、DEAD 更改为-
17.
18. # cells 和 nextCells 是表示游戏中的状态的字典
19. # 它们的键是 (x, y) 元组，它们的值是 ALIVE 或 DEAD
20. nextCells = {}
21. # 将随机死细胞和活细胞放入 nextCells
22. for x in range(WIDTH):   # 循环遍历每个可能的列
23.     for y in range(HEIGHT):   # 循环遍历每个可能的行
24.         # 细胞存活或死亡的概率各占 50%
25.         if random.randint(0, 1) == 0:
26.             nextCells[(x, y)] = ALIVE   # 添加一个活细胞
27.         else:
28.             nextCells[(x, y)] = DEAD    # 添加一个死细胞
29.
30. while True:  # 主循环
31.     # 该循环的每次迭代都是模拟的一个步骤
32.
33.     print('\n' * 50)   # 用换行符分隔每个步骤
34.     cells = copy.deepcopy(nextCells)
35.
36.     # 在屏幕上输出单元格
37.     for y in range(HEIGHT):
38.         for x in range(WIDTH):
39.             print(cells[(x, y)], end='')   # 输出#或空格
40.         print()   # 在行尾输出添加一行空行
41.     print('Press Ctrl-C to quit.')
42.
43.     # 根据当前步骤的单元格计算下一步的单元格
44.     for x in range(WIDTH):
45.         for y in range(HEIGHT):
46.             # 获取 (x, y) 的相邻坐标，即使它们环绕在边缘
47.             left  = (x - 1) % WIDTH
48.             right = (x + 1) % WIDTH
49.             above = (y - 1) % HEIGHT
50.             below = (y + 1) % HEIGHT
51.
52.             # 计算四周处于存活状态的细胞的数量
53.             numNeighbors = 0
54.             if cells[(left, above)] == ALIVE:
55.                 numNeighbors += 1   # 左上角存在活细胞
56.             if cells[(x, above)] == ALIVE:
```

```
57.            numNeighbors += 1  # 正上方存在活细胞
58.         if cells[(right, above)] == ALIVE:
59.            numNeighbors += 1  # 右上角存在活细胞
60.         if cells[(left, y)] == ALIVE:
61.            numNeighbors += 1  # 左边存在活细胞
62.         if cells[(right, y)] == ALIVE:
63.            numNeighbors += 1  # 右边存在活细胞
64.         if cells[(left, below)] == ALIVE:
65.            numNeighbors += 1  # 左下角存在活细胞
66.         if cells[(x, below)] == ALIVE:
67.            numNeighbors += 1  # 正下方存在活细胞
68.         if cells[(right, below)] == ALIVE:
69.            numNeighbors += 1  # 右下角存在活细胞
70.
71.         # 根据康威生命游戏的规则设置单元格
72.         if cells[(x, y)] == ALIVE and (numNeighbors == 2
73.            or numNeighbors == 3):
74.             # 有 2 个或 3 个邻居的活细胞保持活动状态
75.             nextCells[(x, y)] = ALIVE
76.         elif cells[(x, y)] == DEAD and numNeighbors == 3:
77.             # 有 3 个邻居的死细胞变成活细胞
78.             nextCells[(x, y)] = ALIVE
79.         else:
80.             # 其他一切情况细胞都会死亡状态或保持死亡状态
81.             nextCells[(x, y)] = DEAD
82.
83.    try:
84.        time.sleep(1)  # 添加 1 秒暂停以减少闪烁
85.    except KeyboardInterrupt:
86.        print("Conway's Game of Life")
87.        print('By Al Sweigart al@inventwithpython.com')
88.        sys.exit()  # 按 Ctrl-C, 程序结束
```

输入源代码并运行几次后，请尝试对其进行更改。你可以参考标有!的注释对上述程序进行修改，也可以尝试以下操作。

❏ 调整开始时活细胞的百分比，而非总是使用 50%。
❏ 添加从文本文件读取初始状态的功能，以便用户可以手动编辑起始单元格的状态。

探索程序

请尝试找出以下问题的答案。你需要尝试对代码进行一些修改，再次运行程序，查看修改后的效果。

1. 如果将第 8 行的 WIDTH = 79 更改为 WIDTH = 7，会发生什么？
2. 如果删除或注释掉第 33 行的 print('\n' * 50)，会发生什么？
3. 如果将第 25 行的 random.randint(0, 1) 改为 random.randint(0, 10)，会发生什么？
4. 如果将第 81 行的 nextCells[(x, y)] = DEAD 改为 nextCells[(x, y)] = ALIVE，会发生什么？

项目 14　倒计时

本项目实现一个倒计时到零的数字计时器。这里不是直接渲染数字字符，而是使用项目 64 7 段显示模块中的 sevseg.py 模块为每个数字生成图形。请务必先创建该模块文件，这样倒计时程序才能运行。然后，你可以将倒计时器设置为自己喜欢的时、分、秒形式。这个程序类似于项目 19 数字时钟。

运行程序

运行 countdown.py，输出如下所示：

```
 _  _     _  _      _ 
| || | * | || | *  _| |
|_||_| * |_||_| * |_  _|

Press Ctrl-C to quit.
```

工作原理

运行 import sevseg 后，我们就可以调用 sevseg.getSevSegStr() 函数来获取 7 段数字的多行字符串。注意，倒计时程序需要在时、分、秒之间显示由星号组成的冒号。这就需要使用 splitlines() 方法将这些数字的多行字符串的 3 行拆分为 3 个单独的字符串。

```
 1. """倒计时，作者：Al Sweigart al@inventwithpython.com
 2. 使用 7 段显示模块显示倒计时动画
 3. 按 Ctrl-C 停止
 4. 要求 sevseg.py 位于同一文件夹中
 5. 标签：　小，艺术"""
 6.
 7. import sys, time
 8. import sevseg  # 导入 sevseg.py 程序
 9.
10. # (!) 将其更改为任意秒数
11. secondsLeft = 30
12.
13. try:
14.     while True:  # 主循环
```

```
15.        # 通过输出几个换行符来清空屏幕
16.        print('\n' * 60)
17.
18.        # 根据 secondsLeft 设定的值获取小时/分钟/秒
19.        # 例如: 7265 表示的是 2 小时 1 分 5 秒
20.        # 所以 7265 // 3600 表示的是 2 小时
21.        hours = str(secondsLeft // 3600)
22.        # 7265 % 3600 结果是 65, 而 65 // 60 结果是 1, 即 1 分钟
23.        minutes = str((secondsLeft % 3600) // 60)
24.        # 7265 % 60 结果是 5, 即 5s
25.        seconds = str(secondsLeft % 60)
26.
27.        # 利用 sevseg 模块获取数字字符串
28.        hDigits = sevseg.getSevSegStr(hours, 2)
29.        hTopRow, hMiddleRow, hBottomRow = hDigits.splitlines()
30.
31.        mDigits = sevseg.getSevSegStr(minutes, 2)
32.        mTopRow, mMiddleRow, mBottomRow = mDigits.splitlines()
33.
34.        sDigits = sevseg.getSevSegStr(seconds, 2)
35.        sTopRow, sMiddleRow, sBottomRow = sDigits.splitlines()
36.
37.        # 显示数字
38.        print(hTopRow    + '     ' + mTopRow    + '     ' + sTopRow)
39.        print(hMiddleRow + '  *  ' + mMiddleRow + '  *  ' + sMiddleRow)
40.        print(hBottomRow + '  *  ' + mBottomRow + '  *  ' + sBottomRow)
41.
42.        if secondsLeft == 0:
43.            print()
44.            print('    * * * * BOOM * * * *')
45.            break
46.
47.        print()
48.        print('Press Ctrl-C to quit.')
49.
50.        time.sleep(1)   # 插入 1s 的停顿
51.        secondsLeft -= 1
52. except KeyboardInterrupt:
53.     print('Countdown, by Al Sweigart al@inventwithpython.com')
54.     sys.exit()   # 按 Ctrl-C, 程序结束
```

输入源代码并运行几次后,请尝试对其进行更改。你也可以尝试以下操作。

❏ 提示用户输入开始倒计时的时间。

❏ 让用户输入在倒计时结束时显示的消息。

探索程序

请尝试找出以下问题的答案。你需要尝试对代码进行一些修改,再次运行程序,查看修改后的效果。

1. 如果将第 11 行的 `secondsLeft = 30` 改为 `secondsLeft = 30.5`,会发生什么?
2. 如果将第 28、31 和 34 行的 2 改为 1,会发生什么?
3. 如果将第 50 行的 `time.sleep(1)` 改为 `time.sleep(0.1)`,会发生什么?
4. 如果将第 51 行的 `secondsLeft -= 1` 改为 `secondsLeft -= 2`,会发生什么?
5. 如果删除或注释掉第 16 行的 `print('\n' * 60)`,会发生什么?
6. 如果删除或注释掉第 8 行的 `import sevseg`,会得到什么错误消息?

项目 15 地穴冒险

本项目程序的运行效果就像不断向地心深处探索的动画。尽管程序很短，但是能利用计算机屏幕内容的滚动特性来生成有趣且"无休止"的可视化效果，这可以证明不需要太多代码即可生成有趣的内容。本项目类似于项目 58。

运行程序

运行 deepcave.py，输出如下所示：

```
Deep Cave, by Al Sweigart al@inventwithpython.com
Press Ctrl-C to stop.
###################          ##########################################
###################          ##########################################
###################          ##########################################
###################          ##########################################
####################         ##########################################
#####################         #########################################
####################          #########################################
###################           #########################################
##################            #########################################
--snip--
```

工作原理

本项目利用了"输出新行会导致前一行在屏幕上上移"这一特性。通过在每一行输出略微不同的间隙，程序实现了滚动动画。观察者会觉得它好像在向下移动。

左侧的 `#` 字符数由 `leftWidth` 变量跟踪。中间的空格数由 `gapWidth` 变量跟踪。右侧的 `#` 字符数由 `WIDTH - gapWidth - leftWidth` 计算得出，以确保每条线始终保持相同的宽度。

```
1. """地穴冒险，作者：Al Sweigart al@inventwithpython.com
2. 一个不断"深入地球"的动画
3. 标签：小，初学者，动画，艺术"""
4.
5.
6. import random, sys, time
```

```
 7.
 8.    # 创建常量
 9.    WIDTH = 70    # (!) 尝试将其更改为 10 或 30
10.    PAUSE_AMOUNT = 0.05    # (!) 尝试将其更改为 0 或 1.0
11.
12.    print('Deep Cave, by Al Sweigart al@inventwithpython.com')
13.    print('Press Ctrl-C to stop.')
14.    time.sleep(2)
15.
16.    leftWidth = 20
17.    gapWidth = 10
18.
19.    while True:
20.        # 显示深洞
21.        rightWidth = WIDTH - gapWidth - leftWidth
22.        print(('#' * leftWidth) + (' ' * gapWidth) + ('#' * rightWidth))
23.
24.        # 在短暂的暂停期间检查是否按了 Ctrl-C
25.        try:
26.            time.sleep(PAUSE_AMOUNT)
27.        except KeyboardInterrupt:
28.            sys.exit()    # 按 Ctrl-C，程序结束
29.
30.        # 调整左侧宽度
31.        diceRoll = random.randint(1, 6)
32.        if diceRoll == 1 and leftWidth > 1:
33.            leftWidth = leftWidth - 1    # 减小左侧宽度
34.        elif diceRoll == 2 and leftWidth + gapWidth < WIDTH - 1:
35.            leftWidth = leftWidth + 1    # 增大左侧宽度
36.        else:
37.            pass    # 不做调整，左侧宽度没有变化
38.
39.        # 调整间隙宽度
40.        # (!) 尝试取消注释以下所有代码
41.        #diceRoll = random.randint(1, 6)
42.        #if diceRoll == 1 and gapWidth > 1:
43.        #    gapWidth = gapWidth - 1    # 减小间隙宽度
44.        #elif diceRoll == 2 and leftWidth + gapWidth < WIDTH - 1:
45.        #    gapWidth = gapWidth + 1    # 增大间隙宽度
46.        #else:
47.        #    pass    # 不做调整，间隙宽度没有变化
```

输入源代码并运行几次后，请尝试对其进行修改。你可以按照标有!的注释对代码进行修改。

探索程序

请尝试找出以下问题的答案。你需要尝试对代码进行一些修改，再次运行程序，查看修改后的效果。

1. 如果将第 22 行的(' ' * gapWidth)改为('.' * gapWidth)，会发生什么？
2. 如果将第 31 行的 random.randint(1, 6)改为 random.randint(1, 1)，会发生什么？
3. 如果将第 31 行的 random.randint(1, 6)改为 random.randint(2, 2)，会发生什么？
4. 如果删除或注释掉第 16 行的 leftWidth = 20，会得到什么错误消息？
5. 如果将第 9 行的 WIDTH = 70 改为 WIDTH = -70，会发生什么？
6. 如果将第 10 行的 PAUSE_AMOUNT = 0.05 改为 PAUSE_AMOUNT = -0.05，会得到什么错误消息？

项目 16

钻石

本项目将绘制各种尺寸的用ASCII字符串表示的"钻石"(菱形)图案,可以用于绘制指定尺寸的钻石轮廓或具有填充样式的钻石。实现这些功能对初学者来说是非常好的练习。请努力去理解随着尺寸的增加钻石绘画背后的模式。

运行程序

运行diamonds.py,输出如下所示:

```
Diamonds, by Al Sweigart al@inventwithpython.com
/\
\/

 /\
 \/

  /\
 /  \
 \  /
  \/

   /\
  //\\
  \\//
   \/

    /\
   /  \
  /    \
  \    /
   \  /
    \/

     /\
    //\\
   ///\\\
   \\\///
    \\//
     \/
--snip--
```

工作原理

创建这个程序的一个有效方法是先在编辑器中"绘制"几种尺寸的钻石，然后发现钻石变大过程中遵循的规律。此方法会让你意识到，钻石图案轮廓的每一行都包含 4 个部分：前导空格数、外部正斜线数、内部空格数和外部反斜线数；填充样式的钻石图案包含几个内部正斜线和反斜线，而不是内部空间。了解了上述模式，你就可以完成编写 diamonds.py 程序。

```
 1. """钻石，作者：Al Sweigart al@inventwithpython.com
 2. 绘制各种尺寸的钻石图案
 3.                           /\      /\
 4.                          /  \    //\\
 5.            /\     /\    /    \  ///\\\
 6.           /  \   //\\  /      \ ////\\\\
 7.   /\     /    \ ///\\\ \      / \\\\////
 8.  /  \   //\\\   \\\///  \    /   \\\///
 9.  \  /   \\//     \\//    \  /     \\//
10.   \/     \/       \/      \/       \/
11. 标签：小，初学者，艺术"""
12.
13. def main():
14.     print('Diamonds, by Al Sweigart al@inventwithpython.com')
15.
16.     # 显示大小为 0 到 6 的钻石图案
17.     for diamondSize in range(0, 6):
18.         displayOutlineDiamond(diamondSize)
19.         print()  # 输出一个空行
20.         displayFilledDiamond(diamondSize)
21.         print()  # 输出一个空行
22.
23.
24. def displayOutlineDiamond(size):
25.     # 显示钻石图案的上半部分
26.     for i in range(size):
27.         print(' ' * (size - i - 1), end='')  # 左侧空间
28.         print('/', end='')  # 左半部分
29.         print(' ' * (i * 2), end='')  # 内部空间
30.         print('\\')  # 右半部分
31.
32.     # 显示钻石图案的下半部分
33.     for i in range(size):
34.         print(' ' * i, end='')  # 左侧空格
35.         print('\\', end='')  # 左半部分
36.         print(' ' * ((size - i - 1) * 2), end='')  # 内部空格
37.         print('/')  # 右侧半部分
38.
39.
40. def displayFilledDiamond(size):
41.     # 显示钻石图案的上半部分
42.     for i in range(size):
43.         print(' ' * (size - i - 1), end='')  # 左侧空间
44.         print('/' * (i + 1), end='')  # 左半部分
45.         print('\\' * (i + 1))  # 右半部分
46.
47.     # 显示钻石图案的下半部分
48.     for i in range(size):
49.         print(' ' * i, end='')  # 左侧空间
50.         print('\\' * (size - i), end='')  # 左半部分
51.         print('/' * (size - i))  # 右半部分
52.
```

```
53.
54. # 程序运行入口（如果不是被作为模块导入的话）
55. if __name__ == '__main__':
56.     main()
```

输入源代码并运行几次后，请尝试对其进行更改。你也可以尝试以下操作。

❏ 创建其他形状，例如三角形、矩形等。

❏ 将形状输出到文本文件而不是屏幕。

探索程序

请尝试找出以下问题的答案。你需要尝试对代码进行一些修改，再次运行程序，查看修改后的效果。

1. 如果将第 30 行的 print('\\') 改为 print('@')，会发生什么？
2. 如果将第 29 行的 print(' ' * (i * 2), end='') 改为 print('@' * (i * 2), end='')，会发生什么？
3. 如果将第 17 行的 range(0, 6) 改为 range(0, 30)，会发生什么？
4. 如果删除或注释掉第 33 行或第 48 行的 for i in range(size):，会发生什么？

项目 17

骰子数学

本项目将实现一个数学测验程序：掷出 2～6 枚骰子，快速算出这些骰子的点数总和。这个程序不仅会自动闪现卡片，还会将骰子的面绘制到屏幕的随机位置。ASCII 艺术画为我们练习算术方面的程序开发增添了几分乐趣。

运行程序

运行 dicemath.py，输出如下所示：

```
Dice Math, by Al Sweigart al@inventwithpython.com

Add up the sides of all the dice displayed on the screen. You have
30 seconds to answer as many as possible. You get 4 points for each
correct answer and lose 1 point for each incorrect answer.

Press Enter to begin...
```

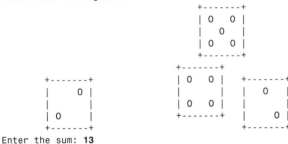

```
Enter the sum: 13
--snip--
```

工作原理

屏幕上的"骰子"由存储在 `canvas` 变量中的字典表示。在 Python 中，元组类似于列表，但它们的内容不能改变。这个字典的键是(x, y)元组，标记了骰子左上角的位置，而其值是 `ALL_DICE` 中的"骰子元组"之一。如第 28 行～80 行代码所示，每个骰子元组包含一个字符串列表，以图形方式表示一个可能的骰子面和一个表示骰子面上有多少点数的整数。程序将据此显示骰子并计算点数的总和。

第 173 行到第 176 行代码类似于项目 13，以单元格的方式在屏幕上显示 canvas 字典中的数据。

```
 1. """骰子数学，作者：Al Sweigart al@inventwithpython.com
 2. 一种闪现卡片加法游戏，你可以在其中计算随机骰子的总和
 3. 标签：大，艺术，游戏，数学"""
 4.
 5. import random, time
 6.
 7. # 创建常量
 8. DICE_WIDTH = 9
 9. DICE_HEIGHT = 5
10. CANVAS_WIDTH = 79
11. CANVAS_HEIGHT = 24 - 3  # 减 3 是为了留出在底部输入总和的空间
12.
13. # 持续时间以秒为单位
14. QUIZ_DURATION = 30   # (!) 尝试将 30 更改为 10 或 60
15. MIN_DICE = 2  # (!) 尝试将 2 更改为 1 或 5
16. MAX_DICE = 6  # (!) 尝试将 6 更改为 14
17.
18. # (!) 尝试将下面的 4 和 1 更改为不同的数字
19. REWARD = 4   # (!) 回答正确可得积分
20. PENALTY = 1  # (!) 回答错误扣除积分
21. # (!) 尝试将 PENALTY 设置为负数以给出错误答案的分数
22.
23. # 如果骰子都无法显示在屏幕上，程序就会挂起
24. assert MAX_DICE <= 14
25.
26. D1 = (['+-------+',
27.        '|       |',
28.        '|   O   |',
29.        '|       |',
30.        '+-------+'], 1)
31.
32. D2a = (['+-------+',
33.         '| O     |',
34.         '|       |',
35.         '|     O |',
36.         '+-------+'], 2)
37.
38. D2b = (['+-------+',
39.         '|     O |',
40.         '|       |',
41.         '| O     |',
42.         '+-------+'], 2)
43.
44. D3a = (['+-------+',
45.         '| O     |',
46.         '|   O   |',
47.         '|     O |',
48.         '+-------+'], 3)
49.
50. D3b = (['+-------+',
51.         '|     O |',
52.         '|   O   |',
53.         '| O     |',
54.         '+-------+'], 3)
55.
```

```
 56. D4 = (['+------+',
 57.        '| O  O |',
 58.        '|      |',
 59.        '| O  O |',
 60.        '+------+'], 4)
 61.
 62. D5 = (['+------+',
 63.        '| O  O |',
 64.        '|   O  |',
 65.        '| O  O |',
 66.        '+------+'], 5)
 67.
 68. D6a = (['+------+',
 69.         '| O  O |',
 70.         '| O  O |',
 71.         '| O  O |',
 72.         '+------+'], 6)
 73.
 74. D6b = (['+------+',
 75.         '| O O O |',
 76.         '|      |',
 77.         '| O O O |',
 78.         '+------+'], 6)
 79.
 80. ALL_DICE = [D1, D2a, D2b, D3a, D3b, D4, D5, D6a, D6b]
 81.
 82. print('''Dice Math, by Al Sweigart al@inventwithpython.com
 83.
 84. Add up the sides of all the dice displayed on the screen. You have
 85. {} seconds to answer as many as possible. You get {} points for each
 86. correct answer and lose {} point for each incorrect answer.
 87. '''.format(QUIZ_DURATION, REWARD, PENALTY))
 88. input('Press Enter to begin...')
 89.
 90. # 记录正确答案和错误答案的数量
 91. correctAnswers = 0
 92. incorrectAnswers = 0
 93. startTime = time.time()
 94. while time.time() < startTime + QUIZ_DURATION:   # 主循环
 95.     # 选择要显示的骰子
 96.     sumAnswer = 0
 97.     diceFaces = []
 98.     for i in range(random.randint(MIN_DICE, MAX_DICE)):
 99.         die = random.choice(ALL_DICE)
100.         # die[0]是骰子一个面的字符串列表
101.         diceFaces.append(die[0])
102.         # die[1]是骰子一个面上的点数
103.         sumAnswer += die[1]
104.
105.     # 包含每个骰子左上角的 (x, y) 元组
106.     topLeftDiceCorners = []
107.
108.     # 计算出骰子的位置
109.     for i in range(len(diceFaces)):
110.         while True:
111.             # 在画布上随机找一个地方放置骰子
112.             left = random.randint(0, CANVAS_WIDTH  - 1 - DICE_WIDTH)
113.             top  = random.randint(0, CANVAS_HEIGHT - 1 - DICE_HEIGHT)
114.
115.             # 获取 4 个角的（x, y）坐标
116.             #       left
117.             #        v
```

```
118.        #top  > +-------+ ^
119.        #     | O     | |
120.        #     |   O   | | DICE_HEIGHT (5)
121.        #     |     O | |
122.        #     +-------+ v
123.        #     <------->
124.        #      DICE_WIDTH (9)
125.        topLeftX = left
126.        topLeftY = top
127.        topRightX = left + DICE_WIDTH
128.        topRightY = top
129.        bottomLeftX = left
130.        bottomLeftY = top + DICE_HEIGHT
131.        bottomRightX = left + DICE_WIDTH
132.        bottomRightY = top + DICE_HEIGHT
133.
134.        # 检查此骰子是否与之前的骰子重叠
135.        overlaps = False
136.        for prevDieLeft, prevDieTop in topLeftDiceCorners:
137.            prevDieRight = prevDieLeft + DICE_WIDTH
138.            prevDieBottom = prevDieTop + DICE_HEIGHT
139.            # 检查这个骰子的每个角，看看它是否在前一个骰子的区域里
140.            for cornerX, cornerY in ((topLeftX, topLeftY),
141.                                     (topRightX, topRightY),
142.                                     (bottomLeftX, bottomLeftY),
143.                                     (bottomRightX, bottomRightY)):
144.                if (prevDieLeft <= cornerX < prevDieRight
145.                    and prevDieTop <= cornerY < prevDieBottom):
146.                    overlaps = True
147.        if not overlaps:
148.            # 它与之前的骰子不重叠，则将骰子放在当前位置
149.            topLeftDiceCorners.append((left, top))
150.            break
151.
152. # 在画布上绘制骰子
153.
154. # 键是整数的 (x, y) 元组，值是该字符在画布上的位置
155. canvas = {}
156. # 循环遍历每个骰子
157. for i, (dieLeft, dieTop) in enumerate(topLeftDiceCorners):
158.     # 循环遍历骰子每个面上的点数
159.     dieFace = diceFaces[i]
160.     for dx in range(DICE_WIDTH):
161.         for dy in range(DICE_HEIGHT):
162.             # 将此字符复制到画布上的正确位置
163.             canvasX = dieLeft + dx
164.             canvasY = dieTop + dy
165.             # 注意在 dieFace 字符串列表中，x 和 y 被交换
166.             canvas[(canvasX, canvasY)] = dieFace[dy][dx]
167.
168. # 在屏幕上显示画布
169. for cy in range(CANVAS_HEIGHT):
170.     for cx in range(CANVAS_WIDTH):
171.         print(canvas.get((cx, cy), ' '), end='')
172.     print()  # 输出一行空行
173.
174. # 让玩家输入答案
175. response = input('Enter the sum: ').strip()
176. if response.isdecimal() and int(response) == sumAnswer:
177.     correctAnswers += 1
178. else:
179.     print('Incorrect, the answer is', sumAnswer)
```

```
180.            time.sleep(2)
181.            incorrectAnswers += 1
182.
183. # 显示最终得分
184. score = (correctAnswers * REWARD) - (incorrectAnswers * PENALTY)
185. print('Correct:  ', correctAnswers)
186. print('Incorrect:', incorrectAnswers)
187. print('Score:    ', score)
```

输入源代码并运行几次后，请尝试对其进行更改。你可以根据标有!的注释对程序进行修改。你也可以自己尝试弄清楚如何执行以下操作。

- ❏ 重新设计 ASCII 艺术骰子面。
- ❏ 添加具有 7、8 或 9 个点的骰子面。

探索程序

请尝试找出以下问题的答案。你需要尝试对代码进行一些修改，再次运行程序，查看修改后的效果。

1. 如果将第 80 行代码改为 `ALL_DICE = [D1]`，会发生什么？
2. 如果将第 171 行的 `get((cx, cy), ' ')` 改为 `get((cx, cy), '.')` 会发生什么？
3. 如果将第 177 行的 `correctAnswers += 1` 改为 `correctAnswers += 0`，会发生什么？
4. 如果删除或注释掉第 91 行的 `correctAnswers = 0`，会得到什么错误消息？

项目 18

掷骰子

在 Dungeons & Dragons 以及其他角色扮演桌游中，玩家会用到特殊的骰子——这些骰子可以有 4、8、10、12 甚至 20 面。这些游戏还使用特定的符号来指示要掷哪个骰子，例如，3d6 意味着掷 3 个 6 面骰子，而 1d10+2 意味着掷 1 个 10 面骰子并增加两点奖励。本项目中的程序将模拟掷这种骰子，还可以模拟物理上不存在的骰子，例如 38 面的骰子。

运行程序

运行 diceroller.py，输出如下所示：

```
Dice Roller, by Al Sweigart al@inventwithpython.com
--snip--
> 3d6
7 (3, 2, 2)
> 1d10+2
9 (7, +2)
> 2d38-1
32 (20, 13, -1)
> 100d6
364 (3, 3, 2, 4, 2, 1, 4, 2, 4, 6, 4, 5, 4, 3, 3, 3, 2, 5, 1, 5, 6, 6, 6, 4,
5, 5, 1, 5, 2, 2, 2, 5, 1, 1, 2, 1, 4, 5, 6, 2, 4, 3, 4, 3, 5, 2, 2, 1, 1, 5,
1, 3, 6, 6, 6, 6, 5, 2, 6, 5, 4, 4, 5, 1, 6, 6, 6, 4, 2, 6, 2, 6, 2, 2, 4, 3,
6, 4, 6, 4, 2, 4, 3, 3, 1, 6, 3, 3, 4, 4, 5, 5, 5, 6, 2, 3, 6, 1, 1, 1)
--snip--
```

工作原理

本项目中的大部分代码专门用于确保用户输入格式正确。实际上，随机掷骰子只是对 `random.randint()` 的简单调用。`random.randint()` 函数没有偏差，传给它的每个整数返回的概率都相同，因此非常适合用来模拟掷骰子。

```
1. """掷骰子，作者：Al Sweigart al@inventwithpython.com
2. 使用 Dungeons & Dragons 中的掷骰子符号模拟掷骰子"""
3. 标签：简短，模拟"""
4.
5. import random, sys
```

```
 6.
 7. print('''Dice Roller, by Al Sweigart al@inventwithpython.com
 8.
 9. Enter what kind and how many dice to roll. The format is the number of
10. dice, followed by "d", followed by the number of sides the dice have.
11. You can also add a plus or minus adjustment.
12.
13. Examples:
14.     3d6 rolls three 6-sided dice
15.     1d10+2 rolls one 10-sided die, and adds 2
16.     2d38-1 rolls two 38-sided die, and subtracts 1
17.     QUIT quits the program
18. ''')
19.
20. while True:    # 主循环
21.     try:
22.         diceStr = input('> ')    # 提示输入骰子字符串
23.         if diceStr.upper() == 'QUIT':
24.             print('Thanks for playing!')
25.             sys.exit()
26.
27.         # 清理骰子字符串
28.         diceStr = diceStr.lower().replace(' ', '')
29.
30.         # 在输入的骰子字符串输入中找到"d"
31.         dIndex = diceStr.find('d')
32.         if dIndex == -1:
33.             raise Exception('Missing the "d" character.')
34.
35.         # 获取骰子的数量 ( "3d6+1" 中的 "3" )
36.         numberOfDice = diceStr[:dIndex]
37.         if not numberOfDice.isdecimal():
38.             raise Exception('Missing the number of dice.')
39.         numberOfDice = int(numberOfDice)
40.
41.         # 查找修饰量是否有加号或减号
42.         modIndex = diceStr.find('+')
43.         if modIndex == -1:
44.             modIndex = diceStr.find('-')
45.
46.         # 查找骰子的面数 ( "3d6+1" 中的 "6" )
47.         if modIndex == -1:
48.             numberOfSides = diceStr[dIndex + 1 :]
49.         else:
50.             numberOfSides = diceStr[dIndex + 1 : modIndex]
51.         if not numberOfSides.isdecimal():
52.             raise Exception('Missing the number of sides.')
53.         numberOfSides = int(numberOfSides)
54.
55.         # 找到修改量 ( "3d6+1" 中的 "1" )
56.         if modIndex == -1:
57.             modAmount = 0
58.         else:
59.             modAmount = int(diceStr[modIndex + 1 :])
60.             if diceStr[modIndex] == '-':
61.                 # 将修改量改为负数
62.                 modAmount = -modAmount
63.
64.         # 模拟掷骰子
65.         rolls = []
66.         for i in range(numberOfDice):
67.             rollResult = random.randint(1, numberOfSides)
```

```
68.            rolls.append(rollResult)
69.
70.        # 显示总点数
71.        print('Total:', sum(rolls) + modAmount, '(Each die:', end='')
72.
73.        # 显示单个点数
74.        for i, roll in enumerate(rolls):
75.            rolls[i] = str(roll)
76.        print(', '.join(rolls), end='')
77.
78.        # 显示修改量
79.        if modAmount != 0:
80.            modSign = diceStr[modIndex]
81.            print(', {}{}'.format(modSign, abs(modAmount)), end='')
82.        print(')')
83.
84.    except Exception as exc:
85.        # 捕获任何异常并向玩家显示消息
86.        print('Invalid input. Enter something like "3d6" or "1d10+2".')
87.        print('Input was invalid because: ' + str(exc))
88.        continue    # 返回骰子字符串的输入提示
```

输入源代码并运行几次后，请尝试对其进行更改。你也可以尝试以下操作。

❑ 除了加法和减法修饰符，你还可以添加乘法修饰符。
❑ 添加自动移除掷出最小骰子的功能。

探索程序

请尝试找出以下问题的答案。你需要尝试对代码进行一些修改，再次运行程序，查看修改后的效果。

1. 如果删除或注释掉第 68 行的 `rolls.append(rollResult)`，会发生什么？
2. 如果将第 68 行的 `rolls.append(rollResult)` 改为 `rolls.append(-rollResult)`，会发生什么？
3. 如果删除或注释掉第 76 行的 `print(', '.join(rolls), end='')`，会发生什么？
4. 如果什么都不输入，掷一次骰子，会发生什么？

项目 19

数字时钟

在本项目中，我们将编写程序，以显示带有当前时间的数字时钟。这里不是直接渲染数字字符，而是用项目 64 中的 sevseg.py 模块为每个数字生成图形。这个程序与项目 14 的程序有异曲同工之妙。

运行程序

运行 digitalclock.py，输出如下所示：

```
 _|  |_| *  _|  _| *  _|  _|
|_|  _| * |_  _| * |_| |_|
```

Press Ctrl-C to quit.

工作原理

数字时钟程序有点类似于项目 14 的倒计时程序——不仅都要导入 sevseg.py 模块，还都必须使用 splitlines() 方法拆分 sevseg.getSevSegStr() 返回的多行字符串。这让我们可以在时钟的时、分、秒数字之间放置一个由星号组成的冒号。你可以将本项目的代码与项目 14 倒计时中的代码加以比较，以了解它们的异同点。

```
 1. """数字时钟，作者：Al Sweigart al@inventwithpython.com
 2. "7 段显示"一个数字时钟的当前时间按 Ctrl-C，程序结束运行
 3. 要求 sevseg.py 也位于同一文件夹中
 4. 标签：小，艺术
 5. """
 6. import sys, time
 7. import sevseg  # 导入 sevseg.py 模块
 8.
 9. try:
10.     while True:  # 主循环
11.         # 通过输出换行符来清空屏幕
12.         print('\n' * 60)
13.
```

```
14.     # 从计算机的时钟获取当前时间
15.     currentTime = time.localtime()
16.     # 用 12 进行取模,表示使用 12 小时制,而不是 24 小时制
17.     hours = str(currentTime.tm_hour % 12)
18.     if hours == '0':
19.         hours = '12'   # 12 小时制显示的是 12:00,而不是 00:00
20.     minutes = str(currentTime.tm_min)
21.     seconds = str(currentTime.tm_sec)
22.
23.     # 从 sevseg 模块中获取数字字符串
24.     hDigits = sevseg.getSevSegStr(hours, 2)
25.     hTopRow, hMiddleRow, hBottomRow = hDigits.splitlines()
26.
27.     mDigits = sevseg.getSevSegStr(minutes, 2)
28.     mTopRow, mMiddleRow, mBottomRow = mDigits.splitlines()
29.
30.     sDigits = sevseg.getSevSegStr(seconds, 2)
31.     sTopRow, sMiddleRow, sBottomRow = sDigits.splitlines()
32.
33.     # 显示数字
34.     print(hTopRow    + '   ' + mTopRow    + '   ' + sTopRow)
35.     print(hMiddleRow + ' * ' + mMiddleRow + ' * ' + sMiddleRow)
36.     print(hBottomRow + ' * ' + mBottomRow + ' * ' + sBottomRow)
37.     print()
38.     print('Press Ctrl-C to quit.')
39.
40.     # 继续循环,直到发生第二次变化
41.     while True:
42.         time.sleep(0.01)
43.         if time.localtime().tm_sec != currentTime.tm_sec:
44.             break
45. except KeyboardInterrupt:
46.     print('Digital Clock, by Al Sweigart al@inventwithpython.com')
47.     sys.exit()   # 按 Ctrl-C,程序结束
```

探索程序

请尝试找出以下问题的答案。你需要尝试对代码进行一些修改,再次运行程序,查看修改后的效果。

1. 如果将第 42 行的 `time.sleep(0.01)` 改为 `time.sleep(2)`,会发生什么?
2. 如果将第 24 行、第 27 行和第 30 行中的 2 改为 1,会发生什么?
3. 如果删除或注释掉第 12 行的 `print('\n' * 60)`,会发生什么?
4. 如果删除或注释掉第 7 行的 `import sevseg`,会得到什么错误消息?

项目 20

数字流

在本项目中，我们将编写程序，实现模仿科幻电影《黑客帝国》中的"数字流"可视化效果。随机的二进制"数字流"从计算机屏幕底部倾泻而出，创造出炫酷的视觉效果。鉴于屏幕内容向下滚动时文本移动的方式，如果不使用诸如 bext 之类的模块，我们很难使其流向向下。

运行程序

运行 digitalstream.py，输出如下所示：

```
Digital Stream Screensaver, by Al Sweigart al@inventwithpython.com
Press Ctrl-C to quit.
                  0                     0
                  0                     0
    1             0     0     1         1 0                         1
    0             0     0     1     0   0 0       0             0
    0             1     0     0     0   1 0 0     1         0   1
    0             1     0     0     1   011 1     1         0     1 0
    0             1     0     0     0   000 11    0         0   1 1 0
    1   1         0 1   0     1     1   110 10  1 0         1   0 1 0
        1       101 0         0     1   000 11  1 1        11   1 1 1
        0       100 1         0         11  00  0 1        01     0
    1 1         001 1         1           0 1   10 0       10     0
    0 0         010 0         1                 1  11 11    0     0
--snip--
```

工作原理

与项目 15 的程序一样，本项目程序通过调用 print() 让数字滚动起来，以实现"数字流"动画效果。每一列由 columns 列表中的一个整数来表示：columns[0] 是最左侧列的整数，columns[1] 是该列右侧列的整数，以此类推。程序最初将这些整数设置为 0，这意味着它会在该列中输出 ' '（空格字符）而不是流。接下来，程序将每个整数更改为介于 MIN_STREAM_LENGTH 和 MAX_STREAM_LENGTH 之间的随机值。每输出一行后该整数减 1。只要一列的整数大于 0，程序就会在该列中随机输出 1 或 0，最终实现我们在屏幕上看到的"数字流"效果。

```
 1. """数字流,作者:Al Sweigart al@inventwithpython.com
 2. 《黑客帝国》电影视觉风格的屏幕保护动画程序
 3. 标签:小,艺术,初学者,动画"""
 4.
 5. import random, shutil, sys, time
 6.
 7. # 创建常量
 8. MIN_STREAM_LENGTH = 6   # (!) 尝试将 6 更改为 1 或 50
 9. MAX_STREAM_LENGTH = 14  # (!) 尝试将 14 更改为 100
10. PAUSE = 0.1  # (!) 尝试将 0.1 更改为 0.0 或 2.0
11. STREAM_CHARS = ['0', '1']  # (!) 尝试将其更改为其他字符
12.
13. # 密度范围为 0.0~1.0
14. DENSITY = 0.02  # (!) 尝试将 0.02 更改为 0.10 或 0.30
15.
16. # 获取终端窗口的大小
17. WIDTH = shutil.get_terminal_size()[0]
18. # 如果不自动添加换行符,我们无法输出 Windows 上的最后一列
19. # 所以宽度减 1
20. WIDTH -= 1
21.
22. print('Digital Stream, by Al Sweigart al@inventwithpython.com')
23. print('Press Ctrl-C to quit.')
24. time.sleep(2)
25.
26. try:
27.     # 对于每一列,如果计数器的值为 0,则不显示流;如果计数器的值不为 0,则其值表示 1 或 0 在列中显示的次数
28.     columns = [0] * WIDTH
29.     while True:
30.         # 为每一列设置计数器
31.         for i in range(WIDTH):
32.             if columns[i] == 0:
33.                 if random.random() <= DENSITY:
34.                     # 重新启动此列上的流
35.                     columns[i] = random.randint(MIN_STREAM_LENGTH,
36.                                                 MAX_STREAM_LENGTH)
37.
38.             # 显示空格或 1/0 字符
39.             if columns[i] > 0:
40.                 print(random.choice(STREAM_CHARS), end='')
41.                 columns[i] -= 1
42.             else:
43.                 print(' ', end='')
44.         print()  # 在 columns 的行末尾输出一个空行
45.         sys.stdout.flush()  # 确保文本出现在屏幕上
46.         time.sleep(PAUSE)
47. except KeyboardInterrupt:
48.     sys.exit()  # 按 Ctrl-C,程序结束
```

输入源代码并运行几次后,请尝试对其进行更改。你可以参考标有!的注释对上述程序进行修改,也可以尝试以下操作。

❑ 引入除 1 和 0 之外的字符。
❑ 引入除线条以外的图形,包括矩形、三角形和菱形等。

探索程序

请尝试找出以下问题的答案。你需要尝试对代码进行一些修改,再次运行程序,查看修改后的效果。

1. 如果将第 43 行的 print(' ', end='') 改为 print('.', end='')，会发生什么？
2. 如果将第 10 行的 PAUSE = 0.1 改为 PAUSE = -0.1，会得到什么错误消息？
3. 如果将第 39 行的 columns[i] > 0 改为 columns[i] < 0，会发生什么？
4. 如果将第 39 行的 columns[i] > 0 改为 columns[i] <= 0，会发生什么？
5. 如果将第 41 行的 columns[i] -= 1 改为 columns[i] += 1，会发生什么？

项目 21

DNA可视化

脱氧核糖核酸（deoxyribonucleic acid，DNA）是一种微小的分子，存在于人体的每个细胞中，蕴含着人体如何生长的"蓝图"。DNA 看上去像一对核苷酸分子构成的双螺旋结构（像一种扭曲的梯子），包含鸟嘌呤、胞嘧啶、腺嘌呤和胸腺嘧啶（分别用字母 G、C、A 和 T 表示）。在显微镜下，DNA 是一个长分子，但如果把它拉长，它的 30 多亿个碱基对可长达 2 米！在本项目中，我们将编写程序，使之可以生成简单的 DNA 动画。

运行程序

运行 dna.py，输出如下所示：

```
DNA Animation, by Al Sweigart al@inventwithpython.com
Press Ctrl-C to quit...
     #G-C#
     #C---G#
    #T-----A#
    #T------A#
   #A------T#
    #G-----C#
     #G---C#
      #C-G#
       ##
      #T-A#
     #C---G#
    #G-----C#
    #G------C#
    #T------A#
    #A-----T#
     #C---G#
      #G-C#
       ##
      #T-A#
      #T---A#
     #A-----T#
--snip--
```

工作原理

这个程序与项目 15 和项目 20 的程序类似，也是通过输出 ROWS 列表中的字符串来创建滚动动画,并使用 format()字符串方法将"AT"和"CG"对插入每个字符串。

```
 1. """DNA 可视化,作者:Al Sweigart al@inventwithpython.com
 2. DNA 双螺旋的简单动画。 按 Ctrl-C 停止
 3. 标签:简短,艺术,动画,科学"""
 4.
 5. import random, sys, time
 6.
 7. PAUSE = 0.15   # (!) 尝试将其更改为 0.5 或 0.0
 8.
 9. # 这些是 DNA 动画的各个行
10. ROWS = [
11.     #123456789 <- 用它来测量空格数
12.     '        ##',  # 索引 0 没有{}
13.     '       #{}-{}#',
14.     '      #{}---{}#',
15.     '     #{}-----{}#',
16.     '    #{}------{}#',
17.     '   #{}------{}#',
18.     '    #{}-----{}#',
19.     '     #{}---{}#',
20.     '      #{}-{}#',
21.     '       ##',   # 索引 9 没有{}
22.     '      #{}-{}#',
23.     '     #{}---{}#',
24.     '    #{}-----{}#',
25.     '   #{}------{}#',
26.     '    #{}------{}#',
27.     '     #{}-----{}#',
28.     '      #{}---{}#',
29.     '       #{}-{}#']
30.     #123456789 <- 用它来测量空格数
31.
32. try:
33.     print('DNA Animation, by Al Sweigart al@inventwithpython.com')
34.     print('Press Ctrl-C to quit...')
35.     time.sleep(2)
36.     rowIndex = 0
37.
38.     while True:   # 主循环
39.         # rowIndex 加 1 以绘制下一行
40.         rowIndex = rowIndex + 1
41.         if rowIndex == len(ROWS):
42.             rowIndex = 0
43.
44.         # 行索引 0 和 9 没有核苷酸
45.         if rowIndex == 0 or rowIndex == 9:
46.             print(ROWS[rowIndex])
47.             continue
48.
49.         # 选择随机核苷酸对(鸟嘌呤-胞嘧啶和腺嘌呤-胸腺嘧啶)
50.         randomSelection = random.randint(1, 4)
51.         if randomSelection == 1:
52.             leftNucleotide, rightNucleotide = 'A', 'T'
53.         elif randomSelection == 2:
54.             leftNucleotide, rightNucleotide = 'T', 'A'
55.         elif randomSelection == 3:
56.             leftNucleotide, rightNucleotide = 'C', 'G'
57.         elif randomSelection == 4:
58.             leftNucleotide, rightNucleotide = 'G', 'C'
59.
60.         # 输出行
```

```
61.          print(ROWS[rowIndex].format(leftNucleotide, rightNucleotide))
62.          time.sleep(PAUSE)    # 添加短暂的停顿
63. except KeyboardInterrupt:
64.     sys.exit()   # 按 Ctrl-C，程序结束
```

探索程序

请尝试找出以下问题的答案。你需要尝试对代码进行一些修改，再次运行程序，查看修改后的效果。

1. 如果将第 40 行的 rowIndex = rowIndex + 1 改为 rowIndex = rowIndex + 2，会发生什么？
2. 如果将第 50 行的 random.randint(1, 4) 改为 random.randint(1, 2)，会发生什么？
3. 如果将第 7 行的 PAUSE = 0.15 设置为 PAUSE = -0.15，会得到什么错误消息？

项目 22

小鸭子

本项目创建一群不断向上移动的小鸭子（用符号表示）。小鸭子有轻微的区别，它们可以面向左或右，并且有2种不同的体型、4种眼睛、2种嘴巴和3种翅膀位置。这为我们提供了96种不同的组合效果，而小鸭子程序会不断产生这些组合。

运行程序

运行 ducklings.py，输出如下所示：

```
Duckling Screensaver, by Al Sweigart al@inventwithpython.com
Press Ctrl-C to quit...
                                                              =" )
   =" )                                             ( v)=" )
  ( ^)                                          ^ ^ ( v) >''')
   ^^                                                ^^     (  ^
                                      >" )                   ^  ^
                                     ( v)         =^^)
  ("< ("<              >" ) ^^                  ( >)
  ( ^) (< )            ( ^)                      ^  ^
   ^^   ^^            ("< ^^                              (``<>^^)
  (^^=                ( ^)                                (< )( ^)
  (v ) ( "<            ^^                                  ^  ^ ^
--snip--
```

工作原理

以下程序用 Duckling 类表示小鸭子。小鸭子的随机身体特征在该类的 __init__() 方法中选择，小鸭子的身体部位分别通过 getHeadStr()、getBodyStr() 和 getFeetStr() 方法返回。

```
1. """小鸭子，作者：Al Sweigart al@inventwithpython.com
2. 许多小鸭子的屏幕保护动画程序
3.
4. >" )    =^^)    (``=   ("=  >")    ("=
```

```
 5. ( >)   ( ^)   (v )   (^ )   ( >)   (v )
 6.  ^ ^    ^ ^    ^ ^    ^^     ^^     ^^
 7.
 8. 标签：大、艺术、面向对象、动画"""
 9.
10. import random, shutil, sys, time
11.
12. # 创建常量
13. PAUSE = 0.2    # (!) 尝试将 0.2 更改为 1.0 或 0.0
14. DENSITY = 0.10   # (!) 尝试将 0.10 更改为范围为 0.0 和 1.0 之间的任意值
15.
16. DUCKLING_WIDTH = 5
17. LEFT = 'left'
18. RIGHT = 'right'
19. BEADY = 'beady'
20. WIDE = 'wide'
21. HAPPY = 'happy'
22. ALOOF = 'aloof'
23. CHUBBY = 'chubby'
24. VERY_CHUBBY = 'very chubby'
25. OPEN = 'open'
26. CLOSED = 'closed'
27. OUT = 'out'
28. DOWN = 'down'
29. UP = 'up'
30. HEAD = 'head'
31. BODY = 'body'
32. FEET = 'feet'
33.
34. # 获取终端窗口的大小
35. WIDTH = shutil.get_terminal_size()[0]
36. # 如果不自动添加换行符，我们无法输出 Windows 上的最后一列，所以宽度减 1
37. WIDTH -= 1
38.
39.
40. def main():
41.     print('Duckling Screensaver, by Al Sweigart')
42.     print('Press Ctrl-C to quit...')
43.     time.sleep(2)
44.
45.     ducklingLanes = [None] * (WIDTH // DUCKLING_WIDTH)
46.
47.     while True:    # 主循环
48.         for laneNum, ducklingObj in enumerate(ducklingLanes):
49.             # 判断我们是否应该在这条"小路"上创建一只小鸭子
50.             if (ducklingObj == None and random.random() <= DENSITY):
51.                 # 在这条"小路"上放一只小鸭子
52.                 ducklingObj = Duckling()
53.                 ducklingLanes[laneNum] = ducklingObj
54.
55.             if ducklingObj != None:
56.                 # 如果这条"小路"上有一只小鸭子，就画一只小鸭子
57.                 print(ducklingObj.getNextBodyPart(), end='')
58.                 # 如果我们画完了小鸭子，就删除它
59.                 if ducklingObj.partToDisplayNext == None:
60.                     ducklingLanes[laneNum] = None
61.             else:
62.                 # 因为这里没有小鸭子，所以我们画 5 个空格
63.                 print(' ' * DUCKLING_WIDTH, end='')
64.
65.         print()    # 输出一行空行
66.         sys.stdout.flush()    # 确保文本显示在屏幕上
```

```
 67.         time.sleep(PAUSE)
 68.
 69.
 70. class Duckling:
 71.     def __init__(self):
 72.         """创建一只身体特征随机的新的小鸭子"""
 73.         self.direction = random.choice([LEFT, RIGHT])
 74.         self.body = random.choice([CHUBBY, VERY_CHUBBY])
 75.         self.mouth = random.choice([OPEN, CLOSED])
 76.         self.wing = random.choice([OUT, UP, DOWN])
 77.
 78.         if self.body == CHUBBY:
 79.             # 胖嘟嘟的小鸭子长着亮晶晶的眼睛
 80.             self.eyes = BEADY
 81.         else:
 82.             self.eyes = random.choice([BEADY, WIDE, HAPPY, ALOOF])
 83.
 84.         self.partToDisplayNext = HEAD
 85.
 86.     def getHeadStr(self):
 87.         """返回小鸭子头的字符串"""
 88.         headStr = ''
 89.         if self.direction == LEFT:
 90.             # 得到小鸭子的嘴
 91.             if self.mouth == OPEN:
 92.                 headStr += '>'
 93.             elif self.mouth == CLOSED:
 94.                 headStr += '='
 95.
 96.             # 得到小鸭子的眼睛
 97.             if self.eyes == BEADY and self.body == CHUBBY:
 98.                 headStr += '"'
 99.             elif self.eyes == BEADY and self.body == VERY_CHUBBY:
100.                 headStr += '" '
101.             elif self.eyes == WIDE:
102.                 headStr += "'''"
103.             elif self.eyes == HAPPY:
104.                 headStr += '^^'
105.             elif self.eyes == ALOOF:
106.                 headStr += '``'
107.
108.             headStr += ')'   # 得到小鸭子的后脑勺
109.
110.         if self.direction == RIGHT:
111.             headStr += ' ('   # 得到小鸭子的后脑勺
112.
113.             # 得到小鸭子的眼睛
114.             if self.eyes == BEADY and self.body == CHUBBY:
115.                 headStr += '"'
116.             elif self.eyes == BEADY and self.body == VERY_CHUBBY:
117.                 headStr += ' "'
118.             elif self.eyes == WIDE:
119.                 headStr += "'''"
120.             elif self.eyes == HAPPY:
121.                 headStr += '^^'
122.             elif self.eyes == ALOOF:
123.                 headStr += '``'
124.
125.             # 得到小鸭子的嘴
126.             if self.mouth == OPEN:
127.                 headStr += '<'
128.             elif self.mouth == CLOSED:
```

```
129.                headStr += '='
130.
131.        if self.body == CHUBBY:
132.            # 给胖嘟嘟的小鸭子留出足够的空间
133.            # 使之看起来和那些非常胖的小鸭子一样宽
134.            headStr += ' '
135.
136.        return headStr
137.
138.    def getBodyStr(self):
139.        """返回小鸭子身体的字符串"""
140.        bodyStr = '('    # 得到小鸭子身体的左半部分
141.        if self.direction == LEFT:
142.            # 得到小鸭子身体的内部
143.            if self.body == CHUBBY:
144.                bodyStr += ' '
145.            elif self.body == VERY_CHUBBY:
146.                bodyStr += '  '
147.
148.            # 得到小鸭子的翅膀
149.            if self.wing == OUT:
150.                bodyStr += '>'
151.            elif self.wing == UP:
152.                bodyStr += '^'
153.            elif self.wing == DOWN:
154.                bodyStr += 'v'
155.
156.        if self.direction == RIGHT:
157.            # 得到小鸭子的翅膀
158.            if self.wing == OUT:
159.                bodyStr += '<'
160.            elif self.wing == UP:
161.                bodyStr += '^'
162.            elif self.wing == DOWN:
163.                bodyStr += 'v'
164.
165.            # 得到小鸭子身体的内部
166.            if self.body == CHUBBY:
167.                bodyStr += ' '
168.            elif self.body == VERY_CHUBBY:
169.                bodyStr += '  '
170.
171.        bodyStr += ')'   # 得到小鸭子身体的右半部分
172.
173.        if self.body == CHUBBY:
174.            # 给胖嘟嘟的胖小鸭子留出足够的空间
175.            # 使之看起来和那些非常胖的小鸭子一样宽
176.            bodyStr += ' '
177.
178.        return bodyStr
179.
180.    def getFeetStr(self):
181.        """返回小鸭子脚的字符串"""
182.        if self.body == CHUBBY:
183.            return ' ^^ '
184.        elif self.body == VERY_CHUBBY:
185.            return ' ^ ^ '
186.
187.    def getNextBodyPart(self):
188.        """为需要显示的下一个身体部位调用适当的显示方法
189.        完成后，将 partToDisplayNext 设置为 None
190.        """
```

```
191.            if self.partToDisplayNext == HEAD:
192.                self.partToDisplayNext = BODY
193.                return self.getHeadStr()
194.            elif self.partToDisplayNext == BODY:
195.                self.partToDisplayNext = FEET
196.                return self.getBodyStr()
197.            elif self.partToDisplayNext == FEET:
198.                self.partToDisplayNext = None
199.                return self.getFeetStr()
200.
201.
202.
203. # 程序运行入口（如果不是作为模块导入的话）
204. if __name__ == '__main__':
205.     try:
206.         main()
207.     except KeyboardInterrupt:
208.         sys.exit()    # 按Ctrl-C，程序结束
```

输入源代码并运行几次后，请尝试对其进行更改。你可以根据标有(!)的注释对程序进行修改。

探索程序

请尝试找出以下问题的答案。你需要尝试对代码进行一些修改，再次运行程序，查看修改后的效果。

1. 如果将第 73 行的 random.choice([LEFT, RIGHT])改为 random.choice([LEFT])，则会发生什么？
2. 如果将第 192 行的 self.partToDisplayNext = BODY 改为 self.partToDisplayNext = None，则会发生什么？
3. 如果将第 195 行的 self.partToDisplayNext = FEET 改为 self.partToDisplayNext = BODY，则会发生什么？
4. 如果将第 193 行的 return self.getHeadStr()改为 return self.getFeetStr()，则会发生什么？

项目 23

蚀刻绘图器

在本项目中，我们将编写程序，以实现这样的效果：当你通过按 W、A、S 和 D 键在计算机屏幕上移动笔尖图标时，蚀刻绘图器通过描绘一条连续的线来生成图像，就像 Etch A Sketch 玩具一样。将你在艺术方面的天赋爆发出来吧，看看能创造出什么样的图像！请将图像保存到文本文件中，以便以后输出。你也可以将其他图像的 WASD 命令复制并粘贴到这个项目的程序中，就像源代码第 7～29 行中针对希尔伯特曲线的 WASD 命令。

运行程序

运行 etchingdrawer.py，输出如下：

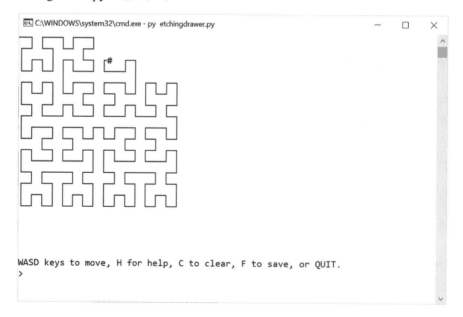

工作原理

与项目 17 的程序一样，在本项目中，我们用存储在名为 canvas 的变量中的字典来记录绘图线条。键是(x，y)元组，值是表示要在屏幕上的"x""y"坐标处绘制的线条形状的字符串。附录 B 给出了 Python 程序中可用的 Unicode 字符的完整列表。

第 139 行代码有一个对 open() 函数的调用，传入一个 encoding='utf-8' 参数。本书对此不多解释，这是 Windows 将行字符写入文本文件所必需的。

```
 1. """蚀刻绘图器，作者：Al Sweigart al@inventwithpython.com
 2. 一个艺术程序，它可实现通过使用 W.A.S.D 键在屏幕周围画一条连续的线
 3. 灵感来自 Etch A Sketch 玩具
 4.
 5. 例如，你可以绘制希尔伯特曲线
 6.
 7. SDWDDSASDSAAWASSDSASSDWDSDWWAWDDDSASSDWDSDWWAWDWWASAAWDWAW
 8. DDSDW
 9.
10. 也可以绘制更大的希尔伯特曲线
11.
12. DDSAASSDDWDDSDDWWAAWDDDDSDDWDDDDSAASDDSAAAAWAASSSDDWDDDDSAASDDSA
13. AAAWA
14.
15. ASAAAAWDDWWAASAAWAASSDDSAASSDDWDDDDSAASDDSAAAAWAASSDDSAASSDDWDDSD
16. DWWA
17.
18. AWDDDDDSAASSDDWDDSDDWWAAWDDWWAASAAAAWDDWAAWDDDDSDDWDDSDDWDDD
19. DSAASD
20. DS
21.
22. AAAAWAASSDDSAASSDDWDDSDDWWAAWDDDDDSAASSDDWDDSDDWWAAWDDWWAASAA
23. AAWDD
24. WA
25.
26. AWDDDDSDDWWAAWDDWWAASAAWAASSDDSAAAAWAASAAAAWDDWAAWDDDDSDDWWW
27. AASAAA
28. AWD
29. DWAAWDDDDSDDWDDDDSAASSDDWDDSDDWWAAWDD
30.
31. 标签：大，艺术"""
32.
33. import shutil, sys
34.
35. # 为行字符创建常量
36. UP_DOWN_CHAR           = chr(9474)   # 字符 9474 是 │
37. LEFT_RIGHT_CHAR        = chr(9472)   # 字符 9472 是 ─
38. DOWN_RIGHT_CHAR        = chr(9484)   # 字符 9484 是 ┌
39. DOWN_LEFT_CHAR         = chr(9488)   # 字符 9488 是 ┐
40. UP_RIGHT_CHAR          = chr(9492)   # 字符 9492 是 └
41. UP_LEFT_CHAR           = chr(9496)   # 字符 9496 是 ┘
42. UP_DOWN_RIGHT_CHAR     = chr(9500)   # 字符 9500 是 ├
43. UP_DOWN_LEFT_CHAR      = chr(9508)   # 字符 9508 是 ┤
44. DOWN_LEFT_RIGHT_CHAR   = chr(9516)   # 字符 9516 是 ┬
45. UP_LEFT_RIGHT_CHAR     = chr(9524)   # 字符 9524 是 ┴
46. CROSS_CHAR             = chr(9532)   # 字符 9532 是 ┼
47.
48. # 获取终端窗口的大小
49. CANVAS_WIDTH, CANVAS_HEIGHT = shutil.get_terminal_size()
50. # 如果不自动添加换行符，我们无法输出 Windows 上的最后一列，故将宽度减 1
51. CANVAS_WIDTH -= 1
52. # 在底部留出几行，用于输出命令行
53. CANVAS_HEIGHT -= 5
54.
55. """canvas 字典的键是整数坐标（x, y）元组，值是字母 W、A、S、D 的集合，可确定所绘制的是什么样的线条"""
```

```
 56. canvas = {}
 57. cursorX = 0
 58. cursorY = 0
 59.
 60.
 61. def getCanvasString(canvasData, cx, cy):
 62.     """返回在 canvasData 中绘制的线条的多行字符串"""
 63.     canvasStr = ''
 64.
 65.     """canvasData 是一个字典,键是(x,y)元组,值是字母 W、A、S 和/或 D 的集合,
 66.        用于表示在每个(x,y)点所绘制的线条的方向"""
 67.     for rowNum in range(CANVAS_HEIGHT):
 68.         for columnNum in range(CANVAS_WIDTH):
 69.             if columnNum == cx and rowNum == cy:
 70.                 canvasStr += '#'
 71.                 continue
 72.
 73.             # 将线条添加到 canvasStr 中
 74.             cell = canvasData.get((columnNum, rowNum))
 75.             if cell in (set(['W', 'S']), set(['W']), set(['S'])):
 76.                 canvasStr += UP_DOWN_CHAR
 77.             elif cell in (set(['A', 'D']), set(['A']), set(['D'])):
 78.                 canvasStr += LEFT_RIGHT_CHAR
 79.             elif cell == set(['S', 'D']):
 80.                 canvasStr += DOWN_RIGHT_CHAR
 81.             elif cell == set(['A', 'S']):
 82.                 canvasStr += DOWN_LEFT_CHAR
 83.             elif cell == set(['W', 'D']):
 84.                 canvasStr += UP_RIGHT_CHAR
 85.             elif cell == set(['W', 'A']):
 86.                 canvasStr += UP_LEFT_CHAR
 87.             elif cell == set(['W', 'S', 'D']):
 88.                 canvasStr += UP_DOWN_RIGHT_CHAR
 89.             elif cell == set(['W', 'S', 'A']):
 90.                 canvasStr += UP_DOWN_LEFT_CHAR
 91.             elif cell == set(['A', 'S', 'D']):
 92.                 canvasStr += DOWN_LEFT_RIGHT_CHAR
 93.             elif cell == set(['W', 'A', 'D']):
 94.                 canvasStr += UP_LEFT_RIGHT_CHAR
 95.             elif cell == set(['W', 'A', 'S', 'D']):
 96.                 canvasStr += CROSS_CHAR
 97.             elif cell == None:
 98.                 canvasStr += ' '
 99.         canvasStr += '\n'  # 在每一行的末尾添加换行符
100.     return canvasStr
101.
102.
103. moves = []
104. while True:  # 主循环
105.     # 根据 canvas 中的数据绘制线条
106.     print(getCanvasString(canvas, cursorX, cursorY))
107.
108.     print('WASD keys to move, H for help, C to clear, '
109.         + 'F to save, or QUIT.')
110.     response = input('> ').upper()
111.
112.     if response == 'QUIT':
113.         print('Thanks for playing!')
114.         sys.exit()  # 退出程序
115.     elif response == 'H':
116.         print('Enter W, A, S, and D characters to move the cursor and')
117.         print('draw a line behind it as it moves. For example, ddd')
```

```
118.            print('draws a line going right and sssdddwwwaaa draws a box.')
119.            print()
120.            print('You can save your drawing to a text file by entering F.')
121.            input('Press Enter to return to the program...')
122.            continue
123.        elif response == 'C':
124.            canvas = {}  # 消除 canvas 中的数据
125.            moves.append('C')  # 记录当前操作
126.        elif response == 'F':
127.            # 将 canvas 字符串保存为一个文本文件
128.            try:
129.                print('Enter filename to save to:')
130.                filename = input('> ')
131.
132.                # 确保文件名以 .txt 为扩展名
133.                if not filename.endswith('.txt'):
134.                    filename += '.txt'
135.                with open(filename, 'w', encoding='utf-8') as file:
136.                    file.write(''.join(moves) + '\n')
137.                    file.write(getCanvasString(canvas, None, None))
138.            except:
139.                print('ERROR: Could not save file.')
140.
141.        for command in response:
142.            if command not in ('W', 'A', 'S', 'D'):
143.                continue  # 忽略这个字母,继续处理下一个字母
144.            moves.append(command)  # 记录当前操作
145.
146.            # 所添加的第一条线需要形成完整的行
147.            if canvas == {}:
148.                if command in ('W', 'S'):
149.                    # 把第一条线变成水平的
150.                    canvas[(cursorX, cursorY)] = set(['W', 'S'])
151.                elif command in ('A', 'D'):
152.                    # 把第一条线变成垂直的
153.                    canvas[(cursorX, cursorY)] = set(['A', 'D'])
154.
155.            # 更新 x 和 y
156.            if command == 'W' and cursorY > 0:
157.                canvas[(cursorX, cursorY)].add(command)
158.                cursorY = cursorY - 1
159.            elif command == 'S' and cursorY < CANVAS_HEIGHT - 1:
160.                canvas[(cursorX, cursorY)].add(command)
161.                cursorY = cursorY + 1
162.            elif command == 'A' and cursorX > 0:
163.                canvas[(cursorX, cursorY)].add(command)
164.                cursorX = cursorX - 1
165.            elif command == 'D' and cursorX < CANVAS_WIDTH - 1:
166.                canvas[(cursorX, cursorY)].add(command)
167.                cursorX = cursorX + 1
168.            else:
169.                # 如果光标没有移动,则说明它已移出了画布的边缘,请不要更改画布的设置[(cursorX, cursorY)]
170.                continue
171.
172.            # 如果没有设置(cursorX,cursorY),则添加一个空集
173.            if (cursorX, cursorY) not in canvas:
174.                canvas[(cursorX, cursorY)] = set()
175.
176.            # 将方向字符串添加到 x 和 y 点的集合中
177.            if command == 'W':
178.                canvas[(cursorX, cursorY)].add('S')
179.            elif command == 'S':
```

```
180.            canvas[(cursorX, cursorY)].add('W')
181.        elif command == 'A':
182.            canvas[(cursorX, cursorY)].add('D')
183.        elif command == 'D':
184.            canvas[(cursorX, cursorY)].add('A')
```

探索程序

请尝试找出以下问题的答案。你需要尝试对代码进行一些修改，再次运行程序，查看修改后的效果。

1. 如果将第 110 行的 `response = input('> ').upper()` 改为 `response = input('> ')`，会发生什么？
2. 如果将第 70 行的 `canvasStr += '#'` 改为 `canvasStr += '@'`，会发生什么？
3. 如果将第 98 行的 `canvasStr += ' '` 改为 `canvasStr += '.'`，会发生什么？
4. 如果将第 103 行的 `moves = []` 改为 `moves = list('SDWDDSASDSAAWASSDSAS')`，会发生什么？

项目 24

因数查找器

整数 a 除以整数 b（b≠0）的商正好是整数而没有余数，我们就说 b 是 a 的因数。也可以说，一个数的因数是任意两个整数乘积等于该数的数。例如，2×13 = 26，故称 2 和 13 是 26 的因数。另外，1×26 = 26，所以 1 和 26 也是 26 的因数。因此，我们说 26 有 4 个因数：1、2、13 和 26。

如果一个数只有两个因数（1 和它本身），则称之为素数，否则，称之为合数。可以使用因数查找器来发现一些新的素数！（提示：素数多以不是 5 的奇数结尾。）你也可以在计算机上运行项目 56 的程序，以发现素数。这个程序所涉及的数学知识并不难，非常适合初学者。

运行程序

运行 factorfinder.py，输出如下所示：

```
Factor Finder, by Al Sweigart al@inventwithpython.com
--snip--
Enter a number to factor (or "QUIT" to quit):
> 26
1, 2, 13, 26
Enter a number to factor (or "QUIT" to quit):
> 4352784
1, 2, 3, 4, 6, 8, 12, 16, 24, 29, 48, 53, 58, 59, 87, 106, 116, 118, 159,
174, 177, 212, 232, 236, 318, 348, 354, 424, 464, 472, 636, 696, 708, 848,
944, 1272, 1392, 1416, 1537, 1711, 2544, 2832, 3074, 3127, 3422, 4611, 5133,
6148, 6254, 6844, 9222, 9381, 10266, 12296, 12508, 13688, 18444, 18762, 20532,
24592, 25016, 27376, 36888, 37524, 41064, 50032, 73776, 75048, 82128, 90683,
150096, 181366, 272049, 362732, 544098, 725464, 1088196, 1450928, 2176392,
4352784
Enter a number to factor (or "QUIT" to quit):
> 9787
1, 9787
Enter a number to factor (or "QUIT" to quit):
> quit
```

工作原理

我们可以通过检查第二个数是否能整除第一个数，来判断一个数是否是另一个数的因数。例如，7 是 21 的因数，因为 21÷7 的结果为 3。这也为我们提供了 21 的另一个因数，即 3。但是，8不是21的因数，因为21÷8 = 2.625，有余数则说明该运算不是整除运算。

%取模运算符执行除法运算并告诉我们是否有余数。21 % 7 的计算结果为0，意味着没有余数，7是21的因数。21 % 8 的计算结果为5，是一个非零值，意味着8不是21的因数。本项目中的程序用第34行的代码来确定哪些数字是因数。

math.sqrt()函数返回接收到的数字的平方根。例如，math.sqrt(25) 返回 5.0，因为5乘其自身的结果为25，所以5是25的平方根。

```
1. """因数查找器，Al Sweigart al@inventwithpython.com
2. 求一个数的所有因数
3. 标签：小，初学者，数学"""
4.
5. import math, sys
6.
7. print('''Factor Finder, by Al Sweigart al@inventwithpython.com
8.
9. A number's factors are two numbers that, when multiplied with each
10. other, produce the number. For example, 2 x 13 = 26, so 2 and 13 are
11. factors of 26. 1 x 26 = 26, so 1 and 26 are also factors of 26. We
12. say that 26 has four factors: 1, 2, 13, and 26.
13.
14. If a number only has two factors (1 and itself), we call that a prime
15. number. Otherwise, we call it a composite number.
16.
17. Can you discover some prime numbers?
18. ''')
19.
20. while True:   # 主循环
21.     print('Enter a positive whole number to factor (or QUIT):')
22.     response = input('> ')
23.     if response.upper() == 'QUIT':
24.         sys.exit()
25.
26.     if not (response.isdecimal() and int(response) > 0):
27.         continue
28.     number = int(response)
29.
30.     factors = ['hello']
31.
32.     # 查找 number 的因数
33.     for i in range(1, int(math.sqrt(number)) + 1):
34.         if number % i == 0:   # 如果没有余数，它就是因数
35.             factors.append(i)
36.             factors.append(number // i)
37.
38.     # 转换为集合，以去除重复的因数
39.     factors = list(set(factors))
40.     factors.sort()
41.
42.     # 显示结果
43.     for i, factor in enumerate(factors):
44.         factors[i] = str(factor)
45.     print(', '.join(factors))
```

探索程序

请尝试找出以下问题的答案。你需要尝试对代码进行一些修改，再次运行程序，查看修改

后的效果。
1. 如果删除或注释掉第 35 行的 `factors.append(i)`，则会发生什么？
2. 如果删除或注释掉第 39 行的 `factors = list(set(factors))`，则会发生什么？（提示：输入一个平方数，例如 25、36 或 49。）
3. 如果删除或注释掉第 40 行的 `factors.sort()`，则会发生什么？
4. 如果将第 30 行的 `factors = []` 改为 `factors = ''`，则会得到什么错误消息？
5. 如果将第 30 行的 `factors = []` 改为 `factors = [-42]`，则会发生什么？
6. 如果将第 30 行的 `factors = []` 改为 `factors = ['hello']`，则会得到什么错误消息？

项目 25

"快速拔枪"

本项目实现的这个程序能测试你的反应速度。一旦看到 DRAW 一词，请你立即按 Enter 键。不过要小心！如果你在 DRAW 出现之前就按了 Enter 键，就算你输。快来试一下，看看你是不是最快的"西部键盘手"！

运行程序

运行 fastdraw.py，输出如下所示：

```
Fast Draw, by Al Sweigart al@inventwithpython.com

Time to test your reflexes and see if you are the fastest
draw in the west!
When you see "DRAW", you have 0.3 seconds to press Enter.
But you lose if you press Enter before "DRAW" appears.

Press Enter to begin...

It is high noon...
DRAW!

You took 0.3485 seconds to draw. Too slow!
Enter QUIT to stop, or press Enter to play again.
> quit
Thanks for playing!
```

工作原理

input()函数可以在等待用户输入字符串时暂停程序。这种简单的行为意味着我们无法仅使用 input()函数来创建实时游戏。但是，程序将缓存键盘输入，这意味着如果你在调用 input()函数之前按 C、A 和 T 键，对应的字符将被保存，一旦 input()函数被执行，它们将立即出现。

通过记录第 22 行 input()函数调用之前的时间和 input()函数调用之后（第 24 行）的

时间，我们可以确定玩家按 Enter 键的时间。不过，如果玩家在调用 `input()` 函数之前按 Enter 键，则缓存的键会导致 `input()` 函数立即返回（通常用时 3ms 左右）。这就是为什么要用第 25 行代码检查时间是否小于 0.01s（10ms），以确定玩家是否过早按下 Enter 键。

```
1.  """"快速拔枪"，作者：Al Sweigart al@inventwithpython.com
2.  测试你的反应能力，看看你是否是最快的 "西部键盘手"
3.  标签：小，初学者，游戏"""
4.
5.  import random, sys, time
6.
7.  print('Fast Draw, by Al Sweigart al@inventwithpython.com')
8.  print()
9.  print('Time to test your reflexes and see if you are the fastest')
10. print('draw in the west!')
11. print('When you see "DRAW", you have 0.3 seconds to press Enter.')
12. print('But you lose if you press Enter before "DRAW" appears.')
13. print()
14. input('Press Enter to begin...')
15.
16. while True:
17.     print()
18.     print('It is high noon...')
19.     time.sleep(random.randint(20, 50) / 10.0)
20.     print('DRAW!')
21.     drawTime = time.time()
22.     input()   # 在按下 Enter 键之前，此函数调用不会返回
23.     timeElapsed = time.time() - drawTime
24.
25.     if timeElapsed < 0.01:
26.         # 如果玩家在 DRAW 出现前按下 Enter 键，input() 方法会立即返回
27.         print('You drew before "DRAW" appeared! You lose.')
28.     elif timeElapsed > 0.3:
29.         timeElapsed = round(timeElapsed, 4)
30.         print('You took', timeElapsed, 'seconds to draw. Too slow!')
31.     else:
32.         timeElapsed = round(timeElapsed, 4)
33.         print('You took', timeElapsed, 'seconds to draw.')
34.         print('You are the fastest draw in the west! You win!')
35.
36.     print('Enter QUIT to stop, or press Enter to play again.')
37.     response = input('> ').upper()
38.     if response == 'QUIT':
39.         print('Thanks for playing!')
40.         sys.exit()
```

探索程序

请尝试找出以下问题的答案。你需要尝试对代码进行一些修改，并再次运行程序，查看修改后的效果。

1. 如果将第 21 行的 `drawTime = time.time()` 改为 `drawTime = 0`，会发生什么？
2. 如果将第 28 行的 `timeElapsed > 0.3` 改为 `timeElapsed < 0.3`，会发生什么？
3. 如果将第 23 行的 `time.time() - drawTime` 改为 `time.time() + drawTime`，会发生什么？
4. 如果删除或注释掉第 14 行的 `input('Press Enter to begin...')`，会发生什么？

项目 26

斐波那契数列

斐波那契数列是 13 世纪意大利数学家斐波那契提出的著名数学模式。序列从 0 和 1 开始，下一个数字总是前两个数字之和。该序列无限继续：

0,1,1,2,3,5,8,13,21,34,55,89,144,233,377,610,987…

斐波那契数列在音乐创作、股市预测、绘制向日葵的花盘图案以及许多其他领域都有应用。本项目中实现的程序可以让我们计算出自己想要的任意长度的数列。

运行程序

运行 fibonacci.py，输出如下所示：

```
Fibonacci Sequence, by Al Sweigart al@inventwithpython.com
--snip--
Enter the Nth Fibonacci number you wish to
calculate (such as 5, 50, 1000, 9999), or QUIT to quit:
> 50

0, 1, 1, 2, 3, 5, 8, 13, 21, 34, 55, 89, 144, 233, 377, 610, 987, 1597, 2584,
4181, 6765, 10946, 17711, 28657, 46368, 75025, 121393, 196418, 317811, 514229,
832040, 1346269, 2178309, 3524578, 5702887, 9227465, 14930352, 24157817,
39088169, 63245986, 102334155, 165580141, 267914296, 433494437, 701408733,
1134903170, 1836311903, 2971215073, 4807526976, 7778742049
```

工作原理

由于斐波那契数很快会变得非常大，第 45~49 行代码检查用户输入的数字是否大于或等于 10000，并显示警告信息，提示整个输出结果在屏幕上完成显示可能需要一些时间。虽然程序可以快速完成数百万次计算，但将文本显示到屏幕上的速度相对较慢，可能需要几秒。程序中的警告信息提醒用户总是可以通过按 Ctrl+C 键终止程序。

```
1. """斐波那契数列, 作者: Al Sweigart al@inventwithpython.com
2. 计算指定长度的斐波那契数列: 0, 1, 1, 2, 3, 5, 8, 13…
3. 标签: 简短, 数学"""
4.
5. import sys
6.
7. print('''Fibonacci Sequence, by Al Sweigart al@inventwithpython.com
```

```
  8.
  9. The Fibonacci sequence begins with 0 and 1, and the next number is the
 10. sum of the previous two numbers. The sequence continues forever:
 11.
 12. 0, 1, 1, 2, 3, 5, 8, 13, 21, 34, 55, 89, 144, 233, 377, 610, 987...
 13. '''
 14.
 15. while True:  # 主循环
 16.     while True:  # 持续询问，直到玩家输入有效数字
 17.         print('Enter the Nth Fibonacci number you wish to')
 18.         print('calculate (such as 5, 50, 1000, 9999), or QUIT to quit:')
 19.         response = input('> ').upper()
 20.
 21.         if response == 'QUIT':
 22.             print('Thanks for playing!')
 23.             sys.exit()
 24.
 25.         if response.isdecimal() and int(response) != 0:
 26.             nth = int(response)
 27.             break  # 当玩家输入有效数字时，循环结束
 28.
 29.         print('Please enter a number greater than 0, or QUIT.')
 30.     print()
 31.
 32.     # 处理用户输入 1 或 2 的特殊情况
 33.     if nth == 1:
 34.         print('0')
 35.         print()
 36.         print('The #1 Fibonacci number is 0.')
 37.         continue
 38.     elif nth == 2:
 39.         print('0, 1')
 40.         print()
 41.         print('The #2 Fibonacci number is 1.')
 42.         continue
 43.
 44.     # 如果玩家输入了较大的数字，则显示警告信息
 45.     if nth >= 10000:
 46.         print('WARNING: This will take a while to display on the')
 47.         print('screen. If you want to quit this program before it is')
 48.         print('done, press Ctrl-C.')
 49.         input('Press Enter to begin...')
 50.
 51.     # 计算第 N 个斐波那契数
 52.     secondToLastNumber = 0
 53.     lastNumber = 1
 54.     fibNumbersCalculated = 2
 55.     print('0, 1, ', end='')  # 显示前两个斐波那契数
 56.
 57.     # 显示斐波那契数列中其余的数字
 58.     while True:
 59.         nextNumber = secondToLastNumber + lastNumber
 60.         fibNumbersCalculated += 1
 61.
 62.         # 显示数列中的下一个数字
 63.         print(nextNumber, end='')
 64.
 65.         # 检查是否找到了玩家想要的第 N 个数字
 66.         if fibNumbersCalculated == nth:
 67.             print()
 68.             print()
 69.             print('The #', fibNumbersCalculated, ' Fibonacci ',
```

```
70.                     'number is ', nextNumber, sep='')
71.             break
72.
73.         # 在数字之间输出逗号
74.         print(', ', end='')
75.
76.         # 移动最后两个数字
77.         secondToLastNumber = lastNumber
78.         lastNumber = nextNumber
```

输入源代码并运行几次后,请尝试对其进行更改。你也可以尝试以下操作。

❑ 使用不同于 0 和 1 的起始数字。

❑ 通过累加前 3 个数字而不是前 2 个数字来创建下一个数字。

探索程序

以上是一个很基础的程序,没有太多的自定义空间。不妨想一想,如何使用这个程序呢?还有哪些其他有用的序列可以通过编程实现呢?

项目 27

虚拟水族箱

在本项目中，你可以在安装了氧气泵并养有海带植物的虚拟水族箱中观赏虚拟鱼。每次运行，程序都会用不同的类型和颜色随机生成鱼。让我们放松一下，享受这个虚拟水族箱所带来的宁静、美好氛围。当然，我们也可以尝试在程序中加入虚拟鲨鱼，用来吓唬其他"社区居民"！我们无法在 IDE 或编辑器中运行程序。这个程序使用了 bext 模块，因此必须从命令提示符窗口或终端窗口运行，才能得以正确显示。

运行程序

运行 fishtank.py，输出如下图所示。

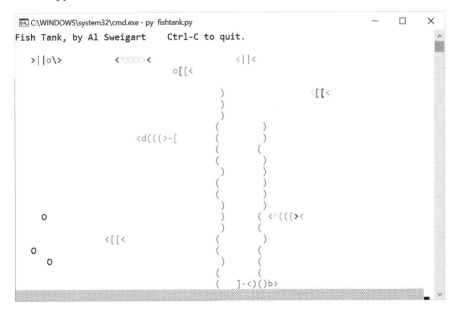

工作原理

通常，现代图形程序通过擦除整个窗口中的内容并每秒实现 30 或 60 次重绘来生成动画。

也就是说，帧速率为 30 帧/秒或 60 帧/秒。帧速率越高，动画动作就越流畅。

实现到终端窗口的绘制要慢得多。如果我们使用 bext 模块擦除整个终端窗口中的内容并加以重绘，通常只能达到大约 3 或 4 帧/秒，这会导致窗口出现明显的闪烁。

我们可以通过仅在终端窗口已更改的部分绘制字符来加快速度。本项目程序的大部分输出是空白空间，所以要让元素移动，clearAquarium() 只需要在鱼、海带和气泡当前所在的位置绘制空格字符。这样可以提高帧速率，减少闪烁，从而制作出更为赏心悦目的水族箱动画。

```
 1. """水族箱，作者：Al Sweigart al@inventwithpython.com
 2. 静态的水族箱。 按 Ctrl-C 停止
 3. 类似于 ASCIIQuarium 或 @EmojiAquarium，但本项目是一个基于较旧的 DOS 版 ASCII 艺术画效果的水族箱
 4. 标签：特大，艺术，bext 模块"""
 5.
 6. import random, sys, time
 7.
 8. try:
 9.     import bext
10. except ImportError:
11.     print('This program requires the bext module, which you')
12.     print('can install by following the instructions at')
13.     print('https://pypi.org/project/Bext/')
14.     sys.exit()
15.
16. # 创建常量
17. WIDTH, HEIGHT = bext.size()
18. # 如果不自动添加换行符，那么我们无法输出到 Windows 上的最后一列，所以将宽度减 1
19. WIDTH -= 1
20.
21. NUM_KELP = 2  # (!) 尝试将 2 改为 10
22. NUM_FISH = 10  # (!) 尝试将 10 改为 2 或 100
23. NUM_BUBBLERS = 1  # (!) 尝试将 1 改为 0 或 10
24. FRAMES_PER_SECOND = 4  # (!) 尝试将 4 改为 1 或 60
25. # (!) 尝试更改常量，以创建一个只有海带的水族箱
26. # 或仅有氧气泵的水族箱
27.
28. # 注意：FISH_TYPES 字典中每个字符串的长度应该相同
29. FISH_TYPES = [
30.   {'right': ['><>'],          'left': ['<><']},
31.   {'right': ['>||>'],         'left': ['<||<']},
32.   {'right': ['>))>'],         'left': ['<[[<']},
33.   {'right': ['>||o', '>||.'], 'left': ['o||<', '.||<']},
34.   {'right': ['>))o', '>)).'], 'left': ['o[[<', '.[[<']},
35.   {'right': ['>-==>'],        'left': ['<==-<']},
36.   {'right': [r'>\\>'],        'left': ['<//<']},
37.   {'right': ['><)))*>'],      'left': ['<*(((><']},
38.   {'right': ['}-[[[*>'],      'left': ['<*]]]-{']},
39.   {'right': [']-<)))b>'],     'left': ['<d(((>-[']},
40.   {'right': ['><XXX*>'],      'left': ['<*XXX><']},
41.   {'right': ['_.-._.-^=>',    '.-._.-.-^=>',
42.              '-._.-._^=>',    '_.-._.-.^=>'],
43.    'left':  ['<=^-._.-._',    '<=^.-._.-.',
44.              '<=^_.-._.-',    '<=^_.-._.']},
45. ]  # (!) 尝试将鱼添加到 FISH_TYPES 中
46. LONGEST_FISH_LENGTH = 10  # FISH_TYPES 中最长的单个字符串
47.
48. # 鱼跑到屏幕边缘的 x 和 y 位置
49. LEFT_EDGE = 0
50. RIGHT_EDGE = WIDTH - 1 - LONGEST_FISH_LENGTH
51. TOP_EDGE = 0
```

```
52. BOTTOM_EDGE = HEIGHT - 2
53.
54.
55. def main():
56.     global FISHES, BUBBLERS, BUBBLES, KELPS, STEP
57.     bext.bg('black')
58.     bext.clear()
59.
60.     # 生成全局变量
61.     FISHES = []
62.     for i in range(NUM_FISH):
63.         FISHES.append(generateFish())
64.
65.     # 注意：气泡是绘制出来的，而不是氧气泵生成的
66.     BUBBLERS = []
67.     for i in range(NUM_BUBBLERS):
68.         # 每个气泡是从随机位置开始的
69.         BUBBLERS.append(random.randint(LEFT_EDGE, RIGHT_EDGE))
70.     BUBBLES = []
71.
72.     KELPS = []
73.     for i in range(NUM_KELP):
74.         kelpx = random.randint(LEFT_EDGE, RIGHT_EDGE)
75.         kelp = {'x': kelpx, 'segments': []}
76.         # 生成海带
77.         for i in range(random.randint(6, HEIGHT - 1)):
78.             kelp['segments'].append(random.choice(['(', ')']))
79.         KELPS.append(kelp)
80.
81.     # 进行模拟
82.     STEP = 1
83.     while True:
84.         simulateAquarium()
85.         drawAquarium()
86.         time.sleep(1 / FRAMES_PER_SECOND)
87.         clearAquarium()
88.         STEP += 1
89.
90.
91. def getRandomColor():
92.     """返回一个随机颜色的字符串"""
93.     return random.choice(('black', 'red', 'green', 'yellow', 'blue',
94.                           'purple', 'cyan', 'white'))
95.
96.
97. def generateFish():
98.     """返回代表一条鱼的字典"""
99.     fishType = random.choice(FISH_TYPES)
100.
101.     # 为鱼文本中的每个字符设置颜色
102.     colorPattern = random.choice(('random', 'head-tail', 'single'))
103.     fishLength = len(fishType['right'][0])
104.     if colorPattern == 'random':  # 鱼的所有部件是随机着色的
105.         colors = []
106.         for i in range(fishLength):
107.             colors.append(getRandomColor())
108.     if colorPattern == 'single' or colorPattern == 'head-tail':
109.         colors = [getRandomColor()] * fishLength  # 都是一样的颜色
110.     if colorPattern == 'head-tail':  # 鱼的头部/尾部与其身体部分颜色不同
111.         headTailColor = getRandomColor()
112.         colors[0] = headTailColor   # 设置鱼头部的颜色
113.         colors[-1] = headTailColor  # 设置鱼尾部的颜色
```

```
114.
115.        # 设置其余的 fish 数据结构
116.        fish = {'right':              fishType['right'],
117.                'left':                fishType['left'],
118.                'colors':              colors,
119.                'hSpeed':              random.randint(1, 6),
120.                'vSpeed':              random.randint(5, 15),
121.                'timeToHDirChange':    random.randint(10, 60),
122.                'timeToVDirChange':    random.randint(2, 20),
123.                'goingRight':          random.choice([True, False]),
124.                'goingDown':           random.choice([True, False])}
125.
126.        # 'x' 总是在鱼体的最左边
127.        fish['x'] = random.randint(0, WIDTH - 1 - LONGEST_FISH_LENGTH)
128.        fish['y'] = random.randint(0, HEIGHT - 2)
129.        return fish
130.
131.
132.    def simulateAquarium():
133.        """模拟一帧水族箱动画"""
134.        global FISHES, BUBBLERS, BUBBLES, KELP, STEP
135.
136.        # 模拟一帧鱼的动画
137.        for fish in FISHES:
138.            # 让鱼水平移动
139.            if STEP % fish['hSpeed'] == 0:
140.                if fish['goingRight']:
141.                    if fish['x'] != RIGHT_EDGE:
142.                        fish['x'] += 1    # 让鱼向右移动
143.                    else:
144.                        fish['goingRight'] = False    # 让鱼掉头
145.                        fish['colors'].reverse()    # 重新着色
146.                else:
147.                    if fish['x'] != LEFT_EDGE:
148.                        fish['x'] -= 1    # 让鱼向左移动
149.                    else:
150.                        fish['goingRight'] = True    # 让鱼掉头
151.                        fish['colors'].reverse()    # 重新着色
152.
153.            #可以随机改变鱼的水平方向
154.            fish['timeToHDirChange'] -= 1
155.            if fish['timeToHDirChange'] == 0:
156.                fish['timeToHDirChange'] = random.randint(10, 60)
157.                # 让鱼掉头
158.                fish['goingRight'] = not fish['goingRight']
159.
160.            # 让鱼垂直移动
161.            if STEP % fish['vSpeed'] == 0:
162.                if fish['goingDown']:
163.                    if fish['y'] != BOTTOM_EDGE:
164.                        fish['y'] += 1    # 让鱼往下移动
165.                    else:
166.                        fish['goingDown'] = False    # 让鱼掉头
167.                else:
168.                    if fish['y'] != TOP_EDGE:
169.                        fish['y'] -= 1    # 让鱼向上移动
170.                    else:
171.                        fish['goingDown'] = True    # 让鱼掉头
172.
173.            #可以随机改变鱼的垂直方向
174.            fish['timeToVDirChange'] -= 1
175.            if fish['timeToVDirChange'] == 0:
```

```
176.            fish['timeToVDirChange'] = random.randint(2, 20)
177.            # 让鱼掉头
178.            fish['goingDown'] = not fish['goingDown']
179.
180.        # 让氧气泵产生气泡
181.        for bubbler in BUBBLERS:
182.            # 有 1/5 的机会可以生成气泡
183.            if random.randint(1, 5) == 1:
184.                BUBBLES.append({'x': bubbler, 'y': HEIGHT - 2})
185.
186.        # 移动气泡
187.        for bubble in BUBBLES:
188.            diceRoll = random.randint(1, 6)
189.            if (diceRoll == 1) and (bubble['x'] != LEFT_EDGE):
190.                bubble['x'] -= 1    # 让气泡向左移动
191.            elif (diceRoll == 2) and (bubble['x'] != RIGHT_EDGE):
192.                bubble['x'] += 1    # 让气泡向右移动
193.
194.            bubble['y'] -= 1    # 让气泡总是向上移动
195.
196.        # 反向迭代 BUBBLES，因为在迭代气泡时会将其从 BUBBLES 中删除
197.
198.        for i in range(len(BUBBLES) - 1, -1, -1):
199.            if BUBBLES[i]['y'] == TOP_EDGE:    # 删除到达顶部的气泡
200.                del BUBBLES[i]
201.
202.        # 模拟一帧海带摆动的动画
203.        for kelp in KELPS:
204.            for i, kelpSegment in enumerate(kelp['segments']):
205.                # 有 1/20 的概率改变摆动
206.                if random.randint(1, 20) == 1:
207.                    if kelpSegment == '(':
208.                        kelp['segments'][i] = ')'
209.                    elif kelpSegment == ')':
210.                        kelp['segments'][i] = '('
211.
212.
213. def drawAquarium():
214.     """在屏幕上绘出水族箱"""
215.     global FISHES, BUBBLERS, BUBBLES, KELP, STEP
216.
217.     # 绘制退出消息
218.     bext.fg('white')
219.     bext.goto(0, 0)
220.     print('Fish Tank, by Al Sweigart    Ctrl-C to quit.', end='')
221.
222.     # 绘制气泡
223.     bext.fg('white')
224.     for bubble in BUBBLES:
225.         bext.goto(bubble['x'], bubble['y'])
226.         print(random.choice(('o', 'O')), end='')
227.
228.     # 绘制鱼
229.     for fish in FISHES:
230.         bext.goto(fish['x'], fish['y'])
231.
232.         # 获取正确的朝右或朝左的鱼文本
233.         if fish['goingRight']:
234.             fishText = fish['right'][STEP % len(fish['right'])]
235.         else:
236.             fishText = fish['left'][STEP % len(fish['left'])]
237.
```

```
238.        # 用正确的颜色绘制鱼文本的每个字符
239.        for i, fishPart in enumerate(fishText):
240.            bext.fg(fish['colors'][i])
241.            print(fishPart, end='')
242.
243.        # 绘制海带
244.        bext.fg('green')
245.        for kelp in KELPS:
246.            for i, kelpSegment in enumerate(kelp['segments']):
247.                if kelpSegment == '(':
248.                    bext.goto(kelp['x'], BOTTOM_EDGE - i)
249.                elif kelpSegment == ')':
250.                    bext.goto(kelp['x'] + 1, BOTTOM_EDGE - i)
251.                print(kelpSegment, end='')
252.
253.        # 在水族箱底部绘制沙子
254.        bext.fg('yellow')
255.        bext.goto(0, HEIGHT - 1)
256.        print(chr(9617) * (WIDTH - 1), end='')  # 绘制沙子
257.
258.        sys.stdout.flush()
259.
260.
261.    def clearAquarium():
262.        """在屏幕上绘制空白区域"""
263.        global FISHES, BUBBLERS, BUBBLES, KELP
264.
265.        # 绘制气泡
266.        for bubble in BUBBLES:
267.            bext.goto(bubble['x'], bubble['y'])
268.            print(' ', end='')
269.
270.        # 绘制鱼
271.        for fish in FISHES:
272.            bext.goto(fish['x'], fish['y'])
273.
274.            # 用正确的颜色绘制鱼文本的每个字符
275.            print(' ' * len(fish['left'][0]), end='')
276.
277.        # 绘制海带
278.        for kelp in KELPS:
279.            for i, kelpSegment in enumerate(kelp['segments']):
280.                bext.goto(kelp['x'], HEIGHT - 2 - i)
281.                print('  ', end='')
282.
283.        sys.stdout.flush()
284.
285.
286.    # 程序运行入口（如果不是作为模块导入的话）
287.    if __name__ == '__main__':
288.        try:
289.            main()
290.        except KeyboardInterrupt:
291.            sys.exit()  # 按 Ctrl-C，程序结束
```

输入源代码并运行几次后，请尝试对其进行更改。你可以参考标有!的注释对程序加以修改，也可以尝试以下操作。

❏ 在水族箱沙子底部添加移动的螃蟹。

❏ 添加一个随机出现在沙子上的 ASCII 艺术城堡。

❑ 让鱼随机地加快游动速度。

探索程序

请尝试找出以下问题的答案。你需要尝试对代码进行一些修改，再次运行程序，查看修改后的效果。

1. 如果将第 46 行的 LONGEST_FISH_LENGTH = 10 改为 LONGEST_FISH_LENGTH = 50，则会发生什么？
2. 如果将第 116 行的 'right':fishType['right'] 改为 'right':fishType['left']，则会发生什么？
3. 如果将第 244 行的 bext.fg('green') 改为 bext.fg('red')，则会发生什么？
4. 如果删除或注释掉第 87 行的 clearAquarium()，则会发生什么？
5. 如果将第 240 行的 bext.fg(fish['colors'][i]) 改为 bext.fg('random')，则会发生什么？
6. 如果将第 156 行的 random.randint(10, 60) 改为 1，则会发生什么？

项目 28

Flooder游戏

Flooder("漫画家")是一款色彩缤纷的游戏。玩家通过改变界面左上角图块的颜色,来用单一颜色填满整个游戏面板。新颜色将扩散到与原始颜色匹配的所有相邻图块。Flooder类似于Flood It手机游戏,这个程序还有一种色盲模式,是使用形状而不是图块填色。它依赖递归泛洪填充算法来绘制游戏面板。其工作方式类似于许多绘画程序中的"paint bucket"(油漆桶)或"fill"(填充)工具。

Flooder游戏的玩法如下:玩家从界面左顶角的图块开始,选择一种不同于左上角图块颜色的另一种颜色进行颜色变换,仔细观察各个图块的分布情况,选择最有利于颜色扩散的颜色进行变换,在20步以内将所有图块变换成同一种颜色。

运行程序

运行 flooder.py,输出如下所示:

工作原理

无障碍性是电子游戏中的一个大问题，解决办法也有多种。例如，在绿色盲或红绿色盲来看，红色和绿色的阴影看起来是相同的，这导致他们难以区分屏幕上的红色和绿色对象。我们可以使用不同形状（而非不同颜色）的模式来为 Flooder 实现无障碍性。注意，即使是色盲模式，仍然需要使用颜色。这意味着你可以取消"标准"模式，甚至可以让非色盲用户在色盲模式下进行游戏。最好从一开始就将无障碍设计纳入考虑范围，而不是将它们作为单独的模式添加。只有这样，才能减少必须编写的代码量，也能让任何未来的错误修复工作更容易实施。

无障碍性设计还包括确保文字足够大以便为弱视者提供阅读方便，为声音效果提供视觉线索，为听力障碍者提供口语替代提示方式（例如字幕），将控件对应到其他键盘键以便玩家能单手玩游戏。YouTube 频道 Game Maker's Toolkit 有一个名为"Designing for Disability"视频系列，涵盖了设计游戏时无障碍性考虑的许多方面。

```
1. """Flooder, 作者: Al Sweigart al@inventwithpython.com
2. 一款色彩缤纷的游戏，你可以尝试用单一颜色填满游戏
3. 这款游戏设有色盲模式
4. 灵感来自"Flood It" 游戏
5. 标签: 大, Bext, 游戏"""
6.
7. import random, sys
8.
9. try:
10.     import bext
11. except ImportError:
12.     print('This program requires the bext module, which you')
13.     print('can install by following the instructions at')
14.     print('https://pypi.org/project/Bext/')
15.     sys.exit()
16.
17. # 创建常量
18. BOARD_WIDTH = 16    # (!) 尝试将 16 更改为 4 或 40
19. BOARD_HEIGHT = 14   # (!) 尝试将 14 更改为 4 或 20
20. MOVES_PER_GAME = 20 # (!) 尝试将 20 更改为 3 或 300
21.
22. # 色盲模式下使用不同形状的常量
23. HEART      = chr(9829)  # 字符 9829 是 "♥"
24. DIAMOND    = chr(9830)  # 字符 9830 是 "♦"
25. SPADE      = chr(9824)  # 字符 9824 是 "♠"
26. CLUB       = chr(9827)  # 字符 9827 是 "♣"
27. BALL       = chr(9679)  # 字符 9679 是 "●"
28. TRIANGLE   = chr(9650)  # 字符 9650 是 "▲"
29.
30. BLOCK      = chr(9608)  # 字符 9608 是 "█"
31. LEFTRIGHT  = chr(9472)  # 字符 9472 是 "─"
32. UPDOWN     = chr(9474)  # 字符 9474 是 "│"
33. DOWNRIGHT  = chr(9484)  # 字符 9484 是 "┌"
34. DOWNLEFT   = chr(9488)  # 字符 9488 是 "┐"
35. UPRIGHT    = chr(9492)  # 字符 9492 是 "└"
36. UPLEFT     = chr(9496)  # 字符 9496 是 "┘"
37.
38. # 游戏面板上使用的所有颜色/形状图块
39. TILE_TYPES = (0, 1, 2, 3, 4, 5)
40. COLORS_MAP = {0: 'red', 1: 'green', 2:'blue',
```

```
41.                    3:'yellow', 4:'cyan', 5:'purple'}
42. COLOR_MODE = 'color mode'
43. SHAPES_MAP = {0: HEART, 1: TRIANGLE, 2: DIAMOND,
44.                3: BALL, 4: CLUB, 5: SPADE}
45. SHAPE_MODE = 'shape mode'
46.
47.
48. def main():
49.     bext.bg('black')
50.     bext.fg('white')
51.     bext.clear()
52.     print('''Flooder, by Al Sweigart al@inventwithpython.com
53.
54. Set the upper left color/shape, which fills in all the
55. adjacent squares of that color/shape. Try to make the
56. entire board the same color/shape.''')
57.
58.     print('Do you want to play in colorblind mode? Y/N')
59.     response = input('> ')
60.     if response.upper().startswith('Y'):
61.         displayMode = SHAPE_MODE
62.     else:
63.         displayMode = COLOR_MODE
64.
65.     gameBoard = getNewBoard()
66.     movesLeft = MOVES_PER_GAME
67.
68.     while True:  # 主循环
69.         displayBoard(gameBoard, displayMode)
70.
71.         print('Moves left:', movesLeft)
72.         playerMove = askForPlayerMove(displayMode)
73.         changeTile(playerMove, gameBoard, 0, 0)
74.         movesLeft -= 1
75.
76.         if hasWon(gameBoard):
77.             displayBoard(gameBoard, displayMode)
78.             print('You have won!')
79.             break
80.         elif movesLeft == 0:
81.             displayBoard(gameBoard, displayMode)
82.             print('You have run out of moves!')
83.             break
84.
85.
86. def getNewBoard():
87.     """返回一个新的待着色的游戏面板字典"""
88.
89.     # 键是 (x,y) 元组，值是该位置的图块
90.     board = {}
91.
92.     # 为游戏面板创建随机颜色
93.     for x in range(BOARD_WIDTH):
94.         for y in range(BOARD_HEIGHT):
95.             board[(x, y)] = random.choice(TILE_TYPES)
96.
97.     # 制作几个与其邻居相同的图块。这会创建颜色/形状相同的图块组
98.     for i in range(BOARD_WIDTH * BOARD_HEIGHT):
99.         x = random.randint(0, BOARD_WIDTH - 2)
100.        y = random.randint(0, BOARD_HEIGHT - 1)
101.        board[(x + 1, y)] = board[(x, y)]
102.    return board
```

```python
103.
104.
105. def displayBoard(board, displayMode):
106.     """在屏幕上显示游戏面板"""
107.     bext.fg('white')
108.     # 显示游戏面板的上边缘
109.     print(DOWNRIGHT + (LEFTRIGHT * BOARD_WIDTH) + DOWNLEFT)
110.
111.     # 显示每一行
112.     for y in range(BOARD_HEIGHT):
113.         bext.fg('white')
114.         if y == 0:  # 第一行以>开头
115.             print('>', end='')
116.         else:  # 后面的行以白色垂直线开始
117.             print(UPDOWN, end='')
118.
119.         # 显示此行中的每个图块
120.         for x in range(BOARD_WIDTH):
121.             bext.fg(COLORS_MAP[board[(x, y)]])
122.             if displayMode == COLOR_MODE:
123.                 print(BLOCK, end='')
124.             elif displayMode == SHAPE_MODE:
125.                 print(SHAPES_MAP[board[(x, y)]], end='')
126.
127.         bext.fg('white')
128.         print(UPDOWN)  # 行以白色垂直线结束
129.     # 显示游戏面板的下边缘
130.     print(UPRIGHT + (LEFTRIGHT * BOARD_WIDTH) + UPLEFT)
131.
132.
133. def askForPlayerMove(displayMode):
134.     """让玩家选用一种颜色来绘制左上角的图块"""
135.     while True:
136.         bext.fg('white')
137.         print('Choose one of ', end='')
138.
139.         if displayMode == COLOR_MODE:
140.             bext.fg('red')
141.             print('(R)ed ', end='')
142.             bext.fg('green')
143.             print('(G)reen ', end='')
144.             bext.fg('blue')
145.             print('(B)lue ', end='')
146.             bext.fg('yellow')
147.             print('(Y)ellow ', end='')
148.             bext.fg('cyan')
149.             print('(C)yan ', end='')
150.             bext.fg('purple')
151.             print('(P)urple ', end='')
152.         elif displayMode == SHAPE_MODE:
153.             bext.fg('red')
154.             print('(H)eart, ', end='')
155.             bext.fg('green')
156.             print('(T)riangle, ', end='')
157.             bext.fg('blue')
158.             print('(D)iamond, ', end='')
159.             bext.fg('yellow')
160.             print('(B)all, ', end='')
161.             bext.fg('cyan')
162.             print('(C)lub, ', end='')
163.             bext.fg('purple')
164.             print('(S)pade, ', end='')
```

```python
165.            bext.fg('white')
166.            print('or QUIT:')
167.            response = input('> ').upper()
168.            if response == 'QUIT':
169.                print('Thanks for playing!')
170.                sys.exit()
171.            if displayMode == COLOR_MODE and response in tuple('RGBYCP'):
172.                # 根据 response 返回图块类型编号
173.                return {'R': 0, 'G': 1, 'B': 2,
174.                        'Y': 3, 'C': 4, 'P': 5}[response]
175.            if displayMode == SHAPE_MODE and response in tuple('HTDBCS'):
176.                # 根据 response 返回图块类型编号
177.                return {'H': 0, 'T': 1, 'D':2,
178.                        'B': 3, 'C': 4, 'S': 5}[response]
179.
180.
181.    def changeTile(tileType, board, x, y, charToChange=None):
182.        """使用递归泛洪填充算法来更改图块的颜色/形状"""
183.        if x == 0 and y == 0:
184.            charToChange = board[(x, y)]
185.            if tileType == charToChange:
186.                return  # 终止条件：游戏面板上的均为同一种图块
187.
188.        board[(x, y)] = tileType
189.
190.        if x > 0 and board[(x - 1, y)] == charToChange:
191.            # 递归：更改位于左边一格的图块
192.            changeTile(tileType, board, x - 1, y, charToChange)
193.        if y > 0 and board[(x, y - 1)] == charToChange:
194.            # 递归：更改位于上方一格的图块
195.            changeTile(tileType, board, x, y - 1, charToChange)
196.        if x < BOARD_WIDTH - 1 and board[(x + 1, y)] == charToChange:
197.            # 递归：更改位于右边一格的图块
198.            changeTile(tileType, board, x + 1, y, charToChange)
199.        if y < BOARD_HEIGHT - 1 and board[(x, y + 1)] == charToChange:
200.            # 递归：更改位于下方一格的图块
201.            changeTile(tileType, board, x, y + 1, charToChange)
202.
203.
204.    def hasWon(board):
205.        """如果整个游戏面板上的图块是同一种颜色/形状，则返回 True"""
206.        tile = board[(0, 0)]
207.
208.        for x in range(BOARD_WIDTH):
209.            for y in range(BOARD_HEIGHT):
210.                if board[(x, y)] != tile:
211.                    return False
212.        return True
213.
214.
215.    # 程序运行入口（如果不是作为模块导入）
216.    if __name__ == '__main__':
217.        main()
```

输入源代码并运行几次后，请尝试对其进行更改。你可以参考标有!的注释对程序加以修改，也可以尝试以下操作。

❑ 添加其他形状和颜色。

❑ 创建除矩形之外的游戏面板形状。

探索程序

请尝试找出以下问题的答案。你需要尝试对代码进行一些修改，再次运行程序，查看修改后的效果。

1. 如果将第 90 行的 `board = {}` 改为 `board = []`，则会得到什么错误消息？
2. 如果将第 102 行的 `return board` 改为 `return None`，则会得到什么错误消息？
3. 如果将第 74 行的 `movesLeft -= 1` 改为 `movesLeft -= 0`，则会发生什么？

项目 29

森林火灾模拟

运行本项目的程序后,你会看到,界面上显现出一片森林,树不断生长,然后被闪电击中并烧毁。在模拟的每一步,空白区域变成一棵树的概率和一棵树被闪电击中并烧毁的概率都为1%。火灾会蔓延到相邻的树木,因此密集的森林比稀疏的森林更容易遭受更大的火灾。本项目的灵感来自 Nicky Case 的 Emoji Sim,具体请参见其个人网站。

运行程序

运行 forestfiresim.py,输出如下所示:

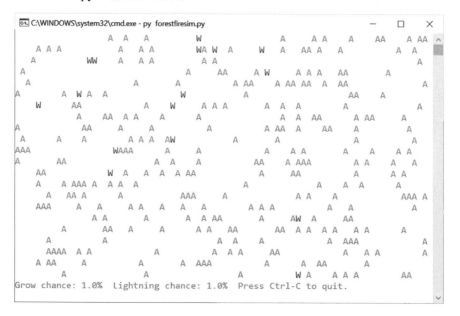

工作原理

这种模拟是突发行为的一个例子——系统中简单部件之间的相互作用产生了复杂的模式。

空白区域长出树，闪电击中并点燃树，火把树烧毁使之变回空白区域，同时火灾蔓延到邻近的树。通过调整树木的生长率和雷击燃烧率，可以使森林显示不同的现象。例如，低雷击燃烧率、高生长率会导致大而持续的森林火灾，因为树木往往彼此靠近并迅速着火；低生长率、高雷击燃烧率会产生几处小火，由于附近缺乏树木，这些火很快就会熄灭。我们没有明确地针对任何这些行为进行编程，而是让它们自然而然地从所创建的系统中"生成"。

```
 1. """森林火灾模拟，作者：Al Sweigart al@inventwithpython.com
 2. 模拟在森林中蔓延的火。 按 Ctrl-C 停止
 3. 标签：简短，bext 模块，模拟"""
 4.
 5. import random, sys, time
 6.
 7. try:
 8.     import bext
 9. except ImportError:
10.     print('This program requires the bext module, which you')
11.     print('can install by following the instructions at')
12.     print('https://pypi.org/project/Bext/')
13.     sys.exit()
14.
15. # 创建常量
16. WIDTH = 79
17. HEIGHT = 22
18.
19. TREE = 'A'
20. FIRE = 'W'
21. EMPTY = ' '
22.
23. # (!) 尝试将以下这些值更改为 0.0 和 1.0 之间的任意值
24. INITIAL_TREE_DENSITY = 0.20  # 森林最初的稠密度
25. GROW_CHANCE = 0.01           # 一个空白区域长出一棵树的概率
26. FIRE_CHANCE = 0.01           # 一棵树被闪电击中并烧毁的概率
27.
28. # (!) 尝试将暂停长度设置为 1.0 或 0.0
29. PAUSE_LENGTH = 0.5
30.
31.
32. def main():
33.     forest = createNewForest()
34.     bext.clear()
35.
36.     while True:   # 主循环
37.         displayForest(forest)
38.
39.         # 运行单个模拟步骤
40.         nextForest = {'width': forest['width'],
41.                       'height': forest['height']}
42.
43.         for x in range(forest['width']):
44.             for y in range(forest['height']):
45.                 if (x, y) in nextForest:
46.                     # 如果我们已经在上一次迭代中设置了 nextForest[(x, y)]
47.                     # 那么在这里什么都不做
48.                     continue
49.
50.                 if ((forest[(x, y)] == EMPTY)
51.                     and (random.random() <= GROW_CHANCE)):
52.                     # 在当前空白区域生成一棵树
53.                     nextForest[(x, y)] = TREE
```

```
54.                    elif ((forest[(x, y)] == TREE)
55.                        and (random.random() <= FIRE_CHANCE)):
56.                        # 闪电击中并点燃了这棵树
57.                        nextForest[(x, y)] = FIRE
58.                    elif forest[(x, y)] == FIRE:
59.                        # 这棵树正在燃烧
60.                        # 循环遍历所有相邻的空间
61.                        for ix in range(-1, 2):
62.                            for iy in range(-1, 2):
63.                                # 火蔓延到邻近的树木
64.                                if forest.get((x + ix, y + iy)) == TREE:
65.                                    nextForest[(x + ix, y + iy)] = FIRE
66.                        # 这棵树已烧毁，则将其删除
67.                        nextForest[(x, y)] = EMPTY
68.                    else:
69.                        # 只需复制现存的对象
70.                        nextForest[(x, y)] = forest[(x, y)]
71.            forest = nextForest
72.
73.            time.sleep(PAUSE_LENGTH)
74.
75.
76.    def createNewForest():
77.        """返回新森林数据结构的字典"""
78.        forest = {'width': WIDTH, 'height': HEIGHT}
79.        for x in range(WIDTH):
80.            for y in range(HEIGHT):
81.                if (random.random() * 100) <= INITIAL_TREE_DENSITY:
82.                    forest[(x, y)] = TREE    # 从树开始
83.                else:
84.                    forest[(x, y)] = EMPTY   # 从空白区域开始
85.        return forest
86.
87.
88.    def displayForest(forest):
89.        """在屏幕上显示森林数据结构"""
90.        bext.goto(0, 0)
91.        for y in range(forest['height']):
92.            for x in range(forest['width']):
93.                if forest[(x, y)] == TREE:
94.                    bext.fg('green')
95.                    print(TREE, end='')
96.                elif forest[(x, y)] == FIRE:
97.                    bext.fg('red')
98.                    print(FIRE, end='')
99.                elif forest[(x, y)] == EMPTY:
100.                    print(EMPTY, end='')
101.            print()
102.        bext.fg('reset')    # 使用默认的字体颜色
103.        print('Grow chance: {}%  '.format(GROW_CHANCE * 100), end='')
104.        print('Lightning chance: {}%  '.format(FIRE_CHANCE * 100), end='')
105.        print('Press Ctrl-C to quit.')
106.
107.
108.    # 程序运行入口（如果不作为模块导入的话）
109.    if __name__ == '__main__':
110.        try:
111.            main()
112.        except KeyboardInterrupt:
113.            sys.exit()    # 按 Ctrl-C 键，程序结束
```

输入源代码并运行几次后,请尝试对其进行更改。你可以根据标有(!)的注释对程序进行修改。你也可以自己尝试弄清楚如何执行以下操作。

- 添加随机创建的湖泊和河流,用作火焰无法穿越的防火带。
- 添加一棵树被相邻的树引燃的概率。
- 添加不同类型的树木,并使它们着火的概率各不相同。
- 为燃烧着的树木添加不同的状态,以模拟树木从被引燃到烧成灰烬的多个步骤。

探索程序

请尝试找出以下问题的答案。你需要尝试对代码进行一些修改,再次运行程序,查看修改后的效果。

1. 如果将第 94 行的 bext.fg('green') 改为 bext.fg('random'),则会发生什么?
2. 如果将第 21 行的 EMPTY = ' ' 改为 EMPTY = '.',则会发生什么?
3. 如果将第 64 行的 forest.get((x + ix, y + iy)) == TREE 改为 forest.get((x + ix, y + iy)) == EMPTY,则会发生什么?
4. 如果将第 67 行的 nextForest[(x, y)] = EMPTY 改为 nextForest[(x, y)] = FIRE,则会发生什么?
5. 如果将第 84 行的 forest[(x, y)] = EMPTY 改为 forest[(x, y)] = TREE,则会发生什么?

项目 30

四子棋

在四子棋这款经典的双人棋盘游戏中,你必须尽快将 4 枚棋子以水平、垂直或对角排列的方式排成一排,同时要阻止对手玩家做到这一点。这有点类似于 Connect Four 游戏。

运行程序

运行 fourinarow.py,输出如下所示:

```
Four in a Row, by Al Sweigart al@inventwithpython.com
--snip--
    1234567
   +-------+
   |.......|
   |.......|
   |.......|
   |.......|
   |.......|
   |.......|
   +-------+
Player X, enter a column or QUIT:
> 3

    1234567
   +-------+
   |.......|
   |.......|
   |.......|
   |.......|
   |.......|
   |..X....|
   +-------+
Player O, enter a column or QUIT:
> 5
--snip--
Player O, enter a column or QUIT:
> 4
```

```
 1234567
+-------+
|.......|
|.......|
|XXX.XO.|
|OOOOXO.|
|OOOXOX.|
|OXXXOXX|
+-------+
Player O has won!
```

工作原理

本书中的棋盘游戏项目遵循类似的程序结构。程序中通常会有用于表示棋盘状态的字典或列表、用于返回棋盘数据结构的 `getNewBoard()` 函数、用于在屏幕上渲染棋盘数据结构的 `displayBoard()` 函数等。你可以依据"棋盘游戏"标签查看本书中的其他项目并加以比较,尤其是当你想编写原创棋盘游戏程序时。

```
 1. """四子棋,作者: Al Sweigart al@inventwithpython.com
 2. 将4枚棋子连成一条线的游戏,类似于 Connect Four 游戏
 3. 标签: 大,游戏,棋盘,双人游戏"""
 4.
 5. import sys
 6.
 7. # 用于显示棋盘的常量
 8. EMPTY_SPACE = '.'  # 句号比空格更容易计数
 9. PLAYER_X = 'X'
10. PLAYER_O = 'O'
11.
12. # 注意: 如果 BOARD_WIDTH 被修改,则更新 displayBoard() 和 COLUMN_LABELS
13. BOARD_WIDTH = 7
14. BOARD_HEIGHT = 6
15. COLUMN_LABELS = ('1', '2', '3', '4', '5', '6', '7')
16. assert len(COLUMN_LABELS) == BOARD_WIDTH
17.
18.
19. def main():
20.     print("""Four in a Row, by Al Sweigart al@inventwithpython.com
21.
22. Two players take turns dropping tiles into one of seven columns, trying
23. to make four in a row horizontally, vertically, or diagonally.
24. """)
25.
26.     # 创建新游戏
27.     gameBoard = getNewBoard()
28.     playerTurn = PLAYER_X
29.
30.     while True:  # 玩家开始走子
31.         # 显示棋盘并获取玩家的下一步走子
32.         displayBoard(gameBoard)
33.         playerMove = askForPlayerMove(playerTurn, gameBoard)
34.         gameBoard[playerMove] = playerTurn
35.
36.         # 判断是否获胜或平局
37.         if isWinner(playerTurn, gameBoard):
38.             displayBoard(gameBoard)   # 最后一次显示棋盘
39.             print('Player ' + playerTurn + ' has won!')
```

```
40.         sys.exit()
41.     elif isFull(gameBoard):
42.         displayBoard(gameBoard)   # 最后一次显示棋盘
43.         print('There is a tie!')
44.         sys.exit()
45.
46.     # 轮到另一位玩家走子
47.     if playerTurn == PLAYER_X:
48.         playerTurn = PLAYER_O
49.     elif playerTurn == PLAYER_O:
50.         playerTurn = PLAYER_X
51.
52.
53. def getNewBoard():
54.     """返回一个 4 枚棋子成一线的字典
55.
56.     键是由两个整数 (columnIndex, rowIndex) 构成的元组
57.     值是 "X" "O" 或 "." （表示空白）字符之一"""
58.     board = {}
59.     for columnIndex in range(BOARD_WIDTH):
60.         for rowIndex in range(BOARD_HEIGHT):
61.             board[(columnIndex, rowIndex)] = EMPTY_SPACE
62.     return board
63.
64.
65. def displayBoard(board):
66.     """在屏幕上显示棋盘及棋子"""
67.
68.     '''Prepare a list to pass to the format() string method for the
69.     board template. The list holds all of the board's tiles (and empty
70.     spaces) going left to right, top to bottom:'''
71.     tileChars = []
72.     for rowIndex in range(BOARD_HEIGHT):
73.         for columnIndex in range(BOARD_WIDTH):
74.             tileChars.append(board[(columnIndex, rowIndex)])
75.
76.     # 显示棋盘
77.     print("""
78.      1234567
79.     +-------+
80.     |{}{}{}{}{}{}{}|
81.     |{}{}{}{}{}{}{}|
82.     |{}{}{}{}{}{}{}|
83.     |{}{}{}{}{}{}{}|
84.     |{}{}{}{}{}{}{}|
85.     |{}{}{}{}{}{}{}|
86.     +-------+""".format(*tileChars))
87.
88.
89. def askForPlayerMove(playerTile, board):
90.     """让玩家选择棋盘上的一列并落子
91.
92.     返回棋子所在的 (column, row) 的元组"""
93.     while True:    # 不断询问玩家，直到他们输入有效的下一步走子
94.         print('Player {}, enter a column or QUIT:'.format(playerTile))
95.         response = input('> ').upper().strip()
96.
97.         if response == 'QUIT':
98.             print('Thanks for playing!')
99.             sys.exit()
100.
101.         if response not in COLUMN_LABELS:
```

```
102.              print('Enter a number from 1 to {}.'.format(BOARD_WIDTH))
103.              continue  # 再次询问玩家的下一步走子
104.
105.          columnIndex = int(response) - 1  # -1 表示以 0 开头的索引
106.
107.          # 如果列已满，再次询问玩家的下一步走子
108.          if board[(columnIndex, 0)] != EMPTY_SPACE:
109.              print('That column is full, select another one.')
110.              continue  # 再次询问玩家的移动
111.
112.          # 从底部开始，找到第一个可落子区域
113.          for rowIndex in range(BOARD_HEIGHT - 1, -1, -1):
114.              if board[(columnIndex, rowIndex)] == EMPTY_SPACE:
115.                  return (columnIndex, rowIndex)
116.
117.
118. def isFull(board):
119.     """如果棋盘上没有可以落子的区域，则返回 True，否则返回 False"""
120.     for rowIndex in range(BOARD_HEIGHT):
121.         for columnIndex in range(BOARD_WIDTH):
122.             if board[(columnIndex, rowIndex)] == EMPTY_SPACE:
123.                 return False  # 找到一个可以落子的区域，所以返回 False
124.     return True  # 没有可以落子的区域
125.
126.
127. def isWinner(playerTile, board):
128.     """如果玩家下的棋子在棋盘上有 4 枚连成一条线，则返回 True，否则返回 False"""
129.
130.     # 遍历整个棋盘，判断是否有 4 枚棋子连成一条线
131.     for columnIndex in range(BOARD_WIDTH - 3):
132.         for rowIndex in range(BOARD_HEIGHT):
133.             # Check for horizontal four-in-a-row going right:   检查水平向右的 4 枚棋子
134.             tile1 = board[(columnIndex, rowIndex)]
135.             tile2 = board[(columnIndex + 1, rowIndex)]
136.             tile3 = board[(columnIndex + 2, rowIndex)]
137.             tile4 = board[(columnIndex + 3, rowIndex)]
138.             if tile1 == tile2 == tile3 == tile4 == playerTile:
139.                 return True
140.
141.     for columnIndex in range(BOARD_WIDTH):
142.         for rowIndex in range(BOARD_HEIGHT - 3):
143.             # 检查垂直向下的 4 枚棋子
144.             tile1 = board[(columnIndex, rowIndex)]
145.             tile2 = board[(columnIndex, rowIndex + 1)]
146.             tile3 = board[(columnIndex, rowIndex + 2)]
147.             tile4 = board[(columnIndex, rowIndex + 3)]
148.             if tile1 == tile2 == tile3 == tile4 == playerTile:
149.                 return True
150.
151.     for columnIndex in range(BOARD_WIDTH - 3):
152.         for rowIndex in range(BOARD_HEIGHT - 3):
153.             # 检查右下对角线的 4 枚棋子
154.             tile1 = board[(columnIndex, rowIndex)]
155.             tile2 = board[(columnIndex + 1, rowIndex + 1)]
156.             tile3 = board[(columnIndex + 2, rowIndex + 2)]
157.             tile4 = board[(columnIndex + 3, rowIndex + 3)]
158.             if tile1 == tile2 == tile3 == tile4 == playerTile:
159.                 return True
160.
161.             # 检查左下对角线的 4 枚棋子
162.             tile1 = board[(columnIndex + 3, rowIndex)]
163.             tile2 = board[(columnIndex + 2, rowIndex + 1)]
```

```
164.                tile3 = board[(columnIndex + 1, rowIndex + 2)]
165.                tile4 = board[(columnIndex, rowIndex + 3)]
166.                if tile1 == tile2 == tile3 == tile4 == playerTile:
167.                    return True
168.     return False
169.
170.
171. # 程序运行入口（如果不作为模块导入的话）
172. if __name__ == '__main__':
173.     main()
```

输入源代码并运行几次后，请尝试对其进行更改。你可以参考标有(!)的注释对程序进行修改，也可以尝试以下操作。

- 创建一个三子棋或五子棋游戏。
- 制作该游戏的 3 人对战版。
- 添加一枚类似"百搭牌"的棋子——该棋子在任意一位玩家落子后随机落子。
- 添加一些任何玩家都无法使用的"障碍"棋子。

探索程序

请尝试找出以下问题的答案。你需要尝试对代码进行一些修改，再次运行程序，查看修改后的效果。

1. 如果将第 10 行的 PLAYER_O = 'O' 改为 PLAYER_O = 'X'，则会发生什么？
2. 如果将第 115 行的 return(columnIndex, rowIndex) 改为 return (columnIndex, 0)，则会发生什么？
3. 如果将第 97 行的 response == 'QUIT' 改为 response != 'QUIT'，则会发生什么？
4. 如果将第 71 行的 tileChars = [] 改为 tileChars = {}，则会得到什么错误消息？

项目 31

猜数字

猜数字是初学者练习基础编程技术的经典游戏。在这个游戏中，计算机给出一个 1~100 的随机数，而玩家有 10 次机会猜出这个数字。玩家每次猜测后，计算机会告诉玩家是猜大了还是猜小了。

运行程序

运行 guess.py，输出如下所示：

```
Guess the Number, by Al Sweigart al@inventwithpython.com

I am thinking of a number between 1 and 100.
You have 10 guesses left. Take a guess.
> 50
Your guess is too high.
You have 9 guesses left. Take a guess.
> 25
Your guess is too low.
--snip--
You have 5 guesses left. Take a guess.
> 42
Yay! You guessed my number!
```

工作原理

本项目将用到几个基本的编程概念：循环、`if-else` 语句、函数、方法调用和随机数。程序用 Python 的 `random` 模块生成伪随机数——这些数字看起来是随机的，但其在技术上是可预测的。对于计算机来说，伪随机数比真正的随机数更容易生成，并且对于电子游戏和一些科学模拟等应用来说，它们被认为是"足够随机的"。

Python 的 `random` 模块根据一个种子值生成伪随机数，并且由同一个种子值生成的每个伪随机数流都是相同的。例如，在交互式 Shell 中输入以下内容：

```
>>> import random
>>> random.seed(42)
>>> random.randint(1, 10); random.randint(1, 10); random.randint(1, 10)
2
1
5
```

如果重新启动交互式 Shell 并再次运行上述代码，那么它将生成相同的伪随机数：2、1、5。三维第一人称沙盒游戏 *Minecraft*（《我的世界》）根据一个初始种子值生成伪随机的虚拟世界，这就是不同玩家可以使用相同种子重新创建相同世界的原因。

```
 1. """猜数字，作者：Al Sweigart al@inventwithpython.com
 2. 尝试根据提示猜测秘密数字
 3. 标签：小，初学者，游戏"""
 4.
 5. import random
 6.
 7.
 8. def askForGuess():
 9.     while True:
10.         guess = input('> ')  # 输入猜测的数字
11.
12.         if guess.isdecimal():
13.             return int(guess)  # 将输入的字符串转换为整数
14.         print('Please enter a number between 1 and 100.')
15.
16.
17. print('Guess the Number, by Al Sweigart al@inventwithpython.com')
18. print()
19. secretNumber = random.randint(1, 100)  # 选择一个随机数
20. print('I am thinking of a number between 1 and 100.')
21.
22. for i in range(10):  # 给玩家 10 次猜测机会
23.     print('You have {} guesses left. Take a guess.'.format(10 - i))
24.
25.     guess = askForGuess()
26.     if guess == secretNumber:
27.         break  # 如果猜测正确，则结束循环
28.
29.     # 给出一个提示
30.     if guess < secretNumber:
31.         print('Your guess is too low.')
32.     if guess > secretNumber:
33.         print('Your guess is too high.')
34.
35. # 显示结果
36. if guess == secretNumber:
37.     print('Yay! You guessed my number!')
38. else:
39.     print('Game over. The number I was thinking of was', secretNumber)
```

输入源代码并运行几次后，请尝试对其进行更改。你也可以尝试以下操作。

❏ 创建一个"猜字母"版本的程序，根据玩家猜测的字母顺序给出提示。
❏ 根据玩家先前的猜测，在其每次猜测后提示说明"更暖"或"更冷"。[1]

---

[1] 程序把所猜测的数字当作温度值。——译者注

探索程序

请尝试找出以下问题的答案。你需要尝试对代码进行一些修改，再次运行程序，查看修改后的效果。

1. 如果将第 10 行的 input('> ') 改为 input(secretNumber)，则会发生什么？
2. 如果将第 13 行的 return int(guess) 改为 return guess，则会得到什么错误消息？
3. 如果将第 19 行的 random.randint(1, 100) 改为 random.randint(1, 1)，则会发生什么？
4. 如果将第 23 行的 format(10 - i) 改为 format(i)，则会发生什么？
5. 如果将第 36 行的 guess == secretNumber 改为 guess = secretNumber，则会得到什么错误消息？

项目 32

"上当受骗"

在这个短小且简单的程序中,你可以学到让一个容易受骗的人忙上几个小时的诀窍和奥妙。这个项目非常适合初学者,你不妨先复制代码来运行程序。

运行程序

运行 gullible.py,输出如下所示:

```
Gullible, by Al Sweigart al@inventwithpython.com
Do you want to know how to keep a gullible person busy for hours? Y/N
> y
Do you want to know how to keep a gullible person busy for hours? Y/N
> y
Do you want to know how to keep a gullible person busy for hours? Y/N
> yes
Do you want to know how to keep a gullible person busy for hours? Y/N
> YES
Do you want to know how to keep a gullible person busy for hours? Y/N
> TELL ME HOW TO KEEP A GULLIBLE PERSON BUSY FOR HOURS
"TELL ME HOW TO KEEP A GULLIBLE PERSON BUSY FOR HOURS" is not a valid yes/no response.
Do you want to know how to keep a gullible person busy for hours? Y/N
> y
Do you want to know how to keep a gullible person busy for hours? Y/N
> y
Do you want to know how to keep a gullible person busy for hours? Y/N
> n
Thank you. Have a nice day!
```

工作原理

为了实现"用户友好",程序应该能够处理来自玩家的一系列可能的输入。例如,向用户询问是/否类型的问题时,玩家只需简单地输入 y 或 n,而不需要输入完整的单词。如果他们的大写键被锁定,程序还可以理解玩家的意图,因为它会对玩家输入的字符串调用 lower() 方法。

这样，程序对 y、yes、Y、Yes 和 YES 的解释都是相同的。如果玩家给出否定的答案，处理方法类似。

```
 1. """"上当受骗"，作者：Al Sweigart al@inventwithpython.com
 2. 如何让一个容易受骗的人忙上几个小时
 3. 标签：小，初学者，诙谐"""
 4.
 5. print('Gullible, by Al Sweigart al@inventwithpython.com')
 6.
 7. while True:    # 主循环
 8.     print('Do you want to know how to keep a gullible person busy for hours? Y/N')
 9.     response = input('> ')  # 获取用户的输入
10.     if response.lower() == 'no' or response.lower() == 'n':
11.         break   # 如果玩家回答'no'，则结束循环
12.     if response.lower() == 'yes' or response.lower() == 'y':
13.         continue  # 如果玩家回答'yes'，则回到循环的开始
14.     print('"{}" is not a valid yes/no response.'.format(response))
15.
16. print('Thank you. Have a nice day!')
```

探索程序

请尝试找出以下问题的答案。你需要尝试对代码进行一些修改，再次运行程序，查看修改后的效果。

1. 如果将第 10 行的 `response.lower() == 'no'` 改为 `response.lower() != 'no'`，则会发生什么？
2. 如果将第 7 行的 `while True:` 改为 `while False:`，则会发生什么？

项目 33

黑客小游戏

在这个游戏中,玩家必须通过猜测用作隐藏密码的 7 个字母的单词来破解计算机。计算机的内存库会显示可能的单词,并提示玩家每次猜测的单词与真实答案的接近程度。例如,如果隐藏密码是 MONITOR,玩家猜到了 CONTAIN,则他们会得到"7 个字母中有 2 个正确"的提示,因为 MONITOR 和 CONTAIN 都将字母 O 和 N 作为第二个和第三个字母。该游戏类似于项目 1 和"辐射"系列电子游戏中的黑客小游戏。

运行程序

运行 hacking.py,输出如下所示:

```
Hacking Minigame, by Al Sweigart al@inventwithpython.com
Find the password in the computer's memory:
0x1150   $],>@|~~RESOLVE^     0x1250   {>+)<!?CHICKEN,%
0x1160   }@%_-:;/$^(|<|!(     0x1260   .][})?#@#ADDRESS
0x1170   _;)][#?<&-$~+&}}     0x1270   ,#=)>{-;/DESPITE
0x1180   %[!]{REFUGEE@?~,     0x1280   }/.}!-DISPLAY%%/
0x1190   _[^%[@}^<_+{_@$~     0x1290   =>>,:*%?_?@+{%#.
0x11a0   )?~/)+PENALTY?-=     0x12a0   >[,?*#IMPROVE@$/
--snip--
Enter password: (4 tries remaining)
> resolve
Access Denied (2/7 correct)
Enter password: (3 tries remaining)
> improve
A C C E S S   G R A N T E D
```

工作原理

这个游戏有一个黑客主题,但并不涉及任何实际的计算机黑客行为。如果我们只是在屏幕上列出可能的单词,那么游戏玩法将是相同的。然而,模拟计算机内存的装饰性元素彰显了一种令人兴奋的计算机黑客感。在本项目中,对细节和用户体验的关注,让平淡乏味的游戏变得趣味盎然。

1. """黑客小游戏,作者:Al Sweigart al@inventwithpython.com
2. 类似于《辐射 3》中的黑客小游戏
3. 根据每次从猜测获得的线索,找出哪 7 个单词是密码

```
  4. 标签：大，艺术，游戏，谜题"""
  5.
  6.
  7. import random, sys
  8.
  9. # 创建常量
 10. # 用于"计算机内存"显示的垃圾填充字符
 11. GARBAGE_CHARS = '~!@#$%^&*()_+-={}[]|;:,.<>?/'
 12.
 13. # 从包含 7 个字母的单词的文本文件中加载 WORDS 列表
 14. with open('sevenletterwords.txt') as wordListFile:
 15.     WORDS = wordListFile.readlines()
 16. for i in range(len(WORDS)):
 17.     # 把每个单词转换为大写形式并删除尾部的换行符
 18.     WORDS[i] = WORDS[i].strip().upper()
 19.
 20.
 21. def main():
 22.     """运行黑客游戏"""
 23.     print('''Hacking Minigame, by Al Sweigart al@inventwithpython.com
 24. Find the password in the computer's memory. You are given clues after
 25. each guess. For example, if the secret password is MONITOR but the
 26. player guessed CONTAIN, they are given the hint that 2 out of 7 letters
 27. were correct, because both MONITOR and CONTAIN have the letter O and N
 28. as their 2nd and 3rd letter. You get four guesses.\n''')
 29.     input('Press Enter to begin...')
 30.
 31.     gameWords = getWords()
 32.     # "计算机内存"虽然只是装饰性元素，但看起来很酷
 33.     computerMemory = getComputerMemoryString(gameWords)
 34.     secretPassword = random.choice(gameWords)
 35.
 36.     print(computerMemory)
 37.     # 从剩余的 4 次猜测开始，并逐渐减少次数
 38.     for triesRemaining in range(4, 0, -1):
 39.         playerMove = askForPlayerGuess(gameWords, triesRemaining)
 40.         if playerMove == secretPassword:
 41.             print('A C C E S S   G R A N T E D')
 42.             return
 43.         else:
 44.             numMatches = numMatchingLetters(secretPassword, playerMove)
 45.             print('Access Denied ({}/7 correct)'.format(numMatches))
 46.     print('Out of tries. Secret password was {}.'.format(secretPassword))
 47.
 48.
 49. def getWords():
 50.     """返回包含可能是密码的 12 个单词的列表
 51.
 52.     密码将是列表中的第一个单词
 53.     为了让游戏公平，我们应尽量保证作为密码的单词中有一些数字"""
 54.     secretPassword = random.choice(WORDS)
 55.
 56.     words = [secretPassword]
 57.
 58.     # 再找两个单词：它们没有匹配的字母
 59.     # 我们使用 "< 3" 是因为密码已经在单词中
 60.     while len(words) < 3:
 61.         randomWord = getOneWordExcept(words)
 62.         if numMatchingLetters(secretPassword, randomWord) == 0:
 63.             words.append(randomWord)
 64.
```

```python
65.     # 找到两个有 3 个匹配字母的单词（如果找不到这么多单词
66.     # 则在 500 次尝试后放弃）
67.     for i in range(500):
68.         if len(words) == 5:
69.             break    # 找到 5 个单词，结束循环
70.
71.         randomWord = getOneWordExcept(words)
72.         if numMatchingLetters(secretPassword, randomWord) == 3:
73.             words.append(randomWord)
74.
75.     # 找到 7 个或 7 个以上至少有一个匹配字母的单词
76.     #（但如果找不到这么多单词，则在 500 次尝试后放弃）
77.     for i in range(500):
78.         if len(words) == 12:
79.             break    # 如果找到 7 个或更多个单词，那么结束循环
80.
81.         randomWord = getOneWordExcept(words)
82.         if numMatchingLetters(secretPassword, randomWord) != 0:
83.             words.append(randomWord)
84.
85.     # 添加所需的任意随机单词，以获得 12 个单词
86.     while len(words) < 12:
87.         randomWord = getOneWordExcept(words)
88.         words.append(randomWord)
89.
90.     assert len(words) == 12
91.     return words
92.
93.
94. def getOneWordExcept(blocklist=None):
95.     """从 WORDS 中返回一个不在 blocklist 中的随机单词"""
96.     if blocklist == None:
97.         blocklist = []
98.
99.     while True:
100.        randomWord = random.choice(WORDS)
101.        if randomWord not in blocklist:
102.            return randomWord
103.
104.
105. def numMatchingLetters(word1, word2):
106.     """返回这两个单词中匹配字母的数量"""
107.     matches = 0
108.     for i in range(len(word1)):
109.         if word1[i] == word2[i]:
110.             matches += 1
111.     return matches
112.
113.
114. def getComputerMemoryString(words):
115.     """返回一个表示"计算机内存"的字符串"""
116.
117.     # 每行字符串包含一个单词
118.     # 字符串共占 16 行，一分为二
119.     linesWithWords = random.sample(range(16 * 2), len(words))
120.     # 起始内存地址（这也是装饰性元素）
121.     memoryAddress = 16 * random.randint(0, 4000)
122.
123.     # 创建"计算机内存"字符串
124.     computerMemory = []    # 将包含 16 个字符串，每行一个
125.     nextWord = 0    # 要放在一行中的单词的索引
```

```
126.    for lineNum in range(16):    # "计算机内存"有16行
127.        # 创建半行垃圾字符
128.        leftHalf = ''
129.        rightHalf = ''
130.        for j in range(16):    # 每半行有16个字符
131.            leftHalf += random.choice(GARBAGE_CHARS)
132.            rightHalf += random.choice(GARBAGE_CHARS)
133.
134.        # 根据单词填写密码
135.        if lineNum in linesWithWords:
136.            # 在半行中找一个随机位置插入单词
137.            insertionIndex = random.randint(0, 9)
138.            # 插入单词
139.            leftHalf = (leftHalf[:insertionIndex] + words[nextWord]
140.                + leftHalf[insertionIndex + 7:])
141.            nextWord += 1    # 更新要放入半行的单词的索引
142.        if lineNum + 16 in linesWithWords:
143.            # 在半行中找一个随机位置插入单词
144.            insertionIndex = random.randint(0, 9)
145.            # 插入单词
146.            rightHalf = (rightHalf[:insertionIndex] + words[nextWord]
147.                + rightHalf[insertionIndex + 7:])
148.            nextWord += 1    # 更新要放入半行的单词的索引
149.
150.        computerMemory.append('0x' + hex(memoryAddress)[2:].zfill(4)
151.                + ' ' + leftHalf + ' '
152.                + '0x' + hex(memoryAddress + (16*16))[2:].zfill(4)
153.                + ' ' + rightHalf)
154.
155.        memoryAddress += 16    # 例如,从0xe680跳转到0xe690
156.
157.    # 将computerMemory列表中的所有字符串拼接成一个大的字符串并返回
158.    return '\n'.join(computerMemory)
159.
160.
161. def askForPlayerGuess(words, tries):
162.    """让玩家输入所猜测的密码"""
163.    while True:
164.        print('Enter password: ({} tries remaining)'.format(tries))
165.        guess = input('> ').upper()
166.        if guess in words:
167.            return guess
168.        print('That is not one of the possible passwords listed above.')
169.        print('Try entering "{}" or "{}".'.format(words[0], words[1]))
170.
171.
172. # 程序运行入口(如果不是作为模块导入的话)
173. if __name__ == '__main__':
174.    try:
175.        main()
176.    except KeyboardInterrupt:
177.        sys.exit()    # 按Ctrl-C,程序结束
```

输入源代码并运行几次后,请尝试对其进行更改。你也可以尝试以下操作。

❏ 请在互联网上查找单词列表,自行创建 Sevenletterwords.txt 文件,使之可以包含由6个或8个字母构成的单词。

❏ 创建"计算机内存"的不同可视化效果。

探索程序

请尝试找出以下问题的答案。你需要尝试对代码进行一些修改，再次运行程序，查看修改后的效果。

1. 如果将第 130 行中的 `j in range(16):` 改为 `for j in range(0):`，则会发生什么？
2. 如果将第 11 行的 `GARBAGE_CHARS = '~!@#$%^&*()_+-={} []|;:,.<>?/'` 改为 `GARBAGE_CHARS = '.'`，则会发生什么？
3. 如果将第 31 行的 `gameWords = getWords()` 改为 `gameWords = ['MALKOVICH'] * 20`，则会发生什么？
4. 如果将第 91 行的 `return words` 改为 `return`，则会得到什么错误消息？
5. 如果将第 100 行的 `randomWord = random.choice(WORDS)` 改为 `secretPassword = 'PASSWORD'`，则会发生什么？

项目 34 "绞刑架"与"断头台"

这个经典的文字游戏是让玩家猜一个秘密单词的字母。玩家每猜错一个字母,"绞刑架"的一部分就会被绘制出来。玩家需要在"绞刑架"绘制完成之前猜出完整的单词。这个版本的秘密单词都是像 RABBIT 和 PIGEON 这样的动物。当然,你也可以用自己的一组单词代替它们。

HANGMAN_PICS 变量包含构成"绞刑架"每一步的 ASCII 艺术字符串:

```
  +--+      +--+      +--+      +--+      +--+      +--+      +--+
  |  |      |  |      |  |      |  |      |  |      |  |      |  |
           O  |      O  |      O  |      O  |      O  |      O  |
            |        |        /|  |     /|\ |     /|\ |     /|\ |
            |        |        |        |        /  |     / \ |
  =====    =====    =====    =====    =====    =====    =====
```

你也可以将 HANGMAN_PICS 变量中的字符串替换为以下样式的内容:

```
 |    |    |    |===|  |===|  |===|  |===|  |===|
 |    |    |    |   |  |   |  |   |  || /|  || /|
 |    |    |    |   |  |   |  |   |  ||/ |  ||/ |
 |    |    |    |   |  |   |  |   |  |   |  |   |
 |    |    |    |   |  |   |  |/-\|  |/-\|  |/-\|
 |    |    |    |   |  |\ /|  |\ /|  |\ /|  |\0/|
|===| |===| |===| |===| |===| |===| |===| |===|
```

运行程序

运行 hangman.py,输出如下所示:

```
Hangman, by Al Sweigart al@inventwithpython.com

  +--+
  |  |
     |
     |
     |
     |
=====
```

项目 34 "绞刑架"与"断头台"

```
The category is: Animals

Missed letters: No missed letters yet.
_ _ _ _ _
Guess a letter.
> e
--snip--
 +--+
 |  |
 O  |
/|  |
    |
    |
=====
The category is: Animals

Missed letters: A I S
O T T E _
Guess a letter.
> r
Yes! The secret word is: OTTER
You have won!
```

工作原理

"绞刑架"与"断头台"遵循相同的游戏机制，只是表现形式有所不同。这使得用"断头台"ASCII 艺术画替换"绞刑架"ASCII 艺术画变得非常容易，无须更改程序所遵循的主要逻辑。程序的视图与逻辑的分离使得更新功能或修改设计变得比较容易。在专业软件开发中，这种策略是软件设计模式或软件架构的典范之一，即注重构建便于理解和修改的程序。尽管这一策略主要应用于大型软件应用研发中，但我们同样可以将其应用于较小的项目中。

```
 1. """"绞刑架"与"断头台"，作者：Al Sweigart al@inventwithpython.com
 2. 在"绞刑架"或"断头台"被绘制出来之前猜出一个秘密单词
 3. 标签：大，游戏，文字，谜题"""
 4.
 5. # 这个游戏的一个版本出现在《Python 游戏快速上手》一书中
 6.
 7. import random, sys
 8.
 9. # 创建常量
10. # (!) 尝试在 HANGMAN_PICS 中添加或更改字符串
11. # 用"断头台"代替"绞刑架"
12. HANGMAN_PICS = [r"""
13.  +--+
14.  |  |
15.     |
16.     |
17.     |
18.     |
19. =====""",
20. r"""
21.  +--+
22.  |  |
23.  O  |
24.     |
25.     |
```

```
26.     |
27. ====="""",
28. r"""
29.  +--+
30.  |  |
31.  O  |
32.  |  |
33.     |
34.     |
35. ====="""",
36. r"""
37.  +--+
38.  |  |
39.  O  |
40. /|  |
41.     |
42.     |
43. ====="""",
44. r"""
45.  +--+
46.  |  |
47.  O  |
48. /|\ |
49.     |
50.     |
51. ====="""",
52. r"""
53.  +--+
54.  |  |
55.  O  |
56. /|\ |
57. /   |
58.     |
59. ====="""",
60. r"""
61.  +--+
62.  |  |
63.  O  |
64. /|\ |
65. / \ |
66.     |
67. ====="""]
68.
69. # (!) 尝试用新的字符串替换 CATEGORY 和 WORDS
70. CATEGORY = 'Animals'
71. WORDS = 'ANT BABOON BADGER BAT BEAR BEAVER CAMEL CAT CLAM COBRA COUGAR COYOTE CROW DEER DOG
         DONKEY DUCK EAGLE FERRET FOX FROG GOAT GOOSE HAWK LION LIZARD LLAMA MOLE MONKEY MOOSE MOUSE MULE NEWT
         OTTER OWL PANDA PARROT PIGEON PYTHON RABBIT RAM RAT RAVEN RHINO SALMON SEAL SHARK SHEEP SKUNK SLOTH
         SNAKE SPIDER STORK SWAN TIGER TOAD TROUT TURKEY TURTLE WEASEL WHALE WOLF WOMBAT ZEBRA'.split()
72.
73.
74. def main():
75.     print('Hangman, by Al Sweigart al@inventwithpython.com')
76.
77.     # 为新游戏创建变量
78.     missedLetters = []    # 猜错的字母列表
79.     correctLetters = []   # 猜对的字母列表
80.     secretWord = random.choice(WORDS)  # 玩家必须猜出的单词
81.
82.     while True:  # 主循环
83.         drawHangman(missedLetters, correctLetters, secretWord)
84.
```

```
85.         # 让玩家输入自己猜测的字母
86.         guess = getPlayerGuess(missedLetters + correctLetters)
87.
88.         if guess in secretWord:
89.             # 将猜对的字母添加到correctLetters中
90.             correctLetters.append(guess)
91.
92.             # 检查玩家是否赢了
93.             foundAllLetters = True   # 假设玩家获胜
94.             for secretWordLetter in secretWord:
95.                 if secretWordLetter not in correctLetters:
96.                     # 若秘密单词里有一个字母没猜对
97.                     # 则玩家并没有获胜
98.                     foundAllLetters = False
99.                     break
100.            if foundAllLetters:
101.                print('Yes! The secret word is:', secretWord)
102.                print('You have won!')
103.                break   # 结束主循环
104.        else:
105.            # 玩家猜错了
106.            missedLetters.append(guess)
107.
108.            # 检查玩家是否因猜了太多次而输了
109.            # 设置为"-1"是因为不应把空绞刑架算进去
110.            # HANGMAN_PICS
111.            if len(missedLetters) == len(HANGMAN_PICS) - 1:
112.                drawHangman(missedLetters, correctLetters, secretWord)
113.                print('You have run out of guesses!')
114.                print('The word was "{}"'.format(secretWord))
115.                break
116.
117.
118. def drawHangman(missedLetters, correctLetters, secretWord):
119.     """绘制出"绞刑架"的当前状态,以及漏掉和猜对了的字母"""
120.     print(HANGMAN_PICS[len(missedLetters)])
121.     print('The category is:', CATEGORY)
122.     print()
123.
124.     # 显示猜错的字母
125.     print('Missed letters: ', end='')
126.     for letter in missedLetters:
127.         print(letter, end=' ')
128.     if len(missedLetters) == 0:
129.         print('No missed letters yet.')
130.     print()
131.
132.     # 显示秘密单词的空格(每个字母一个空格)
133.     blanks = ['_'] * len(secretWord)
134.
135.     # 用正确的字母替换空格
136.     for i in range(len(secretWord)):
137.         if secretWord[i] in correctLetters:
138.             blanks[i] = secretWord[i]
139.
140.     # 显示秘密单词(字母之间有空格)
141.     print(' '.join(blanks))
142.
143.
144. def getPlayerGuess(alreadyGuessed):
145.     """返回玩家输入的字母。这个函数会确保玩家输入了一个他们之前没有猜到的字母"""
146.     while True:   # 继续询问,直到玩家输入有效的字母
```

```
147.        print('Guess a letter.')
148.        guess = input('> ').upper()
149.        if len(guess) != 1:
150.            print('Please enter a single letter.')
151.        elif guess in alreadyGuessed:
152.            print('You have already guessed that letter. Choose again.')
153.        elif not guess.isalpha():
154.            print('Please enter a LETTER.')
155.        else:
156.            return guess
157.
158.
159. # 程序运行入口（如果不作为模块导入的话）
160. if __name__ == '__main__':
161.     try:
162.         main()
163.     except KeyboardInterrupt:
164.         sys.exit()    # 按 Ctrl-C，程序结束
```

输入源代码并运行几次后，请尝试对其进行更改。你可以根据标有!的注释对程序进行修改，也可以尝试以下操作。

- ❑ 添加"类别选择"功能，让玩家选择自己想玩的单词类别。
- ❑ 添加"版本选择"功能，让玩家可以在"绞刑架"和"断头台"两个游戏版本之间进行选择。

探索程序

请尝试找出以下问题的答案。你需要尝试对代码进行一些修改，再次运行程序，查看修改后的效果。

1. 如果删除或注释掉第 106 行的 missedLetters.append(guess)，则会发生什么？
2. 如果将第 83 行的 drawHangman(missedLetters, correctLetters, secretWord) 改为 drawHangman (correctLetters, missLetters, secretWord)，则会发生什么？
3. 如果将第 133 行的 ['_'] 改为 ['*']，则会发生什么？
4. 如果将第 141 行的 print(' '.join(blanks)) 改为 print(secretWord)，则会发生什么？

项目 35

六边形网格

在本项目中，我们将用简短的程序生成六边形网格样式的镶嵌图案（有点像铁丝网）。由此可见，要设计出有趣的东西，并不需要编写大量的代码。关于本项目的一个稍复杂些的版本参见项目 65。

注意，我们将使用原生字符串，在开始引号前加上小写的 r，这样字符串中的反斜杠就不会被解释为转义字符。

运行程序

运行 hexgrid.py，输出结果如下：

工作原理

编程的强大之处在于它可以使计算机快速、无误地执行重复指令。这就是使用十几行代码就可以在屏幕上绘制数百、数千甚至数百万个六边形的原因。

在命令提示符窗口或终端窗口中，你可以将程序的输出从屏幕重定向到文本文件中。在

Windows 上，运行 py hexgrid.py > hextiles.txt 可以创建一个包含六边形图案的文本文件。在 Linux 和 macOS 上，你可以运行 python3 hexgrid.py > hextiles.txt。将绘制内容重定向到文本文件中时，可以不受屏幕大小的限制，我们可以增大 X_REPEAT 和 Y_REPEAT 常量值。在创建文件后，我们就可以轻松地将文件打印在纸上、通过电子邮件发送给他人或将其发布到社交媒体上。这也适用于你创建的其他计算机"艺术品"。

```
1.  """六边形网格，作者：Al Sweigart al@inventwithpython.com
2.  显示一个六边形网格的简单镶嵌图案
3.  标签：小，初学者，艺术"""
4.
5.  # 创建常量
6.  # (!) 试着将这些值改为其他数字
7.  X_REPEAT = 19   # 水平镶嵌的次数
8.  Y_REPEAT = 12   # 垂直镶嵌的次数
9.
10. for y in range(Y_REPEAT):
11.     # 显示六边形的上半部分
12.     for x in range(X_REPEAT):
13.         print(r'/ \_', end='')
14.     print()
15.
16.     # 显示六边形的下半部分
17.     for x in range(X_REPEAT):
18.         print(r'\_/ ', end='')
19.     print()
```

输入源代码并运行几次后，请尝试对其进行更改。你可以根据标有!的注释对程序进行修改，也可以尝试以下操作。

❏ 创建尺寸更大的、平铺的六边形。
❏ 创建平铺的矩形（而不是六边形）的砖块图案。

请以实现更大的六边形网格作为练习来重新创建这个程序，例如使用以下图案模式：

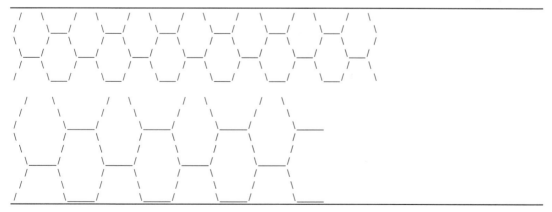

探索程序

这只是一个基础的程序，并没有那么多修改花样，不过你可以思考如何对其他形状的图案进行类似的编程。

项目 36	沙漏

在本项目中，我们将实现一个可视化程序，显示一个简陋的沙漏，以模拟沙子从沙漏的小孔中落下的效果。当沙子全部堆积在沙漏的下半部分后，沙漏会被颠倒过来。这一过程将反复进行。

运行程序

运行 hourglass.py，输出如下所示：

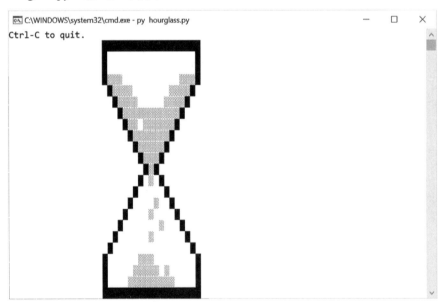

工作原理

本项目中的沙漏程序实现了一个基本的物理引擎。物理引擎是模拟物理对象在重力作用下下落、相互碰撞并根据物理定律移动的软件，常用于电子游戏、计算机动画和科学模拟中。在第90～101行代码中，对于每粒"沙子"会检查其下方的空间是否为空，为空则向下移动，否则检查"沙子"是否可以向左下方移动（第103～111行代码）或向右下方移动（第113～121行代码）。当然，作为研究宏观物体运动的经典物理学分支，运动学的范畴远不止于此。话虽如此，你并不需要拥有物理学学位，也可以对沙漏中的沙子进行原始的模拟，并且可以观察到有趣的沙漏效果。

```
 1. """沙漏,作者: Al Sweigart al@inventwithpython.com
 2. 一个关于沙漏中下落的沙子的动画。 按 Ctrl-C 停止
 3.
 4. 标签: 大，艺术，bext 模块，模拟"""
 5.
 6. import random, sys, time
 7.
 8. try:
 9.     import bext
10. except ImportError:
11.     print('This program requires the bext module, which you')
12.     print('can install by following the instructions at')
13.     print('https://pypi.org/project/Bext/')
14.     sys.exit()
15.
16. # 创建常量
17. PAUSE_LENGTH = 0.2  # (!) 尝试将 0.2 更改为 0.0 或 1.0
18. # (!) 试着把它改成 0 和 100 之间的任意数字
19. WIDE_FALL_CHANCE = 50
20.
21. SCREEN_WIDTH = 79
22. SCREEN_HEIGHT = 25
23. X = 0  # (X, y)元组中 X 值的索引为 0
24. Y = 1  # (x, Y)元组中 Y 值的索引为 1
25. SAND = chr(9617)
26. WALL = chr(9608)
27.
28. #创建沙漏瓶体
29. HOURGLASS = set()  # 用(x, y)元组表示沙漏瓶体所在的位置
30. # (!) 尝试注释一些 HOURGLASS.add()行来擦除瓶壁
31. for i in range(18, 37):
32.     HOURGLASS.add((i, 1))   # 为沙漏的顶部添加瓶壁
33.     HOURGLASS.add((i, 23))  # 为沙漏的底盖添加瓶壁
34. for i in range(1, 5):
35.     HOURGLASS.add((18, i))  # 为左上方的直面添加瓶壁
36.     HOURGLASS.add((36, i))  # 为右上方的直面添加瓶壁
37.     HOURGLASS.add((18, i + 19))  # 在左下方添加瓶壁
38.     HOURGLASS.add((36, i + 19))  # 在右下方添加瓶壁
39. for i in range(8):
40.     HOURGLASS.add((19 + i, 5 + i))  # 在左上角添加斜瓶壁
41.     HOURGLASS.add((35 - i, 5 + i))  # 在右上角添加斜瓶壁
42.     HOURGLASS.add((25 - i, 13 + i))  # 在左下角添加斜瓶壁
43.     HOURGLASS.add((29 + i, 13 + i))  # 在右下角添加斜瓶壁
44.
45. # 在沙漏顶部创建沙子
46. INITIAL_SAND = set()
47. for y in range(8):
48.     for x in range(19 + y, 36 - y):
49.         INITIAL_SAND.add((x, y + 4))
50.
51.
```

```python
52. def main():
53.     bext.fg('yellow')
54.     bext.clear()
55.
56.     #绘制退出消息
57.     bext.goto(0, 0)
58.     print('Ctrl-C to quit.', end='')
59.
60.     # 显示沙漏的瓶壁
61.     for wall in HOURGLASS:
62.         bext.goto(wall[X], wall[Y])
63.         print(WALL, end='')
64.
65.     while True:  # 主循环
66.         allSand = list(INITIAL_SAND)
67.
68.         # 绘制初始的沙子
69.         for sand in allSand:
70.             bext.goto(sand[X], sand[Y])
71.             print(SAND, end='')
72.
73.         runHourglassSimulation(allSand)
74.
75.
76. def runHourglassSimulation(allSand):
77.     """持续模拟沙子下落,直到沙子停止移动"""
78.     while True:  # 继续循环,直到沙子用完
79.         random.shuffle(allSand)  # 模拟沙子的随机顺序
80.
81.         sandMovedOnThisStep = False
82.         for i, sand in enumerate(allSand):
83.             if sand[Y] == SCREEN_HEIGHT - 1:
84.                 # 如果沙子在沙漏最底部,那么它不会移动
85.                 continue
86.
87.             # 如果沙子底下是空的,那么它就会向下移动
88.             noSandBelow = (sand[X], sand[Y] + 1) not in allSand
89.             noWallBelow = (sand[X], sand[Y] + 1) not in HOURGLASS
90.             canFallDown = noSandBelow and noWallBelow
91.
92.             if canFallDown:
93.                 # 在向下一个空间的新位置绘制沙子
94.                 bext.goto(sand[X], sand[Y])
95.                 print(' ', end='')  # 清除旧位置的内容
96.                 bext.goto(sand[X], sand[Y] + 1)
97.                 print(SAND, end='')
98.
99.                 # 把沙子放在下一个空间的新位置上
100.                allSand[i] = (sand[X], sand[Y] + 1)
101.                sandMovedOnThisStep = True
102.            else:
103.                # 检查沙子是否会向左下落
104.                belowLeft = (sand[X] - 1, sand[Y] + 1)
105.                noSandBelowLeft = belowLeft not in allSand
106.                noWallBelowLeft = belowLeft not in HOURGLASS
107.                left = (sand[X] - 1, sand[Y])
108.                noWallLeft = left not in HOURGLASS
109.                notOnLeftEdge = sand[X] > 0
110.                canFallLeft = (noSandBelowLeft and noWallBelowLeft
111.                    and noWallLeft and notOnLeftEdge)
112.
113.                # 检查沙子是否会向右下落
```

```
114.            belowRight = (sand[X] + 1, sand[Y] + 1)
115.            noSandBelowRight = belowRight not in allSand
116.            noWallBelowRight = belowRight not in HOURGLASS
117.            right = (sand[X] + 1, sand[Y])
118.            noWallRight = right not in HOURGLASS
119.            notOnRightEdge = sand[X] < SCREEN_WIDTH - 1
120.            canFallRight = (noSandBelowRight and noWallBelowRight
121.                and noWallRight and notOnRightEdge)
122.
123.            # 设置沙子下落的方向
124.            fallingDirection = None
125.            if canFallLeft and not canFallRight:
126.                fallingDirection = -1  #把沙子置于左边
127.            elif not canFallLeft and canFallRight:
128.                fallingDirection = 1   # 把沙子置于右边
129.            elif canFallLeft and canFallRight:
130.                # 如果两者均有可能，则随机设置
131.                fallingDirection = random.choice((-1, 1))
132.
133.            # 检查沙子是否可以向左或向右"远"落两个空格，而不是只落一个空格
134.            if random.random() * 100 <= WIDE_FALL_CHANCE:
135.                belowTwoLeft = (sand[X] - 2, sand[Y] + 1)
136.                noSandBelowTwoLeft = belowTwoLeft not in allSand
137.                noWallBelowTwoLeft = belowTwoLeft not in HOURGLASS
138.                notOnSecondToLeftEdge = sand[X] > 1
139.                canFallTwoLeft = (canFallLeft and noSandBelowTwoLeft
140.                    and noWallBelowTwoLeft and notOnSecondToLeftEdge)
141.
142.                belowTwoRight = (sand[X] + 2, sand[Y] + 1)
143.                noSandBelowTwoRight = belowTwoRight not in allSand
144.                noWallBelowTwoRight = belowTwoRight not in HOURGLASS
145.                notOnSecondToRightEdge = sand[X] < SCREEN_WIDTH - 2
146.                canFallTwoRight = (canFallRight
147.                    and noSandBelowTwoRight and noWallBelowTwoRight
148.                    and notOnSecondToRightEdge)
149.
150.                if canFallTwoLeft and not canFallTwoRight:
151.                    fallingDirection = -2
152.                elif not canFallTwoLeft and canFallTwoRight:
153.                    fallingDirection = 2
154.                elif canFallTwoLeft and canFallTwoRight:
155.                    fallingDirection = random.choice((-2, 2))
156.
157.            if fallingDirection == None:
158.                # 沙子不会掉下来，所以程序将继续运行
159.                continue
160.
161.            # 在新位置绘制沙子
162.            bext.goto(sand[X], sand[Y])
163.            print(' ', end='')  # 清除旧沙子
164.            bext.goto(sand[X] + fallingDirection, sand[Y] + 1)
165.            print(SAND, end='')  # 绘制新沙子
166.
167.            # 将沙子移到新位置
168.            allSand[i] = (sand[X] + fallingDirection, sand[Y] + 1)
169.            sandMovedOnThisStep = True
170.
171.        sys.stdout.flush()  # 使用 bext 模块需要刷新缓冲区
172.        time.sleep(PAUSE_LENGTH)   # 暂停
173.
174.        # 如果没有沙子在这一步中移动，则重置沙漏
175.        if not sandMovedOnThisStep:
```

```
176.            time.sleep(2)
177.            # Erase all of the sand:
178.            for sand in allSand:
179.                bext.goto(sand[X], sand[Y])
180.                print(' ', end='')
181.            break   # 结束主循环
182.
183.
184. # 程序运行入口（如果不是作为模块导入的话）
185. if __name__ == '__main__':
186.     try:
187.         main()
188.     except KeyboardInterrupt:
189.         sys.exit()   # 按 Ctrl-C，程序结束
```

输入源代码并运行几次后,请尝试对其进行更改。你可以根据标有!的注释对程序进行修改,也可以尝试以下操作。

- 创建不同于沙漏外观的形状。
- 在计算机屏幕上创建不断涌出新沙子的点。

探索程序

请尝试找出以下问题的答案。你需要尝试对代码进行一些修改,再次运行程序,查看修改后的效果。

1. 如果将第 31 行的 range(18, 37) 改为 range(18, 30), 则会发生什么?
2. 如果将第 39 行的 range(8) 改为 range(0), 则会发生什么?
3. 如果将第 81 行的 sandMovedOnThisStep = False 改为 sandMovedOnThisStep = True, 则会发生什么?
4. 如果将第 124 行的 fallingDirection = None 改为 fallingDirection = 1, 则会发生什么?
5. 如果将第 134 行的 random.random() * 100 <= WIDE_FALL_CHANCE 改为 random.random() * 0 <= WIDE_FALL_CHANCE, 则会发生什么?

项目 37

饥饿的机器人

假设你身陷一个迷宫,其中有很多处于饥饿状态的机器人。至于机器人为什么会饥饿,这不是我们要关注的事情。由于程序出错,即使有墙壁遮挡,机器人也会直接向你移动。你必须诱使机器人相互碰撞(或"自相残杀"),才能避免被抓住。

假设有一个私人传送装置,可以将你随机传送到一个新的地方,但它的电池电量只够实现两次传送。此外,你和机器人可以从角落里溜走!

运行程序

运行 hungryrobots.py,输出如下所示:

```
Hungry Robots, by Al Sweigart al@inventwithpython.com
--snip--
▓▓▓▓▓▓▓▓▓▓▓▓▓▓▓▓▓▓▓▓▓▓▓▓▓▓▓▓▓▓▓▓▓▓▓▓
▓   R          R                   ▓
▓▓           R                     ▓
▓                                  ▓
▓     R                            ▓
▓                                  ▓
▓        RX                        ▓
▓      R      R         R          ▓
▓                              R   ▓
▓ R    R      R    R               ▓
▓                                  ▓
▓ @                 R              ▓
▓                                  ▓
▓                     R            ▓
▓X                R   RR    R      ▓
▓RR R     R                 R      ▓
▓      RRR   R                     ▓
▓         R                        ▓
▓▓▓▓▓▓▓▓▓▓▓▓▓▓▓▓▓▓▓▓▓▓▓▓▓▓▓▓▓▓▓▓▓▓▓▓
(T)eleports remaining: 2
                 (Q) (W) ( )
                 (A) (S) (D)
Enter move or QUIT: (Z) (X) ( )
--snip--
```

工作原理

在本项目中，我们用笛卡儿坐标系来表示位置坐标 x 和 y，这样就可以用数学方法来确定机器人应该移动的方向。在编程中，屏幕左上角为坐标原点，x 坐标向右表示增加，y 坐标向下表示增加。也就是说，如果机器人的 x 坐标大于玩家的坐标，它应该向左移动（代码应该减小其当前的 x 坐标），以靠近玩家；如果机器人的 x 坐标较小，则它应该向右移动（代码应该增加其当前的 x 坐标）。这同样适用于基于机器人 y 坐标相对位置的上下移动。

```
 1. """饥饿的机器人，作者：Al Sweigart al@inventwithpython.com
 2. 通过诱使饥饿的机器人相撞，玩家才能逃出生天
 3.
 4. 标签：大，游戏"""
 5.
 6. import random, sys
 7.
 8. # 创建常量
 9. WIDTH = 40              # (!) 试着把 40 改成 70 或 10
10. HEIGHT = 20             # (!) 试着把 20 改成 10
11. NUM_ROBOTS = 10         # (!) 试着把 10 改成 1 或 30
12. NUM_TELEPORTS = 2       # (!) 试着把 2 改成 0 或 9999
13. NUM_DEAD_ROBOTS = 2     # (!) 试着把 2 改成 0 或 20
14. NUM_WALLS = 100         # (!) 试着把 100 改成 0 或 300
15.
16. EMPTY_SPACE = ' '       # (!) 试着把 ' ' 改成 '.'
17. PLAYER = '@'            # (!) 试着把 '@' 改成 'R'
18. ROBOT = 'R'             # (!) 试着把 'R' 改成 '@'
19. DEAD_ROBOT = 'X'        # (!) 试着把 'X' 改成 'R'
20.
21. # (!) 试着把它改成 '#' 或 'O' 或 ' '
22. WALL = chr(9617)  # 字符 9617 表示 '▒'
23.
24.
25. def main():
26.     print('''Hungry Robots, by Al Sweigart al@inventwithpython.com
27.
28. You are trapped in a maze with hungry robots! You don't know why robots
29. need to eat, but you don't want to find out. The robots are badly
30. programmed and will move directly toward you, even if blocked by walls.
31. You must trick the robots into crashing into each other (or dead robots)
32. without being caught. You have a personal teleporter device, but it only
33. has enough battery for {} trips. Keep in mind, you and robots can slip
34. through the corners of two diagonal walls!
35. '''.format(NUM_TELEPORTS))
36.
37.     input('Press Enter to begin...')
38.
39.     # 创建一个新游戏
40.     board = getNewBoard()
41.     robots = addRobots(board)
42.     playerPosition = getRandomEmptySpace(board, robots)
43.     while True:  # 主循环
44.         displayBoard(board, robots, playerPosition)
45.
46.         if len(robots) == 0:  # 检查玩家是否赢了
47.             print('All the robots have crashed into each other and you ')
48.             print('lived to tell the tale! Good job!')
49.             sys.exit()
50.
```

```
51.         # 移动玩家和机器人
52.         playerPosition = askForPlayerMove(board, robots, playerPosition)
53.         robots = moveRobots(board, robots, playerPosition)
54.
55.         for x, y in robots:  # 检查玩家是否输了
56.             if (x, y) == playerPosition:
57.                 displayBoard(board, robots, playerPosition)
58.                 print('You have been caught by a robot!')
59.                 sys.exit()
60.
61.
62. def getNewBoard():
63.     """返回代表该游戏面板的字典。键是(x, y)整数元组,用于表示游戏面板位置,其值是
64.     WALL、EMPTY_SPACE 或 DEAD_ROBOT。字典中也有键"teleports",
65.     表示玩家还剩多少次传送机会。存活的机器人会在与游戏面板的字典分开存储"""
66.     board = {'teleports': NUM_TELEPORTS}
67.
68.     # 创建一个空的游戏面板
69.     for x in range(WIDTH):
70.         for y in range(HEIGHT):
71.             board[(x, y)] = EMPTY_SPACE
72.
73.     # 在游戏面板的边缘添加墙壁
74.     for x in range(WIDTH):
75.         board[(x, 0)] = WALL  # 添加顶部的墙壁
76.         board[(x, HEIGHT - 1)] = WALL  # 添加底部的墙壁
77.     for y in range(HEIGHT):
78.         board[(0, y)] = WALL  # 布置左墙
79.         board[(WIDTH - 1, y)] = WALL  # 布置右墙
80.
81.     # 添加随机墙壁
82.     for i in range(NUM_WALLS):
83.         x, y = getRandomEmptySpace(board, [])
84.         board[(x, y)] = WALL
85.
86.     # 添加处于死亡状态的机器人
87.     for i in range(NUM_DEAD_ROBOTS):
88.         x, y = getRandomEmptySpace(board, [])
89.         board[(x, y)] = DEAD_ROBOT
90.     return board
91.
92.
93. def getRandomEmptySpace(board, robots):
94.     """返回一个(x, y)整数元组,表示游戏面板上的空白空间"""
95.     while True:
96.         randomX = random.randint(1, WIDTH - 2)
97.         randomY = random.randint(1, HEIGHT - 2)
98.         if isEmpty(randomX, randomY, board, robots):
99.             break
100.    return (randomX, randomY)
101.
102.
103. def isEmpty(x, y, board, robots):
104.     """如果游戏面板上的(x, y)为空,并且没有机器人,则返回True"""
105.     return board[(x, y)] == EMPTY_SPACE and (x, y) not in robots
106.
107.
108. def addRobots(board):
109.     """根据NUM_ROBOTS添加机器人到游戏面板上的空白区域,并返回机器人目前所在 (x, y)空间的
110. 列表"""
111.     robots = []
112.     for i in range(NUM_ROBOTS):
```

```
113.          x, y = getRandomEmptySpace(board, robots)
114.          robots.append((x, y))
115.      return robots
116.
117.
118.  def displayBoard(board, robots, playerPosition):
119.      """在屏幕上显示游戏面板、机器人和玩家"""
120.      # 循环遍历游戏面板上的每个区域
121.      for y in range(HEIGHT):
122.          for x in range(WIDTH):
123.              # 绘制出适当的字符
124.              if board[(x, y)] == WALL:
125.                  print(WALL, end='')
126.              elif board[(x, y)] == DEAD_ROBOT:
127.                  print(DEAD_ROBOT, end='')
128.              elif (x, y) == playerPosition:
129.                  print(PLAYER, end='')
130.              elif (x, y) in robots:
131.                  print(ROBOT, end='')
132.              else:
133.                  print(EMPTY_SPACE, end='')
134.          print()  # 输出一行空行
135.
136.
137.  def askForPlayerMove(board, robots, playerPosition):
138.      """根据其当前位置和游戏面板的墙壁，返回玩家下一步移动位置的(x, y)整数元组"""
139.      playerX, playerY = playerPosition
140.
141.      # 找到那些不会被墙挡住的方向
142.      q = 'Q' if isEmpty(playerX - 1, playerY - 1, board, robots) else ' '
143.      w = 'W' if isEmpty(playerX + 0, playerY - 1, board, robots) else ' '
144.      e = 'E' if isEmpty(playerX + 1, playerY - 1, board, robots) else ' '
145.      d = 'D' if isEmpty(playerX + 1, playerY + 0, board, robots) else ' '
146.      c = 'C' if isEmpty(playerX + 1, playerY + 1, board, robots) else ' '
147.      x = 'X' if isEmpty(playerX + 0, playerY + 1, board, robots) else ' '
148.      z = 'Z' if isEmpty(playerX - 1, playerY + 1, board, robots) else ' '
149.      a = 'A' if isEmpty(playerX - 1, playerY + 0, board, robots) else ' '
150.      allMoves = (q + w + e + d + c + x + a + z + 'S')
151.
152.      while True:
153.          # 获取玩家的位置变动信息
154.          print('(T)eleports remaining: {}'.format(board["teleports"]))
155.          print('                ({}) ({}) ({})'.format(q, w, e))
156.          print('                ({})  (S) ({})'.format(a, d))
157.          print('Enter move or QUIT: ({}) ({}) ({})'.format(z, x, c))
158.
159.          move = input('> ').upper()
160.          if move == 'QUIT':
161.              print('Thanks for playing!')
162.              sys.exit()
163.          elif move == 'T' and board['teleports'] > 0:
164.              # 将玩家传送到一个随机的空白区域
165.              board['teleports'] -= 1
166.              return getRandomEmptySpace(board, robots)
167.          elif move != '' and move in allMoves:
168.              # 根据玩家的移动返回其新的位置
169.              return {'Q': (playerX - 1, playerY - 1),
170.                      'W': (playerX + 0, playerY - 1),
171.                      'E': (playerX + 1, playerY - 1),
172.                      'D': (playerX + 1, playerY + 0),
173.                      'C': (playerX + 1, playerY + 1),
174.                      'X': (playerX + 0, playerY + 1),
```

```
175.                        'Z': (playerX - 1, playerY + 1),
176.                        'A': (playerX - 1, playerY + 0),
177.                        'S': (playerX, playerY)}[move]
178.
179.
180.   def moveRobots(board, robotPositions, playerPosition):
181.       """在机器人尝试向玩家移动之后,返回新机器人位置的(x, y)元组列表"""
182.       playerx, playery = playerPosition
183.       nextRobotPositions = []
184.
185.       while len(robotPositions) > 0:
186.           robotx, roboty = robotPositions[0]
187.
188.           # 判断机器人移动的方向
189.           if robotx < playerx:
190.               movex = 1   # 向右移动
191.           elif robotx > playerx:
192.               movex = -1  # 向左移动
193.           elif robotx == playerx:
194.               movex = 0   # 不要水平移动
195.
196.           if roboty < playery:
197.               movey = 1   # 向上移动
198.           elif roboty > playery:
199.               movey = -1  # 向下移动
200.           elif roboty == playery:
201.               movey = 0   # 不要竖直移动
202.
203.           # 检查机器人是否会撞到墙上,并调整路线
204.           if board[(robotx + movex, roboty + movey)] == WALL:
205.               # 机器人会撞到墙,所以调整一个新动作
206.               if board[(robotx + movex, roboty)] == EMPTY_SPACE:
207.                   movey = 0   # 机器人不能水平移动
208.               elif board[(robotx, roboty + movey)] == EMPTY_SPACE:
209.                   movex = 0   # 机器人不能竖直移动
210.               else:
211.                   # 机器人不能移动
212.                   movex = 0
213.                   movey = 0
214.           newRobotx = robotx + movex
215.           newRoboty = roboty + movey
216.
217.           if (board[(robotx, roboty)] == DEAD_ROBOT
218.               or board[(newRobotx, newRoboty)] == DEAD_ROBOT):
219.               # 机器人相撞,把机器人移走
220.               del robotPositions[0]
221.               continue
222.
223.           # 检查机器人是否相撞,如果相撞,那么机器人会损毁
224.           if (newRobotx, newRoboty) in nextRobotPositions:
225.               board[(newRobotx, newRoboty)] = DEAD_ROBOT
226.               nextRobotPositions.remove((newRobotx, newRoboty))
227.           else:
228.               nextRobotPositions.append((newRobotx, newRoboty))
229.
230.           # 如果机器人移动,将其从robotPositions中移除
231.           del robotPositions[0]
232.       return nextRobotPositions
233.
234.
235.   # 程序运行入口(如果不是作为模块导入的话)
236.   if __name__ == '__main__':
237.       main()
```

输入源代码并运行几次后，请试着对其进行更改。你可以根据标有!的注释对程序进行修改，也可以尝试以下操作。

- 创建两种不同类型的机器人：只能沿对角线移动的机器人和只能沿基本方向移动的机器人。
- 给玩家设置一定数量的陷阱，机器人踩到任意一个，都会被阻止追击。
- 给玩家设置一定数量的"瞬时墙"，让他们可以为自己搭建防御装置。

探索程序

请尝试找出以下问题的答案。你需要尝试对代码进行一些修改，再次运行程序，查看修改后的效果。

1. 如果将第 22 行的 WALL = chr(9617) 改为 WALL = 'R'，则会发生什么？
2. 如果将第 232 行的 return nextRobotPositions 改为 return robotsPositions，则会发生什么？
3. 如果删除或注释掉第 44 行的 displayBoard(board, robots, playerPosition)，则会发生什么？
4. 如果删除或注释掉第 53 行的 robots = moveRobots(board, robots, playerPosition)，则会发生什么？

项目 38

"我指证"

假设你是大名鼎鼎的侦探玛蒂尔德·加缪,受人所托找一只名叫佐菲的小猫。假设嫌疑人要么总是说假话,要么总是说真话,你必须根据线索找出真正有罪的一方。你能及时找到小猫佐菲并指证有罪的一方吗?

在本项目的游戏中,你乘坐出租车前往城市各处。所到的每个地点都有一个嫌疑人和一件物品。你可以向嫌疑人询问其他嫌疑人和物品的信息,将其答案与你自己的探索记录加以比较,以确定他们说的是假话还是真话。有些人知道谁绑架了佐菲(或者知道它在哪儿,或者在绑架者所在的位置发现了什么物品),但你必须确定他们说的是否可信。你有 5 分钟的时间找出真正有罪的一方,但如果你做出 3 次错误的指证就输了。该游戏的灵感来自 Homestar Runner 的"找鸡蛋"游戏。

运行程序

运行 jaccuse.py,输出如下所示:

```
J'ACCUSE! (a mystery game)
--snip--
Time left: 5 min, 0 sec
  You are in your TAXI. Where do you want to go?
(A)LBINO ALLIGATOR PIT
(B)OWLING ALLEY
(C)ITY HALL
(D)UCK POND
(H)IPSTER CAFE
(O)LD BARN
(U)NIVERSITY LIBRARY
(V)IDEO GAME MUSEUM
(Z)OO
> a

Time left: 4 min, 48 sec
  You are at the ALBINO ALLIGATOR PIT.
  ESPRESSA TOFFEEPOT with the ONE COWBOY BOOT is here.

(J) "J'ACCUSE!" (3 accusations left)
(Z) Ask if they know where ZOPHIE THE CAT is.
(T) Go back to the TAXI.
(1) Ask about ESPRESSA TOFFEEPOT
(2) Ask about ONE COWBOY BOOT
> z
  They give you this clue: "DUKE HAUTDOG"
Press Enter to continue...
--snip--
```

工作原理

要完全理解这个程序，你应该特别留意 `clues` 字典，在第 50~107 行代码对该字典进行设置。你可以取消对第 149~152 行代码注释，将其显示在计算机屏幕上。该字典用 SUSPECTS 列表中的字符串作为键，用 "clues 字典" 作为值。每个 `clues` 字典包含来自 SUSPECTS 列表和 ITEMS 列表的字符串。当被问及另一个嫌疑人或物品时，最初的嫌疑人将用这些字符串来回答。例如，如果将 `clues['DUKE HAUTDOG']['CANDLESTICK']` 设置为 `'DUCK POND'`，那么当玩家向 DUKE HAUTDOG 询问 CANDLESTICK 时，他们会说它在 DUCK POND。每次玩游戏时，嫌疑人、物品、地点和罪犯都会重置。

本项目的程序的核心数据结构就是上述内容，要看懂程序其他部分，请先理解如下程序。

```
1.  """ "我指证"，  作者：Al Sweigart al@inventwithpython.com
2.  这是关于一只失踪的猫的悬疑类游戏
3.
4.  标签：特大，游戏，诙谐，谜题"""
5.
6.  import time, random, sys
7.
8.  # 创建常量
9.  SUSPECTS = ['DUKE HAUTDOG', 'MAXIMUM POWERS', 'BILL MONOPOLIS', 'SENATOR SCHMEAR',
10. 'MRS. FEATHERTOSS', 'DR. JEAN SPLICER', 'RAFFLES THE CLOWN', 'ESPRESSA TOFFEEPOT',
11. 'CECIL EDGAR VANDERTON']
12. ITEMS = ['FLASHLIGHT', 'CANDLESTICK', 'RAINBOW FLAG', 'HAMSTER WHEEL', 'ANIME VHS TAPE',
13. 'JAR OF PICKLES', 'ONE COWBOY BOOT', 'CLEAN UNDERPANTS', '5 DOLLAR GIFT CARD']
14. PLACES = ['ZOO', 'OLD BARN', 'DUCK POND', 'CITY HALL', 'HIPSTER CAFE', 'BOWLING ALLEY',
15. 'VIDEO GAME MUSEUM', 'UNIVERSITY LIBRARY', 'ALBINO ALLIGATOR PIT']
16. TIME_TO_SOLVE = 300  # 300s（5min）结束游戏
17.
18. # 清单显示需要首字母和最长的地点名字
19. PLACE_FIRST_LETTERS = {}
20. LONGEST_PLACE_NAME_LENGTH = 0
21. for place in PLACES:
22.     PLACE_FIRST_LETTERS[place[0]] = place
23.     if len(place) > LONGEST_PLACE_NAME_LENGTH:
24.         LONGEST_PLACE_NAME_LENGTH = len(place)
25.
26. # 对常量的基本完整性检查
27. assert len(SUSPECTS) == 9
28. assert len(ITEMS) == 9
29. assert len(PLACES) == 9
30. # 首字母必须唯一
31. assert len(PLACE_FIRST_LETTERS.keys()) == len(PLACES)
32.
33.
34. knownSuspectsAndItems = []
35. # visited Places 的键为地点，其值为嫌疑人和物品的字符串
36. visitedPlaces = {}
37. currentLocation = 'TAXI'  # 游戏从租车上开始
38. accusedSuspects = []  # 被指证的嫌疑人不会提供线索
39. liars = random.sample(SUSPECTS, random.randint(3, 4))
40. accusationsLeft = 3  # 玩家最多可以指证 3 个人
41. culprit = random.choice(SUSPECTS)
42.
43. # 使用一致的索引将这些列表元素联系起来，例如 SUSPECTS[0] 和 ITEMS[0] 都在 PLACES[0]
44. random.shuffle(SUSPECTS)
45. random.shuffle(ITEMS)
```

```
 46. random.shuffle(PLACES)
 47.
 48. # 为讲真话的人提供的关于物品和嫌疑人的线索创建数据结构
 49. # clues 的键为被询问线索的嫌疑人，其值为 clues 字典
 50. clues = {}
 51. for i, interviewee in enumerate(SUSPECTS):
 52.     if interviewee in liars:
 53.         continue  # 暂时忽略撒谎的人
 54.
 55.     # 这个 clues 字典的键为物品和嫌疑人，
 56.     # 其值为给出的线索
 57.     clues[interviewee] = {}
 58.     clues[interviewee]['debug_liar'] = False  # 用于调试
 59.     for item in ITEMS:  # 选择与物品相关的线索
 60.         if random.randint(0, 1) == 0:  # 提示物品所在位置
 61.             clues[interviewee][item] = PLACES[ITEMS.index(item)]
 62.         else:  # 提示物品的拥有者是谁
 63.             clues[interviewee][item] = SUSPECTS[ITEMS.index(item)]
 64.     for suspect in SUSPECTS:  # 选择与嫌疑人相关的线索
 65.         if random.randint(0, 1) == 0:  # 提示嫌疑人所在位置
 66.             clues[interviewee][suspect] = PLACES[SUSPECTS.index(suspect)]
 67.         else:  # 提示嫌疑人拥有的物品
 68.             clues[interviewee][suspect] = ITEMS[SUSPECTS.index(suspect)]
 69.
 70. # 为撒谎者的人提供的关于物品和嫌疑人的线索创建数据结构
 71. for i, interviewee in enumerate(SUSPECTS):
 72.     if interviewee not in liars:
 73.         continue  # 已把说真话的人排除了
 74.
 75.     # 这个 clues 字典的键为物品和嫌疑人
 76.     # 其值为给出的线索
 77.     clues[interviewee] = {}
 78.     clues[interviewee]['debug_liar'] = True  # 用于调试
 79.
 80.     # 这个受访者撒谎，给出了错误的线索
 81.     for item in ITEMS:
 82.         if random.randint(0, 1) == 0:
 83.             while True:  # 选择一个随机的（错误的）地点线索
 84.                 # 欺骗物品所在的位置
 85.                 clues[interviewee][item] = random.choice(PLACES)
 86.                 if clues[interviewee][item] != PLACES[ITEMS.index(item)]:
 87.                     # 若选择了错误的线索，则结束循环
 88.                     break
 89.         else:
 90.             while True:  # 选择一个随机的（错误的）嫌疑人线索
 91.                 clues[interviewee][item] = random.choice(SUSPECTS)
 92.                 if clues[interviewee][item] != SUSPECTS[ITEMS.index(item)]:
 93.                     # 若选择了错误的线索，则结束循环
 94.                     break
 95.     for suspect in SUSPECTS:
 96.         if random.randint(0, 1) == 0:
 97.             while True:  # 选择一个随机的（错误的）地点线索
 98.                 clues[interviewee][suspect] = random.choice(PLACES)
 99.                 if clues[interviewee][suspect] != PLACES[ITEMS.index(item)]:
100.                     #若选择了错误的线索，则结束循环
101.                     break
102.         else:
103.             while True:  # 选择一个随机的（错误的）物品线索
104.                 clues[interviewee][suspect] = random.choice(ITEMS)
105.                 if clues[interviewee][suspect] != ITEMS[SUSPECTS.index(suspect)]:
106.                     # 若选择了错误的线索，则结束循环
107.                     break
```

```
108.
109. # 创建关于 zophie Clues 的数据结构
110. zophieClues = {}
111. for interviewee in random.sample(SUSPECTS, random.randint(3, 4)):
112.     kindOfClue = random.randint(1, 3)
113.     if kindOfClue == 1:
114.         if interviewee not in liars:
115.             # 他们会告诉你谁抓了佐菲
116.             zophieClues[interviewee] = culprit
117.         elif interviewee in liars:
118.             while True:
119.                 # 选择一个（错误的）嫌疑人线索
120.                 zophieClues[interviewee] = random.choice(SUSPECTS)
121.                 if zophieClues[interviewee] != culprit:
122.                     # 若选择了错误的线索，则结束循环
123.                     break
124.
125.     elif kindOfClue == 2:
126.         if interviewee not in liars:
127.             # 他们会告诉你佐菲在哪里
128.             zophieClues[interviewee] = PLACES[SUSPECTS.index(culprit)]
129.         elif interviewee in liars:
130.             while True:
131.                 # 选择一个（错误的）地点线索
132.                 zophieClues[interviewee] = random.choice(PLACES)
133.                 if zophieClues[interviewee] != PLACES[SUSPECTS.index(culprit)]:
134.                     # 若选择了错误的线索，则结束循环
135.                     break
136.     elif kindOfClue == 3:
137.         if interviewee not in liars:
138.             # 他们会告诉你佐菲在哪里
139.             zophieClues[interviewee] = ITEMS[SUSPECTS.index(culprit)]
140.         elif interviewee in liars:
141.             while True:
142.                 # 选择一个（错误的）物品线索
143.                 zophieClues[interviewee] = random.choice(ITEMS)
144.                 if zophieClues[interviewee] != ITEMS[SUSPECTS.index(culprit)]:
145.                     # 若选择了错误的线索，则结束循环
146.                     break
147.
148. #进行实验，取消注释该代码，以查看线索数据结构
149. #import pprint
150. #pprint.pprint(clues)
151. #pprint.pprint(zophieClues)
152. #print('culprit =', culprit)
153.
154. # 开始游戏
155. print("""J'ACCUSE! (a mystery game)")
156. By Al Sweigart al@inventwithpython.com
157. Inspired by Homestar Runner\'s "Where\'s an Egg?" game
158.
159. You are the world-famous detective, Mathilde Camus.
160. ZOPHIE THE CAT has gone missing, and you must sift through the clues.
161. Suspects either always tell lies, or always tell the truth. Ask them
162. about other people, places, and items to see if the details they give are
163. truthful and consistent with your observations. Then you will know if
164. their clue about ZOPHIE THE CAT is true or not. Will you find ZOPHIE THE
165. CAT in time and accuse the guilty party?
166. """)
167. input('Press Enter to begin...')
168.
169.
```

```python
170. startTime = time.time()
171. endTime = startTime + TIME_TO_SOLVE
172.
173. while True:    # 主循环
174.     if time.time() > endTime or accusationsLeft == 0:
175.         # 处理"游戏结束"条件
176.         if time.time() > endTime:
177.             print('You have run out of time!')
178.         elif accusationsLeft == 0:
179.             print('You have accused too many innocent people!')
180.         culpritIndex = SUSPECTS.index(culprit)
181.         print('It was {} at the {} with the {} who catnapped her!'.format(culprit,
182.             PLACES[culpritIndex], ITEMS[culpritIndex]))
183.         print('Better luck next time, Detective.')
184.         sys.exit()
185.
186.     print()
187.     minutesLeft = int(endTime - time.time()) // 60
188.     secondsLeft = int(endTime - time.time()) % 60
189.     print('Time left: {} min, {} sec'.format(minutesLeft, secondsLeft))
190.
191.     if currentLocation == 'TAXI':
192.         print('  You are in your TAXI. Where do you want to go?')
193.         for place in sorted(PLACES):
194.             placeInfo = ''
195.             if place in visitedPlaces:
196.                 placeInfo = visitedPlaces[place]
197.             nameLabel = '(' + place[0] + ')' + place[1:]
198.             spacing = " " * (LONGEST_PLACE_NAME_LENGTH - len(place))
199.             print('{} {}{}'.format(nameLabel, spacing, placeInfo))
200.         print('(Q)UIT GAME')
201.         while True:    # 继续询问，直至得到一个有效的回答
202.             response = input('> ').upper()
203.             if response == '':
204.                 continue    # 再问一遍
205.             if response == 'Q':
206.                 print('Thanks for playing!')
207.                 sys.exit()
208.             if response in PLACE_FIRST_LETTERS.keys():
209.                 break
210.         currentLocation = PLACE_FIRST_LETTERS[response]
211.         continue    # 回到主循环的开始部分
212.
213.     # 玩家到达一个地方后，可以询问线索
214.     print('  You are at the {}.'.format(currentLocation))
215.     currentLocationIndex = PLACES.index(currentLocation)
216.     thePersonHere = SUSPECTS[currentLocationIndex]
217.     theItemHere = ITEMS[currentLocationIndex]
218.     print('  {} with the {} is here.'.format(thePersonHere, theItemHere))
219.
220.     # 把这里的嫌疑人和物品加入已知的嫌疑人和物品清单中
221.     if thePersonHere not in knownSuspectsAndItems:
222.         knownSuspectsAndItems.append(thePersonHere)
223.     if ITEMS[currentLocationIndex] not in knownSuspectsAndItems:
224.         knownSuspectsAndItems.append(ITEMS[currentLocationIndex])
225.     if currentLocation not in visitedPlaces.keys():
226.         visitedPlaces[currentLocation] = '({}, {})'.format(thePersonHere.lower(),
227.             theItemHere.lower())
228.
229.     # 如果玩家之前错误地指证了这个人，那么此人就不会提供线索
230.     if thePersonHere in accusedSuspects:
231.         print('They are offended that you accused them,'
```

```
232.         print('and will not help with your investigation.')
233.         print('You go back to your TAXI.')
234.         print()
235.         input('Press Enter to continue...')
236.         currentLocation = 'TAXI'
237.         continue    # 回到循环的开始部分
238.
239.     # 显示已知的嫌疑人和物品的清单,以进行询问
240.     print()
241.     print('(J) "J\'ACCUSE!" ({} accusations left)'.format(accusationsLeft))
242.     print('(Z) Ask if they know where ZOPHIE THE CAT is.')
243.     print('(T) Go back to the TAXI.')
244.     for i, suspectOrItem in enumerate(knownSuspectsAndItems):
245.         print('({}) Ask about {}'.format(i + 1, suspectOrItem))
246.
247.     while True:    # 继续询问,直至得到一个有效的回答
248.         response = input('> ').upper()
249.         if response in 'JZT' or (response.isdecimal() and 0 < int(response) <=
250.         len(knownSuspectsAndItems)):
251.             break
252.
253.     if response == 'J':    # 玩家指证该嫌疑犯
254.         accusationsLeft -= 1    # 用完最后一个指证机会
255.         if thePersonHere == culprit:
256.             # 玩家的指证是正确的
257.             print('You\'ve cracked the case, Detective!')
258.             print('It was {} who had catnapped ZOPHIE THE CAT.'.format(culprit))
259.             minutesTaken = int(time.time() - startTime) // 60
260.             secondsTaken = int(time.time() - startTime) % 60
261.             print('Good job! You solved it in {} min, {} sec.'.format(minutesTaken,
262.                 secondsTaken))
263.             sys.exit()
264.         else:
265.             # 玩家的指证是错误的
266.             accusedSuspects.append(thePersonHere)
267.             print('You have accused the wrong person, Detective!')
268.             print('They will not help you with anymore clues.')
269.             print('You go back to your TAXI.')
270.             currentLocation = 'TAXI'
271.
272.     elif response == 'Z':    # 玩家询问关于佐菲的情况
273.         if thePersonHere not in zophieClues:
274.             print('"I don\'t know anything about ZOPHIE THE CAT."')
275.         elif thePersonHere in zophieClues:
276.             print('  They give you this clue: "{}"'.format(zophieClues[thePersonHere]))
277.             # 在已知事物列表中添加非地点的线索
278.             if zophieClues[thePersonHere] not in knownSuspectsAndItems and zophieClues
279.             [thePersonHere]
280.             not in PLACES:
281.                 knownSuspectsAndItems.append(zophieClues[thePersonHere])
282.
283.     elif response == 'T':    # 玩家回到出租车上
284.         currentLocation = 'TAXI'
285.         continue    # 回到主循环的开始部分
286.
287.     else:    # 玩家询问嫌疑犯或物品的线索
288.         thingBeingAskedAbout = knownSuspectsAndItems[int(response) - 1]
289.         if thingBeingAskedAbout in (thePersonHere, theItemHere):
290.             print('  They give you this clue: "No comment."')
291.         else:
292.             print('  They give you this clue: "{}"'.format(clues[thePersonHere]
293.             [thingBeingAskedAbout]))
```

```
294.            # 在已知事物列表中添加非地点的线索
295.            if clues[thePersonHere][thingBeingAskedAbout] not in knownSuspectsAndItems and
296. clues[thePersonHere][thingBeingAskedAbout] not in PLACES:
297.                knownSuspectsAndItems.append(clues[thePersonHere][thingBeingAskedAbout])
298.
299.      input('Press Enter to continue...')
```

探索程序

请尝试找出以下问题的答案。你需要尝试对代码进行一些修改,再次运行程序,查看修改后的效果。

1. 如果将第 16 行的 TIME_TO_SOLVE = 300 改为 TIME_TO_SOLVE = 0,会发生什么?
2. 如果将第 174 行的 time.time() > endTime or accusationsLeft == 0 改为 time.time() > endTime and accusationsLeft == 0,会发生什么?
3. 如果将第 197 行的 place[1:] 改为 place,会发生什么?
4. 如果将第 171 行的 startTime + TIME_TO_SOLVE 改为 startTime * TIME_TO_SOLVE,会发生什么?

项目 39

朗顿蚂蚁

本项目的朗顿蚂蚁程序是二维网格上的元胞自动机模拟，有点类似于项目 13。在这个项目中，一只"蚂蚁"从具有某种颜色（共有两种颜色）的正方形开始移动。如果所在空间是第一种颜色，蚂蚁将其切换为第二种颜色，顺时针旋转 90°，然后向前移动 1 格。如果所在空间是第二种颜色，蚂蚁将其切换为第一种颜色，逆时针旋转 90°，然后向前移动 1 格。

尽管规则非常简单，但模拟程序显示了复杂的自然行为。该模拟程序可以改为在同一空间中展示多只蚂蚁，当它们彼此相遇时会引发有趣的交互。朗顿蚂蚁是由计算机科学家克里斯·朗顿（Chris Langton）于 1986 年提出的。

运行程序

运行 langtonsant.py，输出如下图所示。

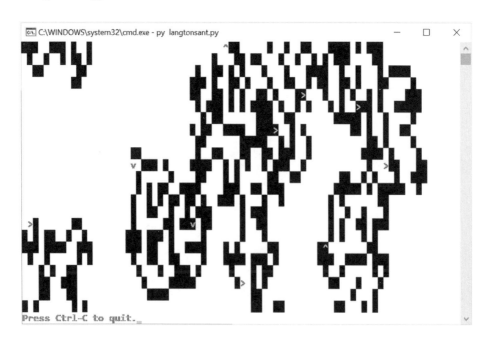

工作原理

本项目中的程序用到了两种"方向"。其一，东、西、南北，代表每只蚂蚁的字典存储基本方向；其二，向左或向右（逆时针或顺时针，因为我们是从上方观察蚂蚁的），表示转动方向。蚂蚁应该根据它们所处位置向左或向右转，因此第73～95行代码根据蚂蚁当前的基本方向和转动方向设置了一个新的基本方向。

```python
1.  """朗顿蚂蚁，作者：Al Sweigart al@inventwithpython.com
2.  元胞自动机的模拟示例。按 Ctrl-C 停止
3.
4.  标签：大，艺术，bext 模块，模拟"""
5.
6.  import copy, random, sys, time
7.
8.  try:
9.      import bext
10. except ImportError:
11.     print('This program requires the bext module, which you')
12.     print('can install by following the instructions at')
13.     print('https://pypi.org/project/Bext/')
14.     sys.exit()
15.
16. # 创建常量
17. WIDTH, HEIGHT = bext.size()
18. # 如果不自动添加换行符，我们就无法输出到 Windows 上的最后一列，所以将宽度减 1
19. WIDTH -= 1
20. HEIGHT -= 1  # 调整底部的退出消息
21.
22. NUMBER_OF_ANTS = 10  # (!) 尝试将 10 更改为 1 或 50
23. PAUSE_AMOUNT = 0.1  # (!) 尝试将 0.1 更改为 1.0 或 0.0
24.
25. # (!) 尝试做出如下更改，让蚂蚁看起来不一样
26. ANT_UP = '^'
27. ANT_DOWN = 'v'
28. ANT_LEFT = '<'
29. ANT_RIGHT = '>'
30.
31. # (!) 尝试将如下颜色更改为'black'、'red'、'green'、'yellow'、'blue'、'purple'、'cyan'或'white'中的一种
32. (这些是 bext 模块支持的颜色)
33. ANT_COLOR = 'red'
34. BLACK_TILE = 'black'
35. WHITE_TILE = 'white'
36.
37. NORTH = 'north'
38. SOUTH = 'south'
39. EAST = 'east'
40. WEST = 'west'
41.
42.
43. def main():
44.     bext.fg(ANT_COLOR)   # 蚂蚁的颜色是前景色
45.     bext.bg(WHITE_TILE)  # 开始前将背景色设置为白色
46.     bext.clear()
47.
48.     # 创建新的 board 数据结构
49.     board = {'width': WIDTH, 'height': HEIGHT}
50.
51.     # 创建 ants 数据结构
```

```
52.    ants = []
53.    for i in range(NUMBER_OF_ANTS):
54.        ant = {
55.            'x': random.randint(0, WIDTH - 1),
56.            'y': random.randint(0, HEIGHT - 1),
57.            'direction': random.choice([NORTH, SOUTH, EAST, WEST]),
58.        }
59.        ants.append(ant)
60.
61.    # 记录发生改变的图块并在屏幕上重新绘制
62.    changedTiles = []
63.
64.    while True:    #主循环
65.        displayBoard(board, ants, changedTiles)
66.        changedTiles = []
67.
68.        # nextBoard 是下一个步骤中的 board。 从当前步骤的 board 副本开始
69.        nextBoard = copy.copy(board)
70.
71.        # 为每只蚂蚁运行一个模拟步骤
72.        for ant in ants:
73.            if board.get((ant['x'], ant['y']), False) == True:
74.                nextBoard[(ant['x'], ant['y'])] = False
75.                #顺时针方向旋转
76.                if ant['direction'] == NORTH:
77.                    ant['direction'] = EAST
78.                elif ant['direction'] == EAST:
79.                    ant['direction'] = SOUTH
80.                elif ant['direction'] == SOUTH:
81.                    ant['direction'] = WEST
82.                elif ant['direction'] == WEST:
83.                    ant['direction'] = NORTH
84.            else:
85.                nextBoard[(ant['x'], ant['y'])] = True
86.                # 逆时针方向旋转
87.                if ant['direction'] == NORTH:
88.                    ant['direction'] = WEST
89.                elif ant['direction'] == WEST:
90.                    ant['direction'] = SOUTH
91.                elif ant['direction'] == SOUTH:
92.                    ant['direction'] = EAST
93.                elif ant['direction'] == EAST:
94.                    ant['direction'] = NORTH
95.            changedTiles.append((ant['x'], ant['y']))
96.
97.            # 让蚂蚁朝它所面对的任意方向前进
98.            if ant['direction'] == NORTH:
99.                ant['y'] -= 1
100.            if ant['direction'] == SOUTH:
101.                ant['y'] += 1
102.            if ant['direction'] == WEST:
103.                ant['x'] -= 1
104.            if ant['direction'] == EAST:
105.                ant['x'] += 1
106.
107.            # 如果蚂蚁越过了屏幕的边缘,它应该绕到另一边
108.            ant['x'] = ant['x'] % WIDTH
109.            ant['y'] = ant['y'] % HEIGHT
110.
111.            changedTiles.append((ant['x'], ant['y']))
112.
113.        board = nextBoard
```

```
114.
115.
116. def displayBoard(board, ants, changedTiles):
117.     """在屏幕上显示挡板和蚂蚁。changedTiles 参数是一个(x, y)元组列表，用于显示屏幕上已更改且需
118. 要重新绘制的图块 """
119.
120.     # 用于绘制 board 的数据结构
121.     for x, y in changedTiles:
122.         bext.goto(x, y)
123.         if board.get((x, y), False):
124.             bext.bg(BLACK_TILE)
125.         else:
126.             bext.bg(WHITE_TILE)
127.
128.         antIsHere = False
129.         for ant in ants:
130.             if (x, y) == (ant['x'], ant['y']):
131.                 antIsHere = True
132.                 if ant['direction'] == NORTH:
133.                     print(ANT_UP, end='')
134.                 elif ant['direction'] == SOUTH:
135.                     print(ANT_DOWN, end='')
136.                 elif ant['direction'] == EAST:
137.                     print(ANT_LEFT, end='')
138.                 elif ant['direction'] == WEST:
139.                     print(ANT_RIGHT, end='')
140.                 break
141.         if not antIsHere:
142.             print(' ', end='')
143.
144.     # 在屏幕底部显示退出信息
145.     bext.goto(0, HEIGHT)
146.     bext.bg(WHITE_TILE)
147.     print('Press Ctrl-C to quit.', end='')
148.
149.     sys.stdout.flush()   # 使用 bext 模块需要刷新缓冲区
150.     time.sleep(PAUSE_AMOUNT)
151.
152.
153. # 程序运行（如果不是作为模块导入的话）
154. if __name__ == '__main__':
155.     try:
156.         main()
157.     except KeyboardInterrupt:
158.         print("Langton's Ant, by Al Sweigart al@inventwithpython.com")
159.         sys.exit()    # 按下 Ctrl-C，程序结束
```

输入源代码并运行几次后，请试着对其进行更改。你可以根据标有!的注释对程序进行修改，也可以尝试以下操作。

❏ 让玩家使用文本文件存储和加载所有图块的状态。
❏ 使用新的移动规则创建其他图块状态，并查看其出现的行为。
❏ 实现维基百科文章中关于朗顿蚂蚁的一些想法。

探索程序

请尝试找出以下问题的答案。你需要尝试对代码进行一些修改，再次运行程序，查看修改

后的效果。

1. 如果将第 142 行的 print(' ', end='') 改为 print('.', end='')，则会发生什么？
2. 如果将第 101 行的 ant['y'] += 1 改为 ant['y'] -= 1，则会发生什么？
3. 如果将第 74 行的 nextBoard[(ant['x'], ant['y'])] = False 改为 nextBoard[(ant['x'], ant['y'])] = True，则会发生什么？
4. 如果将第 19 行的 WIDTH -= 1 改为 WIDTH -= 40，则会发生什么？
5. 如果将第 113 行的 board = nextBoard 改为 board = board，则会发生什么？

项目 40

火星文

要秀黑客技能的话，恐怕没有比用数字替换文本中的字母更酷的方式了！在本项目中，我们把用这种方式形成的文字称为"火星文"，比如"mad hacker skills"的火星文可以是"m4d h4x0r 5k1llz"！本项目的程序会自动将英文转换为火星文，这是非常酷的在线聊天方式。

例如，"It takes a while to get used to, but with some practice, you'll eventually be able to read leetspeak fluently"（习惯火星文需要一段时间，但通过一些练习，你终将能够流利地阅读火星文），其火星文版本为"1t +@]<3s 4 w|-|1le +o g37 |_|s3|) 70, b|_|+ y0u (an 3\/3nt|_|/-\lly r3a|) l33t$peak phl|_|3n+ly"。初次接触火星文，你可能会感到很难读懂，但本项目的程序非常简单，很适合初学者。

运行程序

运行 leetspeak.py，输出如下所示：

```
L3375P34]< (leetspeek)
By Al Sweigart al@inventwithpython.com

Enter your leet message:
> I am a leet hacker. Fear my mad skills. The 90s were over two decades ago.

! @m a l33t h@(]<er. ph3@r my m4|) $k|ll$. +h3 90s w3r3 0ver tw0 d3(ad3$ 4g0.
(Copied leetspeak to clipboard.)
```

工作原理

如下面的程序所示，第 34 行的 charMapping 字典将纯英文字符映射到火星文字符。但是，由于可以有多个可能的火星文字符（例如字母't'对应的'7'或'+'），charMapping 字典中的每个值都是一个字符串列表。当创建新的火星文字符串时，程序有 30%的概率使用原始英文消息中的字符，有 70%的概率使用火星文字符之一。这意味着相同的英文信息有多种可能的火星文版本。

```
1. """火星文，作者：Al Sweigart al@inventwithpython.com
2. 将英文翻译成火星文
3. 标签：小，初学者，文字"""
4.
5. import random
6.
```

```python
 7. try:
 8.     import pyperclip  # pyperclip 模块可以将文本复制到剪贴板
 9. except ImportError:
10.     pass  # pyperclip 模块并不是必需的,不安装也没关系
11.
12.
13. def main():
14.     print('''L3375P34]< (leetspeek)
15. By Al Sweigart al@inventwithpython.com
16.
17. Enter your leet message:''')
18.     english = input('> ')
19.     print()
20.     leetspeak = englishToLeetspeak(english)
21.     print(leetspeak)
22.
23.     try:
24.         #如果没有导入 pyperclip,尝试使用它将引发"名字错误"异常
25.         pyperclip.copy(leetspeak)
26.         print('(Copied leetspeak to clipboard.)')
27.     except NameError:
28.         pass  # 如果未安装 pyperclip,则不用执行任何操作
29.
30.
31. def englishToLeetspeak(message):
32.     """转换消息中的英文字符串并返回对应的火星文"""
33.     # 确保 charMapping 中的所有键都是小写的
34.     charMapping = {
35.     'a': ['4', '@', '/-\\'], 'c': ['(', 'd': ['|)'], 'e': ['3'],
36.     'f': ['ph'], 'h': [']-[', '|-|'], 'i': ['1', '!', '|'], 'k': [']<'],
37.     'o': ['0'], 's': ['$', '5'], 't': ['7', '+'], 'u': ['|_|'],
38.     'v': ['\\/']}
39.     leetspeak = ''
40.     for char in message:  # 检查每个字符
41.         # 我们有 70%的概率把这个字符转换成火星文
42.         if char.lower() in charMapping and random.random() <= 0.70:
43.             possibleLeetReplacements = charMapping[char.lower()]
44.             leetReplacement = random.choice(possibleLeetReplacements)
45.             leetspeak = leetspeak + leetReplacement
46.         else:
47.             # 不转换字符
48.             leetspeak = leetspeak + char
49.     return leetspeak
50.
51.
52. # 程序运行入口(如果不作为模块导入的话)
53. if __name__ == '__main__':
54.     main()
```

输入源代码并运行几次后,请尝试对其进行更改。你也可以尝试以下操作。

❑ 修改 `charMapping` 字典,使其支持新的火星文字符。

❑ 添加可以将火星文转换回英文的功能。

探索程序

请尝试找出以下问题的答案。你需要尝试对代码进行一些修改,再次运行程序,查看修改后的效果。

1. 如果将第 49 行的 return leetspeak 改为 return message，则会发生什么？
2. 如果将第 42 行的 char.lower() 改为 char，则会发生什么？
3. 如果将第 42 行的 char.lower() 改为 char.upper()，则会发生什么？
4. 如果将第 45 行的 leetspeak = leetspeak + leetReplacement 改为 leetspeak = leetReplacement，则会发生什么？

项目 41

幸运星

在本项目中,我们来玩一个游戏。你需要通过掷骰子来收集星星。掷骰子的次数越多,你能得到的星星也就越多,但如果你掷出了 3 个骷髅头,得到的星星就会全部清零!这款快速实现的多人游戏可以支持任意数量的玩家来玩,非常适合在聚会上玩。

轮到你时,你需从骰子筒中随机取出 3 颗骰子并掷出它们。你可以掷出星星、骷髅头和问号。如果你结束游戏,则一颗星星记一分;如果你选择再次掷骰子,则保留问号并取出新骰子来替换星星和骷髅头;如果你掷出 3 个骷髅头,则将失去所有星星,游戏结束。

如果一个玩家率先获得 13 分,那么其他玩家剩 1 次游戏机会,最后,得到最多点数的玩家获胜。

骰子筒里有 6 颗金骰子、4 颗银骰子和 3 颗铜骰子。金骰子上的星星多些,铜骰子上的骷髅头多些,银骰子上的星星和骷髅头数量均等。

运行程序

运行 luckystars.py,输出如下所示:

```
Lucky Stars, by Al Sweigart al@inventwithpython.com
--snip--
SCORES: Alice=0, Bob=0
It is Alice's turn.

+----------+ +----------+ +----------+
|          | |    .     | |          |
|          | |   ,O,    | |          |
|    ?     | | 'ooOOOoo'| |    ?     |
|          | |   `OOO`  | |          |
|          | |   O' 'O  | |          |
+----------+ +----------+ +----------+
    GOLD         GOLD        BRONZE
Stars collected: 1 Skulls collected: 0
Do you want to roll again? Y/N
> y

+----------+ +----------+ +----------+
|    .     | |    ___   | |          |
|   ,O,    | |   /   \  | |          |
| 'ooOOOoo'| |  |() ()| | |    ?     |
```

```
|    `ooo`   | |   \ ^ /   | |           |
|    O'  O   | |    VVV    | |           |
+-----------+ +-----------+ +-----------+
    GOLD         BRONZE        BRONZE
Stars collected: 2 Skulls collected: 1
Do you want to roll again? Y/N
--snip--
```

工作原理

在本项目中，基于文本的图形以字符串形式存储在 `STAR_FACE`、`SKULL_FACE` 和 `QUESTION_FACE` 变量的列表中。这种形式使它们易于在代码编辑器中编写，而第 152～155 行中的代码会将它们显示在屏幕上。注意，因为 3 颗骰子一起显示，代码必须一次性在 3 颗骰子面上输出出各水平文本行。简单运行 `print(STAR_FACE)` 之类的代码会导致 3 颗骰子呈上下排列显示而不是并排显示。

```
 1. """幸运星，作者：Al Sweigart al@inventwithpython.com
 2. 这是一款"碰运气"的游戏，你可以通过掷骰子来收集尽可能多的星星
 3. 你想掷多少次骰子都可以，但如果你掷出 3 个骷髅头，则会失去所有的星星
 4.
 5. 这款游戏的灵感来自僵尸骰子游戏
 6. 标签：大，游戏，多人游戏"""
 7.
 8. import random
 9.
10. # 创建常量
11. GOLD = 'GOLD'
12. SILVER = 'SILVER'
13. BRONZE = 'BRONZE'
14.
15. STAR_FACE = ["+-----------+",
16.              "|     .     |",
17.              "|    ,O,    |",
18.              "| 'ooOOOoo' |",
19.              "|   `OOO`   |",
20.              "|   O' 'O   |",
21.              "+-----------+"]
22. SKULL_FACE = ['+-----------+',
23.               '|    ___    |',
24.               '|   /   \\   |',
25.               '|  |() ()|  |',
26.               '|   \\ ^ /   |',
27.               '|    VVV    |',
28.               '+-----------+']
29. QUESTION_FACE = ['+-----------+',
30.                  '|           |',
31.                  '|           |',
32.                  '|     ?     |',
33.                  '|           |',
34.                  '|           |',
35.                  '+-----------+']
36. FACE_WIDTH = 13
37. FACE_HEIGHT = 7
38.
39. print("""Lucky Stars, by Al Sweigart al@inventwithpython.com
40.
41. A "press your luck" game where you roll dice with Stars, Skulls, and
```

```
 42. Question Marks.
 43.
 44. On your turn, you pull three random dice from the dice cup and roll
 45. them. You can roll Stars, Skulls, and Question Marks. You can end your
 46. turn and get one point per Star. If you choose to roll again, you keep
 47. the Question Marks and pull new dice to replace the Stars and Skulls.
 48. If you collect three Skulls, you lose all your Stars and end your turn.
 49.
 50. When a player gets 13 points, everyone else gets one more turn before
 51. the game ends. Whoever has the most points wins.
 52.
 53. There are 6 Gold dice, 4 Silver dice, and 3 Bronze dice in the cup.
 54. Gold dice have more Stars, Bronze dice have more Skulls, and Silver is
 55. even.
 56. """)
 57.
 58. print('How many players are there?')
 59. while True:  # 继续循环,直到玩家输入一个数字
 60.     response = input('> ')
 61.     if response.isdecimal() and int(response) > 1:
 62.         numPlayers = int(response)
 63.         break
 64.     print('Please enter a number larger than 1.')
 65.
 66. playerNames = []  # 存储玩家名字的列表
 67. playerScores = {}  # 键是玩家的名字,值是整数型的分数
 68. for i in range(numPlayers):
 69.     while True:  # 继续循环,直到玩家输入一个名称
 70.         print('What is player #' + str(i + 1) + '\'s name?')
 71.         response = input('> ')
 72.         if response != '' and response not in playerNames:
 73.             playerNames.append(response)
 74.             playerScores[response] = 0
 75.             break
 76.         print('Please enter a name.')
 77. print()
 78.
 79. turn = 0  # 索引为 0 的玩家将首先开始游戏
 80. # (!) 取消注释,让取名为 "Al" 的玩家以 3 分为底分开始游戏
 81. #playerScores['Al'] = 3
 82. endGameWith = None
 83. while True:  # Main game loop
 84.     # 显示每个玩家的分数
 85.     print()
 86.     print('SCORES: ', end='')
 87.     for i, name in enumerate(playerNames):
 88.         print(name + ' = ' + str(playerScores[name]), end='')
 89.         if i != len(playerNames) - 1:
 90.             # 除了最后一个玩家,所有玩家的名字都用逗号隔开
 91.             print(', ', end='')
 92.     print('\n')
 93.
 94.     # 从 0 开始统计星星和骷髅头的数量
 95.     stars = 0
 96.     skulls = 0
 97.     # 骰子筒里有 6 颗金骰子、4 颗银骰子、3 颗铜骰子
 98.     cup = ([GOLD] * 6) + ([SILVER] * 4) + ([BRONZE] * 3)
 99.     hand = []  # 玩家一开始没有骰子
100.     print('It is ' + playerNames[turn] + '\'s turn.')
101.     while True:  # 该循环的每次迭代都是在掷骰子
102.         print()
103.
```

```
104.        # 检查骰子筒里是否有足够的骰子
105.        if (3 - len(hand)) > len(cup):
106.            # 因为没有足够的骰子，所以结束当前回合
107.            print('There aren\'t enough dice left in the cup to '
108.                  + 'continue ' + playerNames[turn] + '\'s turn.')
109.            break
110.
111.        # 从骰子筒里取骰子，直到有 3 颗骰子在玩家手中
112.        random.shuffle(cup)  # 在杯子中摇骰子
113.        while len(hand) < 3:
114.            hand.append(cup.pop())
115.
116.        # 掷骰子
117.        rollResults = []
118.        for dice in hand:
119.            roll = random.randint(1, 6)
120.            if dice == GOLD:
121.                # 掷出一颗金骰子（3 颗星星，2 个问号，1 个骷髅头）
122.                if 1 <= roll <= 3:
123.                    rollResults.append(STAR_FACE)
124.                    stars += 1
125.                elif 4 <= roll <= 5:
126.                    rollResults.append(QUESTION_FACE)
127.                else:
128.                    rollResults.append(SKULL_FACE)
129.                    skulls += 1
130.            if dice == SILVER:
131.                # 掷出一颗银骰子（2 颗星星，2 个问号，2 个骷髅头）
132.                if 1 <= roll <= 2:
133.                    rollResults.append(STAR_FACE)
134.                    stars += 1
135.                elif 3 <= roll <= 4:
136.                    rollResults.append(QUESTION_FACE)
137.                else:
138.                    rollResults.append(SKULL_FACE)
139.                    skulls += 1
140.            if dice == BRONZE:
141.                # 掷出一颗铜骰子（1 颗星星，2 个问号，3 个骷髅头）
142.                if roll == 1:
143.                    rollResults.append(STAR_FACE)
144.                    stars += 1
145.                elif 2 <= roll <= 4:
146.                    rollResults.append(QUESTION_FACE)
147.                else:
148.                    rollResults.append(SKULL_FACE)
149.                    skulls += 1
150.
151.        # 显示掷骰子的结果
152.        for lineNum in range(FACE_HEIGHT):
153.            for diceNum in range(3):
154.                print(rollResults[diceNum][lineNum] + '  ', end='')
155.            print()  # 输出一行空行
156.
157.        # 展示每颗骰子的类型（金，银，铜）
158.        for diceType in hand:
159.            print(diceType.center(FACE_WIDTH) + '  ', end='')
160.        print()  # 输出一个换行符
161.
162.        print('Stars collected:', stars, '  Skulls collected:', skulls)
163.
164.        # 检查玩家是否收集了 3 个或更多的骷髅头
165.        if skulls >= 3:
```

```python
166.             print('3 or more skulls means you\'ve lost your stars!')
167.             input('Press Enter to continue...')
168.             break
169.
170.         print(playerNames[turn] + ', do you want to roll again? Y/N')
171.         while True:   # 一直继续询问，直到玩家输入 Y 或 N
172.             response = input('> ').upper()
173.             if response != '' and response[0] in ('Y', 'N'):
174.                 break
175.             print('Please enter Yes or No.')
176.
177.         if response.startswith('N'):
178.             print(playerNames[turn], 'got', stars, 'stars!')
179.             # 将星星对应的积分添加到玩家的积分总数中
180.             playerScores[playerNames[turn]] += stars
181.
182.             # 检查玩家是否已经得到 13 或更高的积分
183.             # （！）试着把它改成 5 或 50
184.             if (endGameWith == None
185.                 and playerScores[playerNames[turn]] >= 13):
186.                     # 由于这个玩家获得了 13 分，因此其他玩家只剩 1 次掷骰子机会
187.                     print('\n\n' + ('!' * 60))
188.                     print(playerNames[turn] + ' has reached 13 points!!!')
189.                     print('Everyone else will get one more turn!')
190.                     print(('!' * 60) + '\n\n')
191.                     endGameWith = playerNames[turn]
192.             input('Press Enter to continue...')
193.             break
194.
195.         # 去掉星星和骷髅头，但保留问号
196.         nextHand = []
197.         for i in range(3):
198.             if rollResults[i] == QUESTION_FACE:
199.                 nextHand.append(hand[i])    # 保留问号
200.         hand = nextHand
201.
202.     # 轮到下一个玩家掷骰子
203.     turn = (turn + 1) % numPlayers
204.
205.     # 如果游戏已经结束，则结束当前循环
206.     if endGameWith == playerNames[turn]:
207.         break   # 结束游戏
208.
209. print('The game has ended...')
210.
211. # 显示每个玩家的分数
212. print()
213. print('SCORES: ', end='')
214. for i, name in enumerate(playerNames):
215.     print(name + ' = ' + str(playerScores[name]), end='')
216.     if i != len(playerNames) - 1:
217.         # 除了最后一个玩家，其他玩家的名字之间都用逗号隔开
218.         print(', ', end='')
219. print('\n')
220.
221. # 查看谁是赢家
222. highestScore = 0
223. winners = []
224. for name, score in playerScores.items():
225.     if score > highestScore:
226.         # 此玩家得分最高
227.         highestScore = score
```

```
228.            winners = [name]   # 覆盖以前所有的赢家
229.        elif score == highestScore:
230.            # 这名玩家的得分与最高分持平
231.            winners.append(name)
232.
233.    if len(winners) == 1:
234.        # 只有一个赢家
235.        print('The winner is ' + winners[0] + '!!!')
236.    else:
237.        # 有多个赢家
238.        print('The winners are: ' + ', '.join(winners))
239.
240.    print('Thanks for playing!')
```

输入源代码并运行几次后，请试着对其进行更改。你可以根据标有!的注释对程序进行修改。

探索程序

请尝试找出以下问题的答案。你需要尝试对代码进行一些修改，再次运行程序，查看修改后的效果。

1. 如果删除或注释掉第 112 行的 random.shuffle(cup)，则会发生什么？
2. 如果将第 165 行的 skulls >= 3 改为 skulls > 3，则会发生什么？
3. 如果将第 203 行的 (turn + 1) % numPlayers 改为 (turn + 1)，则会得到什么错误消息？
4. 如果将第 82 行的 endGameWith = None 改为 endGameWith = playerNames[0]，则会发生什么？
5. 如果删除或注释掉第 174 行的 break，则会发生什么？
6. 如果将第 74 行的 playerScores[response] = 0 改为 playerScores[response] = 10，则会发生什么？

项目 42　魔法幸运球

在本项目中，我们会编写魔法幸运球程序，利用 Python 随机数模块来预测未来，并以 100%的准确率对你的是/否问题作答。该程序实现的效果类似于魔力 8 号球玩具，只是你不必摇晃这个魔法幸运球了。它还能实现缓慢输出文本字符串（每个字符之间留有空格）的效果，营造一种秘密气氛。

本项目中的大部分代码特意用于营造秘密的气氛，程序本身只会随机选择一条消息进行显示，作为对玩家所提出问题的响应。

运行程序

运行 magicfortuneball.py，输出如下所示：

```
MAGiC  FORTUNE  BALL,  BY  AL  SWEiGART

ASK  ME  YOUR  YES/NO  QUESTiON.

> Isn't fortune telling just a scam to trick money out of gullible people?
LET  ME  THiNK  ON  THiS...

. . . . . . . .
I  HAVE  AN  ANSWER...

AFFiRMATiVE
```

工作原理

魔法幸运球程序主要用于显示随机选择的字符串。它其实完全忽略了用户的问题。没错，第 26 行调用 input('> ')，但它不会将返回值存储在任何变量中，因为程序实际上并不使用该文本。让玩家输入问题只是为了使他们觉得程序具有厉害的洞察力。

slowSpacePrint()函数会将文本中所有大写字母 I 显示为小写字母 i，使消息看起来颇具个性。该函数还在字符串的每个字符之间插入空格，并且缓慢地显示它们。程序并不需要复杂到真的能够预测未来才有趣！

```
1. """魔法幸运球，作者：Al Sweigart al@inventwithpython.com
2. 这款游戏会对玩家的是/否问题作答，其灵感来自魔力 8 号球玩具
3. 标签：小，初学者，诙谐"""
```

```
 4.
 5. import random, time
 6.
 7.
 8. def slowSpacePrint(text, interval=0.1):
 9.     """缓慢显示文本，每个字符和小写字母 i 之间留有空格"""
10.     for character in text:
11.         if character == 'I':
12.             # I 以小写形式显示
13.             print('i ', end='', flush=True)
14.         else:
15.             # 其他字符正常显示
16.             print(character + ' ', end='', flush=True)
17.         time.sleep(interval)
18.     print()  # 在末尾输出两行空行
19.     print()
20.
21.
22. # 问题提示
23. slowSpacePrint('MAGIC FORTUNE BALL, BY AL SWEiGART')
24. time.sleep(0.5)
25. slowSpacePrint('ASK ME YOUR YES/NO QUESTION.')
26. input('> ')
27.
28. # 显示一个简短的回复
29. replies = [
30.     'LET ME THINK ON THIS...',
31.     'AN INTERESTING QUESTION...',
32.     'HMMM... ARE YOU SURE YOU WANT TO KNOW..?',
33.     'DO YOU THINK SOME THINGS ARE BEST LEFT UNKNOWN..?',
34.     'I MIGHT TELL YOU, BUT YOU MIGHT NOT LIKE THE ANSWER...',
35.     'YES... NO... MAYBE... I WILL THINK ON IT...',
36.     'AND WHAT WILL YOU DO WHEN YOU KNOW THE ANSWER? WE SHALL SEE...',
37.     'I SHALL CONSULT MY VISIONS...',
38.     'YOU MAY WANT TO SIT DOWN FOR THIS...',
39. ]
40. slowSpacePrint(random.choice(replies))
41.
42. # 故弄玄虚的停顿
43. slowSpacePrint('.' * random.randint(4, 12), 0.7)
44.
45. # 给出答案
46. slowSpacePrint('I HAVE AN ANSWER...', 0.2)
47. time.sleep(1)
48. answers = [
49.     'YES, FOR SURE',
50.     'MY ANSWER IS NO',
51.     'ASK ME LATER',
52.     'I AM PROGRAMMED TO SAY YES',
53.     'THE STARS SAY YES, BUT I SAY NO',
54.     'I DUNNO MAYBE',
55.     'FOCUS AND ASK ONCE MORE',
56.     'DOUBTFUL, VERY DOUBTFUL',
57.     'AFFIRMATIVE',
58.     'YES, THOUGH YOU MAY NOT LIKE IT',
59.     'NO, BUT YOU MAY WISH IT WAS SO',
60. ]
61. slowSpacePrint(random.choice(answers), 0.05)
```

输入源代码并运行几次后，请尝试对其进行更改。你也可以尝试以下操作。

❏ 检查玩家的问题是否以问号结尾。

❏ 添加程序可以给出的其他答案。

探索程序

请尝试找出以下问题的答案。你需要尝试对代码进行一些修改，再次运行程序，查看修改后的效果。

1. 如果将第 43 行的 random.randint(4, 12)改为 random.randint(4, 9999)，则会发生什么？
2. 如果将第 47 行的 time.sleep(1)改为 time.sleep(-1)，则会出现什么错误？

项目 43

播棋

播棋（Mancala），又称为"非洲棋"，英文名为 Sowing Game，意为"播种游戏"，是一款二人对弈的棋类游戏。

播棋这种棋类游戏至少有 2000 年的历史，其与项目 63 的游戏一样，都很古老。在游戏中，两名玩家从坑中选择种子，将其分配到棋盘上的其他坑中，同时玩家努力往自己的仓库中收集尽可能多的种子。该游戏在不同的文化背景下有多个版本，其名字起源于阿拉伯语 naqala，意思是"搬运"。

玩游戏时，玩家从棋盘上自己这一侧的一个坑中抓取种子，以逆时针的方向往后续的每个坑（也包括自己的仓库）中各放一个种子，但不向对手的仓库中放种子。如果玩家的最后一颗种子落在自己的空坑中，则将对手坑中的种子放到自己的仓库中。如果玩家最后放的种子在自己的仓库中，则该玩家可以接着玩下一个回合。

当任意一名玩家的坑中都为空时，游戏结束，此时另一名玩家将棋盘上自己这一侧所有坑中的种子都放到自己的仓库中。最后，谁仓库中的种子越多，谁就是最终的获胜者。

运行程序

运行 mancala.py，输出如下所示：

```
Mancala, by Al Sweigart al@inventwithpython.com
--snip--

+------+------+--<<<<<-Player 2----+------+------+------+
2     |G     |H     |I     |J     |K     |L     |     1
      |    4 |    4 |    4 |    4 |    4 |    4 |
S     |      |      |      |      |      |      |     S
T    0+------+------+------+------+------+------+0    T
O     |A     |B     |C     |D     |E     |F     |     O
R     |    4 |    4 |    4 |    4 |    4 |    4 |     R
E     |      |      |      |      |      |      |     E
+------+------+------+-Player 1->>>>>-----+------+------+

Player 1, choose move: A-F (or QUIT)
> f

+------+------+--<<<<<-Player 2----+------+------+------+
2     |G     |H     |I     |J     |K     |L     |     1
      |    4 |    4 |    4 |    5 |    5 |    5 |
S     |      |      |      |      |      |      |     S
```

```
T   O +------+------+------+------+------+------+  1 T
O    |A     |B     |C     |D     |E     |F     |    O
R    |4     |4     |4     |4     |4     |0     |    R
E    |      |      |      |      |      |      |    E
     +------+------+------+-Player 1->>>>>-----+------+
Player 2, choose move: G-L (or QUIT)
```
--snip--

工作原理

本项目以 ASCII 艺术画的形式显示棋盘。注意，每个坑中不仅要有种子数量，还要有一个标签。为避免混淆，使用字母 A~L 作为标签，它们不会被误认为是坑中种子的数量。字典 `NEXT_PIT` 和 `OPPOSITE_PIT` 分别将一个坑的字母映射到它旁边或对面的坑的字母。因此，表达式 `NEXT_PIT['A']` 的计算结果为 B，而表达式 `OPPOSITE_PIT['A']` 的计算结果为 G。要注意代码中是如何使用这些字典的。如果不这么做，则该播棋程序将需要使用一长串 `if` 和 `elif` 语句来实现游戏步骤。

```
 1. """播棋，作者：Al Sweigart al@inventwithpython.com
 2. 古老的"播种游戏"
 3. 标签：大，棋盘，游戏，双人游戏"""
 4.
 5. import sys
 6.
 7. # 存储玩家的坑的元组
 8. PLAYER_1_PITS = ('A', 'B', 'C', 'D', 'E', 'F')
 9. PLAYER_2_PITS = ('G', 'H', 'I', 'J', 'K', 'L')
10.
11. # 一个键为坑、值是其对面坑的字典
12. OPPOSITE_PIT = {'A': 'G', 'B': 'H', 'C': 'I', 'D': 'J', 'E': 'K',
13.                 'F': 'L', 'G': 'A', 'H': 'B', 'I': 'C', 'J': 'D',
14.                 'K': 'E', 'L': 'F'}
15.
16. # 一个键是坑、值是其下一个坑的字典
17. NEXT_PIT = {'A': 'B', 'B': 'C', 'C': 'D', 'D': 'E', 'E': 'F', 'F': '1',
18.             '1': 'L', 'L': 'K', 'K': 'J', 'J': 'I', 'I': 'H', 'H': 'G',
19.             'G': '2', '2': 'A'}
20.
21. # 每个坑标签，从 A 开始按逆时针顺序旋转
22. PIT_LABELS = 'ABCDEF1LKJIHG2'
23.
24. # 在新游戏开始时，每个坑中有多少颗种子
25. STARTING_NUMBER_OF_SEEDS = 4   # (!) 尝试将改成 1 或 10
26.
27.
28. def main():
29.     print('''Mancala, by Al Sweigart al@inventwithpython.com
30.
31. The ancient two-player, seed-sowing game. Grab the seeds from a pit on
32. your side and place one in each following pit, going counterclockwise
33. and skipping your opponent's store. If your last seed lands in an empty
34. pit of yours, move the opposite pit's seeds into your store. The
35. goal is to get the most seeds in your store on the side of the board.
36. If the last placed seed is in your store, you get a free turn.
37.
38. The game ends when all of one player's pits are empty. The other player
39. claims the remaining seeds for their store, and the winner is the one
```

```
40.   with the most seeds.
41.   ''')
42.     input('Press Enter to begin...')
43.
44.     gameBoard = getNewBoard()
45.     playerTurn = '1'  # 玩家 1 先开始游戏
46.
47.     while True:  # 运行一个回合
48.         # 通过输出许多新行来清空屏幕,让原来的游戏界面(棋盘)不再可见
49.         print('\n' * 60)
50.         # 显示棋盘并获取玩家的下一步走子
51.         displayBoard(gameBoard)
52.         playerMove = askForPlayerMove(playerTurn, gameBoard)
53.
54.         # 执行玩家的下一步走子
55.         playerTurn = makeMove(gameBoard, playerTurn, playerMove)
56.
57.         # 检查游戏是否结束以及哪位玩家获胜
58.         winner = checkForWinner(gameBoard)
59.         if winner == '1' or winner == '2':
60.             displayBoard(gameBoard)  # 最后一次显示棋盘
61.             print('Player ' + winner + ' has won!')
62.             sys.exit()
63.         elif winner == 'tie':
64.             displayBoard(gameBoard)  # 最后一次显示棋盘
65.             print('There is a tie!')
66.             sys.exit()
67.
68.
69. def getNewBoard():
70.     """返回表示开始状态下播棋的字典:每个坑中有 4 颗种子,仓库中有 0 颗种子"""
71.
72.     # 美化语法——使用更短的变量名
73.     s = STARTING_NUMBER_OF_SEEDS
74.
75.     # 创建棋盘的数据结构,包括仓库中的 0 颗种子以及游戏开始时坑中种子的数量
76.     return {'1': 0, '2': 0, 'A': s, 'B': s, 'C': s, 'D': s, 'E': s,
77.             'F': s, 'G': s, 'H': s, 'I': s, 'J': s, 'K': s, 'L': s}
78.
79.
80. def displayBoard(board):
81.     """在棋盘字典的基础上以 ASCII 艺术画的形式显示棋盘"""
82.
83.     seedAmounts = []
84.     # 下面这个 GHIJKL21ABCDEF 字符串是 pits 按照从左到右和从上到下的顺序排列的
85.     for pit in 'GHIJKL21ABCDEF':
86.         numSeedsInThisPit = str(board[pit]).rjust(2)
87.         seedAmounts.append(numSeedsInThisPit)
88.
89.     print("""
90. +------+------+--<<<<<-Player 2----+------+------+------+
91. 2      |G     |H     |I     |J     |K     |L     |      1
92.        | {}   | {}   | {}   | {}   | {}   | {}   |
93. S      |      |      |      |      |      |      |      S
94. T {}   +------+------+------+------+------+------+ {}   T
95. O      |A     |B     |C     |D     |E     |F     |      O
96. R      | {}   | {}   | {}   | {}   | {}   | {}   |      R
97. E      |      |      |      |      |      |      |      E
98. +------+------+------+-Player 1->>>>>-----+------+------+
99.
100. """.format(*seedAmounts))
101.
```

```python
102.
103. def askForPlayerMove(playerTurn, board):
104.     """询问玩家选择在棋盘一侧的哪个坑播种。 以字符串形式返回所选坑的大写字母标签 """
105.
106.     while True:  # 继续询问玩家,直到他们输入有效的下一步走子
107.         # 让玩家在自己这边选择一个坑
108.         if playerTurn == '1':
109.             print('Player 1, choose move: A-F (or QUIT)')
110.         elif playerTurn == '2':
111.             print('Player 2, choose move: G-L (or QUIT)')
112.         response = input('> ').upper().strip()
113.
114.         # 检查玩家是否想要退出
115.         if response == 'QUIT':
116.             print('Thanks for playing!')
117.             sys.exit()
118.
119.         # 确保它是一个有效的 pit 以供选择
120.         if (playerTurn == '1' and response not in PLAYER_1_PITS) or (
121.             playerTurn == '2' and response not in PLAYER_2_PITS
122.         ):
123.             print('Please pick a letter on your side of the board.')
124.             continue  # 再次询问玩家的下一步走子
125.         if board.get(response) == 0:
126.             print('Please pick a non-empty pit.')
127.             continue  # 再次询问玩家的下一步走子
128.         return response
129.
130.
131. def makeMove(board, playerTurn, pit):
132.     """修改棋盘的数据结构,让 1 号或 2 号玩家依次选择一个坑作为播种的坑。根据下一回合轮到谁来返回 1 或 2 """
133.
134.     seedsToSow = board[pit]  # 从选择的坑中获得相应数量的种子
135.     board[pit] = 0  # 清空选择的坑
136.
137.     while seedsToSow > 0:  # 继续播种,直到没有种子为止
138.         pit = NEXT_PIT[pit]  # 去到下一个坑
139.         if (playerTurn == '1' and pit == '2') or (
140.             playerTurn == '2' and pit == '1'
141.         ):
142.             continue  # 跳过对手的仓库
143.         board[pit] += 1
144.         seedsToSow -= 1
145.
146.     # 如果最后一颗种子放入玩家的仓库,那么玩家再来一回合
147.     if (pit == playerTurn == '1') or (pit == playerTurn == '2'):
148.         # 最后一颗种子放入玩家的仓库;再来一回合
149.         return playerTurn
150.
151.     # 检查最后一颗种子是否在一个空的坑里;取走对面坑里的种子
152.     if playerTurn == '1' and pit in PLAYER_1_PITS and board[pit] == 1:
153.         oppositePit = OPPOSITE_PIT[pit]
154.         board['1'] += board[oppositePit]
155.         board[oppositePit] = 0
156.     elif playerTurn == '2' and pit in PLAYER_2_PITS and board[pit] == 1:
157.         oppositePit = OPPOSITE_PIT[pit]
158.         board['2'] += board[oppositePit]
159.         board[oppositePit] = 0
160.
161.     # 轮到另一位玩家玩游戏
162.     if playerTurn == '1':
163.         return '2'
```

```
164.        elif playerTurn == '2':
165.            return '1'
166.
167.
168.    def checkForWinner(board):
169.        """观察棋盘,如果有赢家则返回 1 或 2,如果没有则返回平手或没有赢家
170.        如果一个玩家的坑空了,则游戏结束;另一个玩家将剩余的种子放入自己的仓库中。拥有种子最多的玩家获胜
171.        """
172.
173.        player1Total = board['A'] + board['B'] + board['C']
174.        player1Total += board['D'] + board['E'] + board['F']
175.        player2Total = board['G'] + board['H'] + board['I']
176.        player2Total += board['J'] + board['K'] + board['L']
177.
178.        if player1Total == 0:
179.            # 玩家 2 得到自己剩余的所有种子
180.            board['2'] += player2Total
181.            for pit in PLAYER_2_PITS:
182.                board[pit] = 0   # Set all pits to 0
183.        elif player2Total == 0:
184.            # 玩家 1 得到自己剩余的所有种子
185.            board['1'] += player1Total
186.            for pit in PLAYER_1_PITS:
187.                board[pit] = 0   # 设置所有坑为 0
188.        else:
189.            return 'no winner'  # 目前还没有赢家
190.
191.        # 游戏结束,找到得分最高的玩家
192.        if board['1'] > board['2']:
193.            return '1'
194.        elif board['2'] > board['1']:
195.            return '2'
196.        else:
197.            return 'tie'
198.
199.
200.    # 程序运行入口(如果不是被作为模块导入的话)
201.    if __name__ == '__main__':
202.        main()
```

输入源代码并运行几次后,请尝试对其进行更改。你也可以尝试以下操作。

❑ 换一个坑更多的棋盘。
❑ 随机选择一个"奖励坑",如果最后一颗种子落入其中,则玩家可以再玩一个回合。
❑ 为 4 个玩家(而不是 2 个玩家)创建一个正方形棋盘。

探索程序

请尝试找出以下问题的答案。你需要尝试对代码进行一些修改,再次运行程序,查看修改后的效果。

1. 如果将第 163 行的 `return '2'` 改为 `return '1'`,则会发生什么?
2. 如果将第 195 行的 `return '2'` 改为 `return '1'`,则会发生什么?
3. 如果将第 115 行的 `response == 'QUIT'` 改为 `response == 'quit'`,则会发生什么?

4. 如果将第 135 行的 board[pit] = 0 改为 board[pit] = 1，则会发生什么？
5. 如果将第 49 行的 print('\n' * 60) 改为 print('\n' * 0)，则会发生什么？
6. 如果将第 45 行的 playerTurn = '1' 改为 playerTurn = '2'，则会发生什么？
7. 如果将第 125 行的 board.get(response) == 0 改为 board.get(response) == -1，则会发生什么？

项目 44

二维版移动迷宫

在本项目中，我们用二维版移动迷宫程序向玩家展示了在文本编辑器（例如用于编写.py文件的IDE）中创建的迷宫文件俯视图。使用W、A、S和D键，玩家可以分别将@符号向上、向左、向下和向右移动，最后将其移至由X字符标记的出口处。

要制作迷宫文件，请打开文本编辑器并创建以下图案。请勿在图案的顶部和左侧输入数字，因为它们并非图案的构成部分，只是用作绘图参照。

```
 123456789
1#########
2#S# # # #
3#########
4# # # # #
5#########
6# # # # #
7#########
8# # # #E#
9#########
```

在文件中，#字符代表墙壁，S字符标记开始，E字符标记退出[1]。#字符代表可以移除以形成迷宫的墙壁。不要清除奇数列和奇数行的墙壁，也不要清除迷宫的边界。设置完成后，请将迷宫保存为.txt（文本）文件。二维版移动迷宫的效果如下所示：

```
#########
#S    # #
# ### # #
# #   # #
# ##### #
#   #   #
### # # #
#   #E#
#########
```

当然，这只是一个简单的迷宫。你可以制作任意大小的迷宫文件，让它们的行数和列数为奇数即可。此外，请确保其大小与计算机屏幕相匹配！

---

[1] 程序运行后代表出口的E会被替换为X。——译者注

运行程序

运行 mazerunner2d.py，输出如下所示：

```
Maze Runner 2D, by Al Sweigart al@inventwithpython.com

(Maze files are generated by mazemakerrec.py)
Enter the filename of the maze (or LIST or QUIT):
> maze65x11s1.txt
```

```
▒▒▒▒▒▒▒▒▒▒▒▒▒▒▒▒▒▒▒▒▒▒▒▒▒▒▒▒▒▒▒▒▒▒▒▒▒▒▒▒▒▒▒▒▒▒▒▒▒▒▒▒▒▒▒▒▒▒▒▒▒▒▒▒▒
▒@▒   ▒       ▒       ▒   ▒   ▒     ▒             ▒   ▒       ▒ ▒
▒ ▒ ▒ ▒ ▒▒▒ ▒ ▒ ▒▒▒▒▒ ▒ ▒ ▒ ▒ ▒ ▒▒▒ ▒ ▒▒▒▒▒▒▒▒▒ ▒ ▒ ▒ ▒ ▒▒▒▒▒ ▒ ▒
...
                                                              X
▒▒▒▒▒▒▒▒▒▒▒▒▒▒▒▒▒▒▒▒▒▒▒▒▒▒▒▒▒▒▒▒▒▒▒▒▒▒▒▒▒▒▒▒▒▒▒▒▒▒▒▒▒▒▒▒▒▒▒▒▒▒▒▒▒
                           W
Enter direction, or QUIT: ASD
--snip--
```

工作原理

本项目中的程序将迷宫墙壁的数据从文本文件加载到存储在 `maze` 变量中的字典中。该字典用(x，y)元组作为键，用 WALL、EMPTY、START 或 EXIT 常量中的字符串作为值。在项目 45 中，我们用到类似的迷宫字典表示。本项目与项目 45 的区别在于在屏幕上呈现迷宫的代码。由于二维版移动迷宫更简单，建议在学习三维版移动迷宫之前先熟悉二维版移动迷宫的程序。

```
 1. """二维版移动迷宫，作者：Al Sweigart al@inventwithpython.com
 2. 在迷宫中移动并尝试逃跑。迷宫文件由 mazemakerrec.py 生成
 3. 标签：大，游戏，迷宫"""
 4.
 5. import sys, os
 6.
 7. # 迷宫文件常量
 8. WALL = '#'
 9. EMPTY = ' '
10. START = 'S'
11. EXIT = 'E'
12.
13. PLAYER = '@'    # (!) 尝试将其改成'+'或'o'
14. BLOCK = chr(9617)   # 字符 9617 表示的是 "▒"
15.
16.
17. def displayMaze(maze):
18.     # 显示迷宫
19.     for y in range(HEIGHT):
20.         for x in range(WIDTH):
21.             if (x, y) == (playerx, playery):
22.                 print(PLAYER, end='')
23.             elif (x, y) == (exitx, exity):
24.                 print('X', end='')
```

```
25.            elif maze[(x, y)] == WALL:
26.                print(BLOCK, end='')
27.            else:
28.                print(maze[(x, y)], end='')
29.        print()  # 输出一个空行
30.
31.
32. print('''Maze Runner 2D, by Al Sweigart al@inventwithpython.com
33.
34. (Maze files are generated by mazemakerrec.py)''')
35.
36. # 从玩家处获取迷宫文件的文件名
37. while True:
38.     print('Enter the filename of the maze (or LIST or QUIT):')
39.     filename = input('> ')
40.
41.     # 列出当前文件夹中的所有迷宫文件
42.     if filename.upper() == 'LIST':
43.         print('Maze files found in', os.getcwd())
44.         for fileInCurrentFolder in os.listdir():
45.             if (fileInCurrentFolder.startswith('maze') and
46.                 fileInCurrentFolder.endswith('.txt')):
47.                 print('  ', fileInCurrentFolder)
48.         continue
49.
50.     if filename.upper() == 'QUIT':
51.         sys.exit()
52.
53.     if os.path.exists(filename):
54.         break
55.     print('There is no file named', filename)
56.
57. # 从文件中加载迷宫
58. mazeFile = open(filename)
59. maze = {}
60. lines = mazeFile.readlines()
61. playerx = None
62. playery = None
63. exitx = None
64. exity = None
65. y = 0
66. for line in lines:
67.     WIDTH = len(line.rstrip())
68.     for x, character in enumerate(line.rstrip()):
69.         assert character in (WALL, EMPTY, START, EXIT), 'Invalid character at column {},
70. line{}'.format(x + 1, y + 1)
71.         if character in (WALL, EMPTY):
72.             maze[(x, y)] = character
73.         elif character == START:
74.             playerx, playery = x, y
75.             maze[(x, y)] = EMPTY
76.         elif character == EXIT:
77.             exitx, exity = x, y
78.             maze[(x, y)] = EMPTY
79.     y += 1
80. HEIGHT = y
81.
82. assert playerx != None and playery != None, 'No start in maze file.'
83. assert exitx != None and exity != None, 'No exit in maze file.'
84.
85. while True:  # 主循环
86.     displayMaze(maze)
```

```
 87.
 88.        while True:  # 获取玩家的下一步移动
 89.            print('                              W')
 90.            print('Enter direction, or QUIT: ASD')
 91.            move = input('> ').upper()
 92.
 93.            if move == 'QUIT':
 94.                print('Thanks for playing!')
 95.                sys.exit()
 96.
 97.            if move not in ['W', 'A', 'S', 'D']:
 98.                print('Invalid direction. Enter one of W, A, S, or D.')
 99.                continue
100.
101.            # 检查玩家是否可以朝那个方向移动
102.            if move == 'W' and maze[(playerx, playery - 1)] == EMPTY:
103.                break
104.            elif move == 'S' and maze[(playerx, playery + 1)] == EMPTY:
105.                break
106.            elif move == 'A' and maze[(playerx - 1, playery)] == EMPTY:
107.                break
108.            elif move == 'D' and maze[(playerx + 1, playery)] == EMPTY:
109.                break
110.
111.            print('You cannot move in that direction.')
112.
113.        # 继续朝当前方向移动，直至遇到一个分支点
114.        if move == 'W':
115.            while True:
116.                playery -= 1
117.                if (playerx, playery) == (exitx, exity):
118.                    break
119.                if maze[(playerx, playery - 1)] == WALL:
120.                    break  # 碰到墙壁时就结束循环
121.                if (maze[(playerx - 1, playery)] == EMPTY
122.                    or maze[(playerx + 1, playery)] == EMPTY):
123.                    break  # 如果到达一个分支点，就结束循环
124.        elif move == 'S':
125.            while True:
126.                playery += 1
127.                if (playerx, playery) == (exitx, exity):
128.                    break
129.                if maze[(playerx, playery + 1)] == WALL:
130.                    break  # 碰到墙壁时就结束循环
131.                if (maze[(playerx - 1, playery)] == EMPTY
132.                    or maze[(playerx + 1, playery)] == EMPTY):
133.                    break  # 如果到达一个分支点，就结束循环
134.        elif move == 'A':
135.            while True:
136.                playerx -= 1
137.                if (playerx, playery) == (exitx, exity):
138.                    break
139.                if maze[(playerx - 1, playery)] == WALL:
140.                    break  # 碰到墙壁时就结束循环
141.                if (maze[(playerx, playery - 1)] == EMPTY
142.                    or maze[(playerx, playery + 1)] == EMPTY):
143.                    break  # 如果到达一个分支点，就结束循环
144.        elif move == 'D':
145.            while True:
146.                playerx += 1
147.                if (playerx, playery) == (exitx, exity):
148.                    break
```

```
149.            if maze[(playerx + 1, playery)] == WALL:
150.                break  # 碰到墙壁时就结束循环
151.            if (maze[(playerx, playery - 1)] == EMPTY
152.                or maze[(playerx, playery + 1)] == EMPTY):
153.                break  # 如果到达一个分支点，就结束循环
154.
155.    if (playerx, playery) == (exitx, exity):
156.        displayMaze(maze)
157.        print('You have reached the exit! Good job!')
158.        print('Thanks for playing!')
159.        sys.exit()
```

探索程序

请尝试找出以下问题的答案。你需要尝试对代码进行一些修改，再次运行程序，查看修改后的效果。

1. 如果将第 73 行的 character == START 改为 character == EXIT，会得到什么错误消息？
2. 如果将第 104 行的 playery + 1 改为 playery-1，则会发生什么？
3. 如果将第 155 行的 (exitx, exity) 改为 (None, None)，则会发生什么？
4. 如果将第 88 行的 while True: 改为 while False:，则会得到什么错误消息？
5. 如果将第 103 行的 break 改为 continue，则会发生什么？
6. 如果将第 120 行的 break 改为 continue，则会得到什么错误消息？

项目 45

三维版移动迷宫

在本项目中,我们用三维版移动迷宫程序为玩家提供了迷宫内部第一人称视角的效果。请努力寻找出路!你可以按照项目 44 中的说明生成迷宫文件。

运行程序

运行 mazerunner3d.py,输出如下所示:

```
Maze Runner 3D, by Al Sweigart al@inventwithpython.com
(Maze files are generated by mazemakerrec.py)
Enter the filename of the maze (or LIST or QUIT):
> maze75x11s1.txt
```

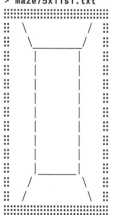

```
Location (1, 1) Direction: NORTH
(W)
Enter direction: (A) (D) or QUIT.
> d
```

```
      \
       _____
       |       |
       |       |
       |       |
       |       |
       |       |
       |       |
       |      /
       |_____/
Location (1, 1) Direction: EAST
--snip--
```

工作原理

本项目中的三维透视 ASCII 艺术画一开始显示路径未被墙壁封闭的场景（用存储在 ALL_OPEN 中的多行字符串描绘），然后基于此绘制墙（相应的字符串）存储在 CLOSED 字典中，以生成由任意可能的封闭路径组成的 ASCII 艺术画。例如，以下视图展示了程序如何生成墙壁并使其位于玩家的左侧：

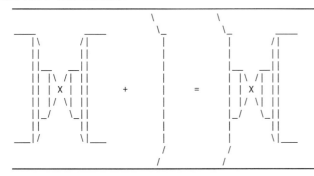

在显示字符串之前，源代码中 ASCII 艺术画中的句点（.）会被删除。句点的存在只是为了让输入代码更容易一些——可以避免插入或遗漏空格。

三维版移动迷宫的源代码如下所示：

```
 1. """三维版移动迷宫，作者：Al Sweigart al@inventwithpython.com
 2. 在三维版移动迷宫中四处移动以逃离迷宫
 3. 标签：特大，艺术，游戏，迷宫"""
 4. import copy, sys, os
 5.
 6. #创建常量
 7. WALL = '#'
 8. EMPTY = ' '
 9. START = 'S'
10. EXIT = 'E'
11. BLOCK = chr(9617)   # 字符 9617 表示的是 "▒"
```

```
12. NORTH = 'NORTH'
13. SOUTH = 'SOUTH'
14. EAST = 'EAST'
15. WEST = 'WEST'
16.
17.
18. def wallStrToWallDict(wallStr):
19.     """获取墙壁的字符串表示形式（像ALL_OPEN或CLOSED中的那些表示形式），并返回字典中的表示形式
20.     其中键为(x, y)元组，值为在该(x, y)位置绘制的字符"""
21.     wallDict = {}
22.     height = 0
23.     width = 0
24.     for y, line in enumerate(wallStr.splitlines()):
25.         if y > height:
26.             height = y
27.         for x, character in enumerate(line):
28.             if x > width:
29.                 width = x
30.             wallDict[(x, y)] = character
31.     wallDict['height'] = height + 1
32.     wallDict['width'] = width + 1
33.     return wallDict
34.
35. EXIT_DICT = {(0, 0): 'E', (1, 0): 'X', (2, 0): 'I',
36.              (3, 0): 'T', 'height': 1, 'width': 4}
37.
38. # 创建所要显示的字符串的方法是：使用wallStrToWallDict()将这些多行字符串中的图片转换为字典
39. # 然后根据玩家的位置和方向，将CLOSED中的字典"粘贴"到ALL_OPEN中的字典之上，从而组成墙壁
40.
41.
42.
43. ALL_OPEN = wallStrToWallDict(r'''
44. ................
45. ____.........____
46. ...|\......./|...
47. ...||.......||...
48. ...||__...__||...
49. ...||.|\./|.||...
50. ...||.|.X.|.||...
51. ...||.|/.\|.||...
52. ...||_/...\_||...
53. ...||.......||...
54. ___|/.......\|___
55. ................
56. ................'''.strip())
57. # strip()用于在这个多行字符串的开头删除换行符
58.
59. CLOSED = {}
60. CLOSED['A'] = wallStrToWallDict(r'''
61. ____
62. .....
63. .....
64. .....
65. ____'''.strip()) # 粘贴到（6,4）处
66.
67. CLOSED['B'] = wallStrToWallDict(r'''
68. .\.
69. ..\
70. ...
71. ...
72. ...
73. ../
74. ./.'''.strip()) # 粘贴到（4,3）处
```

```
 75.
 76. CLOSED['C'] = wallStrToWallDict(r'''
 77. _____
 78. ..........
 79. ..........
 80. ..........
 81. ..........
 82. ..........
 83. ..........
 84. ..........
 85. ..........
 86. _____'''.strip())  # 粘贴到（3，1）处
 87.
 88. CLOSED['D'] = wallStrToWallDict(r'''
 89. ./.
 90. /..
 91. ...
 92. ...
 93. ...
 94. \..
 95. .\.'''.strip())  # 粘贴到（10,3）处
 96.
 97. CLOSED['E'] = wallStrToWallDict(r'''
 98. ..\..
 99. ...\_
100. ....|
101. ....|
102. ....|
103. ....|
104. ....|
105. ....|
106. ....|
107. ....|
108. ....|
109. .../.
110. ../..'''.strip())  # 粘贴到（0,0）处
111.
112. CLOSED['F'] = wallStrToWallDict(r'''
113. ../..
114. _/...
115. |....
116. |....
117. |....
118. |....
119. |....
120. |....
121. |....
122. |....
123. |....
124. .\...
125. ..\..'''.strip())  # 粘贴到（12,0）处
126.
127. def displayWallDict(wallDict):
128.     """在屏幕上显示由wallStrToWallDict()返回的字典"""
129.     print(BLOCK * (wallDict['width'] + 2))
130.     for y in range(wallDict['height']):
131.         print(BLOCK, end='')
132.         for x in range(wallDict['width']):
133.             wall = wallDict[(x, y)]
134.             if wall == '.':
135.                 wall = ' '
136.             print(wall, end='')
137.         print(BLOCK)  # 换行输出BLOCK
```

```
138.        print(BLOCK * (wallDict['width'] + 2))
139.
140.
141. def pasteWallDict(srcWallDict, dstWallDict, left, top):
142.     """将 srcWallDict 中的字典复制到 dstWallDict 中的字典之上
143. 偏移到左上方给定的位置"""
144.     dstWallDict = copy.copy(dstWallDict)
145.     for x in range(srcWallDict['width']):
146.         for y in range(srcWallDict['height']):
147.             dstWallDict[(x + left, y + top)] = srcWallDict[(x, y)]
148.     return dstWallDict
149.
150.
151. def makeWallDict(maze, playerx, playery, playerDirection, exitx, exity):
152.     """根据玩家在迷宫中的位置和方向（在 exitx 和 exity 处有一个出口），
153. 通过将字典粘贴到 ALL_OPEN 的顶部来创建表示墙壁的字典，然后将其返回"""
154.
155.     # 根据 A~F "区域"（与玩家方向相关）确定在迷宫中检查哪些墙壁
156.     # 看看是否需要将它们粘贴到正在创建的字典上
157.
158.     if playerDirection == NORTH:
159.         # 各区域的地图，与 A 相关
160.         # 取决于玩家 @:           BCD  (玩家面向北方)
161.         #                         E@F
162.         offsets = (('A', 0, -2), ('B', -1, -1), ('C', 0, -1),
163.                    ('D', 1, -1), ('E', -1, 0), ('F', 1, 0))
164.     if playerDirection == SOUTH:
165.         # 各区域的地图，与 F@E 相关
166.         # 取决于玩家 @:           DCB  (玩家面向南方)
167.         #                          A
168.         offsets = (('A', 0, 2), ('B', 1, 1), ('C', 0, 1),
169.                    ('D', -1, 1), ('E', 1, 0), ('F', -1, 0))
170.     if playerDirection == EAST:
171.         # 各区域的地图，与 EB 相关
172.         # 取决于玩家 @:           @CA  (玩家面向东方)
173.         #                          FD
174.         offsets = (('A', 2, 0), ('B', 1, -1), ('C', 1, 0),
175.                    ('D', 1, 1), ('E', 0, -1), ('F', 0, 1))
176.     if playerDirection == WEST:
177.         # 各区域的地图，与 DF 相关
178.         # 取决于玩家@:            AC@  (玩家面向西方)
179.         #                          BE
180.         offsets = (('A', -2, 0), ('B', -1, 1), ('C', -1, 0),
181.                    ('D', -1, -1), ('E', 0, 1), ('F', 0, -1))
182.
183.     section = {}
184.     for sec, xOff, yOff in offsets:
185.         section[sec] = maze.get((playerx + xOff, playery + yOff), WALL)
186.         if (playerx + xOff, playery + yOff) == (exitx, exity):
187.             section[sec] = EXIT
188.
189.     wallDict = copy.copy(ALL_OPEN)
190.     PASTE_CLOSED_TO = {'A': (6, 4), 'B': (4, 3), 'C': (3, 1),
191.                        'D': (10, 3), 'E': (0, 0), 'F': (12, 0)}
192.     for sec in 'ABDCEF':
193.         if section[sec] == WALL:
194.             wallDict = pasteWallDict(CLOSED[sec], wallDict,
195.                 PASTE_CLOSED_TO[sec][0], PASTE_CLOSED_TO[sec][1])
196.
197.     # 若有需要，则绘制退出标志
198.     if section['C'] == EXIT:
199.         wallDict = pasteWallDict(EXIT_DICT, wallDict, 7, 9)
```

```
200.        if section['E'] == EXIT:
201.            wallDict = pasteWallDict(EXIT_DICT, wallDict, 0, 11)
202.        if section['F'] == EXIT:
203.            wallDict = pasteWallDict(EXIT_DICT, wallDict, 13, 11)
204.
205.    return wallDict
206.
207.
208. print('Maze Runner 3D, by Al Sweigart al@inventwithpython.com')
209. print('(Maze files are generated by mazemakerrec.py)')
210.
211. # 从玩家处获取迷宫文件的文件名
212. while True:
213.     print('Enter the filename of the maze (or LIST or QUIT):')
214.     filename = input('> ')
215.
216.     #列出当前文件夹中的所有迷宫文件
217.     if filename.upper() == 'LIST':
218.         print('Maze files found in', os.getcwd())
219.         for fileInCurrentFolder in os.listdir():
220.             if (fileInCurrentFolder.startswith('maze')
221.                 and fileInCurrentFolder.endswith('.txt')):
222.                 print('  ', fileInCurrentFolder)
223.         continue
224.
225.     if filename.upper() == 'QUIT':
226.         sys.exit()
227.
228.     if os.path.exists(filename):
229.         break
230.     print('There is no file named', filename)
231.
232. # 从文件中加载迷宫文件
233. mazeFile = open(filename)
234. maze = {}
235. lines = mazeFile.readlines()
236. px = None
237. py = None
238. exitx = None
239. exity = None
240. y = 0
241. for line in lines:
242.     WIDTH = len(line.rstrip())
243.     for x, character in enumerate(line.rstrip()):
244.         assert character in (WALL, EMPTY, START, EXIT), 'Invalid character at column {},
245. line {}'.format(x + 1, y + 1)
246.         if character in (WALL, EMPTY):
247.             maze[(x, y)] = character
248.         elif character == START:
249.             px, py = x, y
250.             maze[(x, y)] = EMPTY
251.         elif character == EXIT:
252.             exitx, exity = x, y
253.             maze[(x, y)] = EMPTY
254.     y += 1
255. HEIGHT = y
256.
257. assert px != None and py != None, 'No start point in file.'
258. assert exitx != None and exity != None, 'No exit point in file.'
259. pDir = NORTH
260.
261.
262. while True: #主循环
```

```
263.      displayWallDict(makeWallDict(maze, px, py, pDir, exitx, exity))
264.
265.      while True:  # 获取玩家的下一步移动
266.          print('Location ({}, {})   Direction: {}'.format(px, py, pDir))
267.          print('                  (W)')
268.          print('Enter direction: (A) (D)   or QUIT.')
269.          move = input('> ').upper()
270.
271.          if move == 'QUIT':
272.              print('Thanks for playing!')
273.              sys.exit()
274.
275.          if (move not in ['F', 'L', 'R', 'W', 'A', 'D']
276.              and not move.startswith('T')):
277.              print('Please enter one of F, L, or R (or W, A, D).')
278.              continue
279.
280.          # 根据玩家的下一步移动来前进
281.          if move == 'F' or move == 'W':
282.              if pDir == NORTH and maze[(px, py - 1)] == EMPTY:
283.                  py -= 1
284.                  break
285.              if pDir == SOUTH and maze[(px, py + 1)] == EMPTY:
286.                  py += 1
287.                  break
288.              if pDir == EAST and maze[(px + 1, py)] == EMPTY:
289.                  px += 1
290.                  break
291.              if pDir == WEST and maze[(px - 1, py)] == EMPTY:
292.                  px -= 1
293.                  break
294.          elif move == 'L' or move == 'A':
295.              pDir = {NORTH: WEST, WEST: SOUTH,
296.                      SOUTH: EAST, EAST: NORTH}[pDir]
297.              break
298.          elif move == 'R' or move == 'D':
299.              pDir = {NORTH: EAST, EAST: SOUTH,
300.                      SOUTH: WEST, WEST: NORTH}[pDir]
301.              break
302.          elif move.startswith('T'):  # 作弊代码: 'T x,y'
303.              px, py = move.split()[1].split(',')
304.              px = int(px)
305.              py = int(py)
306.              break
307.          else:
308.              print('You cannot move in that direction.')
309.
310.      if (px, py) == (exitx, exity):
311.          print('You have reached the exit! Good job!')
312.          print('Thanks for playing!')
313.          sys.exit()
```

探索程序

请尝试找出以下问题的答案。你需要尝试对代码进行一些修改，再次运行程序，查看修改后的效果。

1. 如果将第 271 行的 move == 'QUIT' 改为 move == 'quit'，会导致什么错误？
2. 怎样才能消除作弊行为？（提示：程序运行后，输入 T 73,9，可以直达迷宫出口。）

项目 46 掷100万次骰子结果统计模拟器

当你掷两颗 6 面骰子时,掷出点数和为 7 的概率约为 17%,比掷出点数和为 2 的概率要大得多,后者的概率约为 3%。这是因为只有一种骰子组合可以得到 2(只有两个骰子都掷出 1,才会出现这种结果),但有多个组合的点数加起来为 7,包括 1 和 6、2 和 5、3 和 4 等组合。

那么当掷 3 颗骰子时呢?4 颗呢?1000 颗呢?我们可以用数学方法计算出理论概率,也可以让计算机掷 100 万次骰子来凭经验计算出概率。本项目的程序采用后一种方法。在本程序中,我们让计算机掷 N 个骰子 100 万次并记住结果,然后显示每个点数总和的百分比概率。

本程序做了大量的计算,但其计算逻辑并不难理解。

运行程序

运行 milliondicestats.py,输出如下所示:

```
Million Dice Roll Statistics Simulator
By Al Sweigart al@inventwithpython.com

Enter how many six-sided dice you want to roll:
> 2
Simulating 1,000,000 rolls of 2 dice...
36.2% done...
73.4% done...
TOTAL - ROLLS - PERCENTAGE
   2 - 27590 rolls - 2.8%
   3 - 55730 rolls - 5.6%
   4 - 83517 rolls - 8.4%
   5 - 111526 rolls - 11.2%
   6 - 139015 rolls - 13.9%
   7 - 166327 rolls - 16.6%
   8 - 139477 rolls - 13.9%
   9 - 110268 rolls - 11.0%
  10 - 83272 rolls - 8.3%
  11 - 55255 rolls - 5.5%
  12 - 28023 rolls - 2.8%
```

工作原理

我们在第 29 行通过调用 random.randint(1, 6) 来模拟掷单个 6 面骰子。这将返回一个

1~6 的随机数，无论掷多少个骰子，这个随机数都会被累加到点数总和中。random.randint() 函数具有均匀分布的特征，这意味着每个数字返回的概率相同。

以下程序将一次掷骰子的结果存储在 results 字典中。这个字典的键是每次可能掷出的骰子总数，值则是掷出骰子的次数。为了获得频率（百分比），我们将该总次数除以 1000000（此模拟中掷骰子的次数）并将其乘 100（以获得 0.0～100.0 而不是 0.0～1.0 的百分比）。算一下就会明白这与我们在第 36 行所做的将总次数除以 10000 的效果是相同的。

```
1.  """掷 100 万次骰子结果统计模拟器
2.  作者: Al Sweigart al@inventwithpython.com
3.  模拟掷 100 万次骰子
4.  标签: 小，初学者，数学，模拟"""
5.
6.  import random, time
7.
8.  print('''Million Dice Roll Statistics Simulator
9.  By Al Sweigart al@inventwithpython.com
10.
11. Enter how many six-sided dice you want to roll:''')
12. numberOfDice = int(input('> '))
13.
14. # 设置一个字典来存储每次掷骰子的结果
15. results = {}
16. for i in range(numberOfDice, (numberOfDice * 6) + 1):
17.     results[i] = 0
18.
19. # 模拟掷骰子
20. print('Simulating 1,000,000 rolls of {} dice...'.format(numberOfDice))
21. lastPrintTime = time.time()
22. for i in range(1000000):
23.     if time.time() > lastPrintTime + 1:
24.         print('{}% done...'.format(round(i / 10000, 1)))
25.         lastPrintTime = time.time()
26.
27.     total = 0
28.     for j in range(numberOfDice):
29.         total = total + random.randint(1, 6)
30.     results[total] = results[total] + 1
31.
32. # 显示结果
33. print('TOTAL - ROLLS - PERCENTAGE')
34. for i in range(numberOfDice, (numberOfDice * 6) + 1):
35.     roll = results[i]
36.     percentage = round(results[i] / 10000, 1)
37.     print('  {} - {} rolls - {}%'.format(i, roll, percentage))
```

输入源代码并运行几次后，请尝试对其进行更改。你也可以尝试以下操作。

❏ 尝试掷 8 面、10 面、12 面或 20 面骰子。

❏ 尝试模拟抛双面硬币。

探索程序

请尝试找出以下问题的答案。你需要尝试对代码进行一些修改，再次运行程序，查看修改后的效果。

1. 如果将第 23 行的 `lastPrintTime + 1` 改为 `lastPrintTime + 2`，会发生什么？
2. 如果删除或注释掉第 30 行的 `results[total] = results[total] + 1`，会导致什么错误？
3. 如果用户输入字母而不是数字来表示要掷的 6 面骰子的数量，会发生什么错误？

项目 47　蒙德里安艺术品生成器

彼埃·蒙德里安（Piet Mondrian）是20世纪著名的荷兰画家，也是抽象艺术运动新造型主义的创始人之一。他最具代表性的画作是基于原色（蓝色、黄色、红色）、黑色和白色绘制的。他利用极简主义的方法，通过水平元素和垂直元素将这些颜色分开。

在本项目中，我们将生成遵循蒙德里安风格的随机绘画程序。

运行程序

bext 模块能让 Python 程序在输出文本时显示明亮的原色（本书为黑白印刷，只能显示黑白图像）。运行 mondrian.py，输出如下图所示。

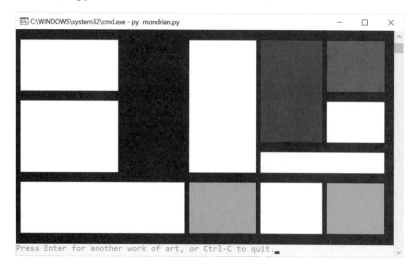

工作原理

在本项目中，我们通过创建具有随机间隔的垂直线和水平线的数据结构（canvas 字典）来实现相应效果：

工作原理 185

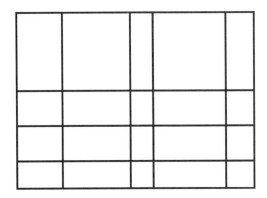

接下来，删除一些线，以创建一些较大的矩形：

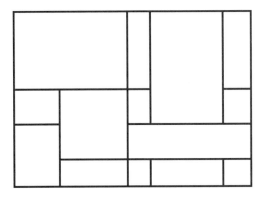

最后，随机选择矩形，用黄色、红色、蓝色或黑色随机进行填充：

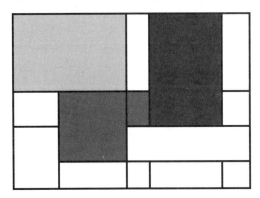

```
1. """蒙德里安艺术品生成器，作者：Al Sweigart al@inventwithpython.com
2. 随机生成蒙德里安风格的艺术品
3. 标签：大，艺术，bext 模块"""
4.
5. import sys, random
6.
7. try:
8.     import bext
9. except ImportError:
```

```
10.     print('This program requires the bext module, which you')
11.     print('can install by following the instructions at')
12.     print('https://pypi.org/project/Bext/')
13.     sys.exit()
14.
15. # 创建常量
16. MIN_X_INCREASE = 6
17. MAX_X_INCREASE = 16
18. MIN_Y_INCREASE = 3
19. MAX_Y_INCREASE = 6
20. WHITE = 'white'
21. BLACK = 'black'
22. RED = 'red'
23. YELLOW = 'yellow'
24. BLUE = 'blue'
25.
26. # 创建屏幕
27. width, height = bext.size()
28. # 如果不自动添加换行符，我们就无法输出到 Windows 上的最后一列，所以将宽度减少 1
29. width -= 1
30.
31. height -= 3
32.
33. while True:    # 主循环
34.     # 在画布上预填充空白
35.     canvas = {}
36.     for x in range(width):
37.         for y in range(height):
38.             canvas[(x, y)] = WHITE
39.
40.     # 生成垂直线
41.     numberOfSegmentsToDelete = 0
42.     x = random.randint(MIN_X_INCREASE, MAX_X_INCREASE)
43.     while x < width - MIN_X_INCREASE:
44.         numberOfSegmentsToDelete += 1
45.         for y in range(height):
46.             canvas[(x, y)] = BLACK
47.         x += random.randint(MIN_X_INCREASE, MAX_X_INCREASE)
48.
49.     # 生成水平线
50.     y = random.randint(MIN_Y_INCREASE, MAX_Y_INCREASE)
51.     while y < height - MIN_Y_INCREASE:
52.         numberOfSegmentsToDelete += 1
53.         for x in range(width):
54.             canvas[(x, y)] = BLACK
55.         y += random.randint(MIN_Y_INCREASE, MAX_Y_INCREASE)
56.
57.     numberOfRectanglesToPaint = numberOfSegmentsToDelete - 3
58.     numberOfSegmentsToDelete = int(numberOfSegmentsToDelete * 1.5)
59.
60.     # 随机选择点并试着将其移除
61.     for i in range(numberOfSegmentsToDelete):
62.         while True:   # 继续选择并试着将其移除
63.             # 在一个现有的段上获得一个随机的起点
64.             startx = random.randint(1, width - 2)
65.             starty = random.randint(1, height - 2)
66.             if canvas[(startx, starty)] == WHITE:
67.                 continue
68.
69.             # 判断是在垂直段还是水平段上
70.             if (canvas[(startx - 1, starty)] == WHITE and
71.                 canvas[(startx + 1, starty)] == WHITE):
```

```
 72.                orientation = 'vertical'
 73.            elif (canvas[(startx, starty - 1)] == WHITE and
 74.                  canvas[(startx, starty + 1)] == WHITE):
 75.                orientation = 'horizontal'
 76.            else:
 77.                # 起点在一个交叉点上，所以得到一个新的随机起点
 78.                continue
 79.
 80.            pointsToDelete = [(startx, starty)]
 81.
 82.            canDeleteSegment = True
 83.            if orientation == 'vertical':
 84.                # 从起点向上"走"一条"路"，看看能否将其移除
 85.                for changey in (-1, 1):
 86.                    y = starty
 87.                    while 0 < y < height - 1:
 88.                        y += changey
 89.                        if (canvas[(startx - 1, y)] == BLACK and
 90.                            canvas[(startx + 1, y)] == BLACK):
 91.                            # 我们发现了一个十字交叉口
 92.                            break
 93.                        elif ((canvas[(startx - 1, y)] == WHITE and
 94.                               canvas[(startx + 1, y)] == BLACK) or
 95.                              (canvas[(startx - 1, y)] == BLACK and
 96.                               canvas[(startx + 1, y)] == WHITE)):
 97.                            # 我们找到了一个T形路口，但不能删除这一段
 98.                            canDeleteSegment = False
 99.                            break
100.                        else:
101.                            pointsToDelete.append((startx, y))
102.
103.            elif orientation == 'horizontal':
104.                # 从起点向上"走"一条"路"，看看能否将其移除
105.                for changex in (-1, 1):
106.                    x = startx
107.                    while 0 < x < width - 1:
108.                        x += changex
109.                        if (canvas[(x, starty - 1)] == BLACK and
110.                            canvas[(x, starty + 1)] == BLACK):
111.                            #我们发现了一个十字交叉口
112.                            break
113.                        elif ((canvas[(x, starty - 1)] == WHITE and
114.                               canvas[(x, starty + 1)] == BLACK) or
115.                              (canvas[(x, starty - 1)] == BLACK and
116.                               canvas[(x, starty + 1)] == WHITE)):
117.                            # 我们找到了一个T形路口，但不能删除这一段
118.                            canDeleteSegment = False
119.                            break
120.                        else:
121.                            pointsToDelete.append((x, starty))
122.            if not canDeleteSegment:
123.                continue  # 随机获取一个新的起点
124.            break  # 继续删除段
125.
126.        # 如果我们可以删除这一段，将所有的点设置为白色
127.        for x, y in pointsToDelete:
128.            canvas[(x, y)] = WHITE
129.
130.    # 添加边框线
131.    for x in range(width):
132.        canvas[(x, 0)] = BLACK  # 上边框
133.        canvas[(x, height - 1)] = BLACK  # 下边框
```

```
134.    for y in range(height):
135.        canvas[(0, y)] = BLACK  # 左侧边框
136.        canvas[(width - 1, y)] = BLACK  # 右侧边框
137.
138.    # 画矩形
139.    for i in range(numberOfRectanglesToPaint):
140.        while True:
141.            startx = random.randint(1, width - 2)
142.            starty = random.randint(1, height - 2)
143.
144.            if canvas[(startx, starty)] != WHITE:
145.                continue  # 获得一个新的随机起点
146.            else:
147.                break
148.
149.        # 泛洪填充算法
150.        colorToPaint = random.choice([RED, YELLOW, BLUE, BLACK])
151.        pointsToPaint = set([(startx, starty)])
152.        while len(pointsToPaint) > 0:
153.            x, y = pointsToPaint.pop()
154.            canvas[(x, y)] = colorToPaint
155.            if canvas[(x - 1, y)] == WHITE:
156.                pointsToPaint.add((x - 1, y))
157.            if canvas[(x + 1, y)] == WHITE:
158.                pointsToPaint.add((x + 1, y))
159.            if canvas[(x, y - 1)] == WHITE:
160.                pointsToPaint.add((x, y - 1))
161.            if canvas[(x, y + 1)] == WHITE:
162.                pointsToPaint.add((x, y + 1))
163.
164.    # 画布数据结构
165.    for y in range(height):
166.        for x in range(width):
167.            bext.bg(canvas[(x, y)])
168.            print(' ', end='')
169.
170.        print()
171.
172.    # 提示用户创建一个新的艺术品
173.    try:
174.        input('Press Enter for another work of art, or Ctrl-C to quit.')
175.    except KeyboardInterrupt:
176.        sys.exit()
```

输入源代码并运行几次后，请试着对其进行更改。你也可以尝试以下操作。

□ 创建具有不同调色板的程序。

□ 使用 Pillow 模块生成蒙德里安艺术品的图像文件。你可以阅读《Python 编程快速上手——让繁琐工作自动化》一书，学习如何使用该模块。

探索程序

请尝试找出以下问题的答案。你需要尝试对代码进行一些修改，再次运行程序，查看修改后的效果。

1. 如果将第 38 行的 `canvas[(x, y)] = WHITE` 改为 `canvas[(x, y)] = RED`，则会发生什么错误？

2. 如果将第 168 行的 `print(' ', end='')` 改为 `print('A', end='')`，则会发生什么？

项目 48

3扇门问题

本项目中的 3 扇门问题揭示了一个令人惊讶的概率事实。该问题大致基于老牌电视游戏节目 *Let's Make a Deal* 及其主持人 Monty Hall。对于 3 扇门问题,你可以选择 3 扇门中的 1 扇门。其中 1 扇门后面有奖品(一辆新车),其他两扇门后各有 1 只山羊(在本项目中,山羊是毫无价值的)。假设你选择了 1 号门,在该门被打开之前,主持人会打开另一扇有山羊的门(2 号门或 3 号门)。现在,你可以选择打开自己最初选择的那扇门,或者改选另一扇未打开的门。

可能看上去是否改变选择没有影响,但如果你改变选择,中奖概率确实会提高!本项目的程序会通过重复实验来论证 3 扇门问题。

要了解为什么中奖概率会提高,不妨考虑有 1000 扇门而不是 3 扇门的"N 扇门"问题。你选择 1 扇门,然后主持人打开 998 扇门,所有门后都有 1 只山羊,未打开的两扇门只有你选择的那扇门和另一扇门。如果你一开始就正确地选择了后面有车的门(1/1000 的概率),那么主持人会随机选择 1 扇后面有山羊的门使其保持关闭。如果你选择了后面有山羊的门(99/1000 的概率),主持人会专门选择后面有车的门使其保持关闭。选择打开哪些门不是随机的;主持人总是会让后面有车的门保持关闭。几乎可以认为,你一开始没有选中后面有车的门,因此你应该改选另一扇门。

另一种思考方式是,现在有 1000 个盒子,其中 1 个盒子中有奖品。你猜奖品在哪个盒子里,主持人会把这个盒子交给你。你认为奖品是在你的盒子里还是在其他 999 个盒子中的其中一个里?你不需要主持人打开 999 个不含奖品的盒子中的 998 个,选中的概率与 1000 扇门问题的相同。你一开始猜中的概率是 1/1000,未猜中的概率(奖品在其他某个盒子里)大约是 99/1000。

运行程序

运行 montyhall.py,输出如下所示:

```
The Monty Hall Problem, by Al Sweigart al@inventwithpython.com
--snip--
+------+ +------+ +------+
|      | |      | |      |
|  1   | |  2   | |  3   |
|      | |      | |      |
|      | |      | |      |
|      | |      | |      |
+------+ +------+ +------+
Pick a door 1, 2, or 3 (or "quit" to stop):
```

```
> 1
+------+ +------+ +------+
|      | |      | |  ((  |
|  1   | |  2   | |  oo  |
|      | |      | | /_/|_|
|      | |      | |      |
|      | |      | ||GOAT|||
+------+ +------+ +------+
Door 3 contains a goat!
Do you want to swap doors? Y/N
> y
+------+ +------+ +------+
|  ((  | | CAR! | |  ((  |
|  oo  | |  __  | |  oo  |
| _/ |_| |  _/  | | /_/|_|
|      | | /_ __| |      |
|GOAT||| |  O   | |GOAT|||
+------+ +------+ +------+
Door 2 has the car!
You won!

Swapping:     1 wins, 0 losses, success rate 100.0%
Not swapping: 0 wins, 0 losses, success rate 0.0%

Press Enter to repeat the experiment...
--snip--
```

工作原理

在本项目中，我们将用于表示 ASCII 艺术画的多行字符串存储在几个常量（例如 ALL_CLOSED、FIRST_GOAT 和 FIRST_CAR_OTHERS_GOAT）中。对于使用了这些常量的代码，如第 120 行的 print(FIRST_GOAT)，即使我们更新了图形，也不需要对其加以修改。通过将多行字符串一起放在源代码文件的顶部，我们可以更轻松地比较它们，以确保图形的一致。

```
 1. """3 扇门问题，作者：Al Sweigart al@inventwithpython.com
 2. 一个模拟 3 扇门游戏的问题
 3. 标签：大，游戏，数学，模拟"""
 4.
 5. import random, sys
 6.
 7. ALL_CLOSED = """
 8. +------+ +------+ +------+
 9. |      | |      | |      |
10. |  1   | |  2   | |  3   |
11. |      | |      | |      |
12. |      | |      | |      |
13. |      | |      | |      |
14. +------+ +------+ +------+"""
15.
16. FIRST_GOAT = """
17. +------+ +------+ +------+
18. |  ((  | |      | |      |
19. |  oo  | |  2   | |  3   |
20. | /_/|_|| |      | |      |
21. |      | |      | |      |
22. |GOAT|||| |      | |      |
23. +------+ +------+ +------+"""
24.
```

```
25. SECOND_GOAT = """
26. +------+  +------+  +------+
27. |      |  |  ((  |  |      |
28. |  1   |  |  oo  |  |  3   |
29. |      |  | /_/|_|  |      |
30. |      |  |   | |   |      |
31. |      |  |GOAT| |  |      |
32. +------+  +------+  +------+"""
33.
34. THIRD_GOAT = """
35. +------+  +------+  +------+
36. |      |  |      |  |  ((  |
37. |  1   |  |  2   |  |  oo  |
38. |      |  |      |  | /_/|_|
39. |      |  |      |  |   | |
40. |      |  |      |  |GOAT| |
41. +------+  +------+  +------+"""
42.
43. FIRST_CAR_OTHERS_GOAT = """
44. +------+  +------+  +------+
45. | CAR! |  |  ((  |  |  ((  |
46. |   __|  |  oo  |  |  oo  |
47. |  _/  |  | /_/|_|  | /_/|_|
48. | /_ __|  |   | |   |   | |
49. |   O  |  |GOAT| |  |GOAT| |
50. +------+  +------+  +------+"""
51.
52. SECOND_CAR_OTHERS_GOAT = """
53. +------+  +------+  +------+
54. |  ((  |  | CAR! |  |  ((  |
55. |  oo  |  |   __|  |  oo  |
56. | /_/|_|  |  _/  |  | /_/|_|
57. |   | |   | /_ __|  |   | |
58. |GOAT| |  |   O  |  |GOAT| |
59. +------+  +------+  +------+"""
60.
61. THIRD_CAR_OTHERS_GOAT = """
62. +------+  +------+  +------+
63. |  ((  |  |  ((  |  | CAR! |
64. |  oo  |  |  oo  |  |   __|
65. | /_/|_|  | /_/|_|  |  _/  |
66. |   | |   |   | |   | /_ __|
67. |GOAT| |  |GOAT| |  |   O  |
68. +------+  +------+  +------+"""
69.
70. print('''The Monty Hall Problem, by Al Sweigart al@inventwithpython.com
71.
72. In the Monty Hall game show, you can pick one of three doors. One door
73. has a new car for a prize. The other two doors have worthless goats:
74. {}
75. Say you pick Door #1.
76. Before the door you choose is opened, another door with a goat is opened:
77. {}
78. You can choose to either open the door you originally picked or swap
79. to the other unopened door.
80.
81. It may seem like it doesn't matter if you swap or not, but your odds
82. do improve if you swap doors! This program demonstrates the Monty Hall
83. problem by letting you do repeated experiments.
84. '''.format(ALL_CLOSED, THIRD_GOAT))
85.
86. input('Press Enter to start...')
```

```
 87.
 88.
 89. swapWins = 0
 90. swapLosses = 0
 91. stayWins = 0
 92. stayLosses = 0
 93. while True:    # 主循环
 94.     # 由计算机选择哪一扇门后面有汽车
 95.     doorThatHasCar = random.randint(1, 3)
 96.
 97.     # 由玩家选择一扇门
 98.     print(ALL_CLOSED)
 99.     while True:    # 不断询问玩家，直到他们选中一扇有效的门
100.         print('Pick a door 1, 2, or 3 (or "quit" to stop):')
101.         response = input('> ').upper()
102.         if response == 'QUIT':
103.             # 结束游戏
104.             print('Thanks for playing!')
105.             sys.exit()
106.
107.         if response == '1' or response == '2' or response == '3':
108.             break
109.     doorPick = int(response)
110.
111.     # 找出该向玩家呈现的后面有山羊的门
112.     while True:
113.         # 选择一扇未被玩家选中的后面有山羊的门
114.         showGoatDoor = random.randint(1, 3)
115.         if showGoatDoor != doorPick and showGoatDoor != doorThatHasCar:
116.             break
117.
118.     # 向玩家展示后面有山羊的门
119.     if showGoatDoor == 1:
120.         print(FIRST_GOAT)
121.     elif showGoatDoor == 2:
122.         print(SECOND_GOAT)
123.     elif showGoatDoor ==3:
124.         print(THIRD_GOAT)
125.
126.     print('Door {} contains a goat!'.format(showGoatDoor))
127.
128.     # 询问玩家是否想换一扇门来选择
129.     while True:    # 继续询问，直到玩家输入 Y 或 N
130.         print('Do you want to swap doors? Y/N')
131.         swap = input('> ').upper()
132.         if swap == 'Y' or swap == 'N':
133.             break
134.
135.     # 如果玩家想交换，则交换玩家选择的门
136.     if swap == 'Y':
137.         if doorPick == 1 and showGoatDoor == 2:
138.             doorPick = 3
139.         elif doorPick == 1 and showGoatDoor == 3:
140.             doorPick = 2
141.         elif doorPick == 2 and showGoatDoor == 1:
142.             doorPick = 3
143.         elif doorPick == 2 and showGoatDoor == 3:
144.             doorPick = 1
145.         elif doorPick == 3 and showGoatDoor == 1:
146.             doorPick = 2
147.         elif doorPick == 3 and showGoatDoor == 2:
148.             doorPick = 1
```

```
149.
150.        # 打开所有的门
151.        if doorThatHasCar == 1:
152.            print(FIRST_CAR_OTHERS_GOAT)
153.        elif doorThatHasCar == 2:
154.            print(SECOND_CAR_OTHERS_GOAT)
155.        elif doorThatHasCar == 3:
156.            print(THIRD_CAR_OTHERS_GOAT)
157.
158.        print('Door {} has the car!'.format(doorThatHasCar))
159.
160.        # 记录交换和不交换的胜与负记录
161.        if doorPick == doorThatHasCar:
162.            print('You won!')
163.            if swap == 'Y':
164.                swapWins += 1
165.            elif swap == 'N':
166.                stayWins += 1
167.        else:
168.            print('Sorry, you lost.')
169.            if swap == 'Y':
170.                swapLosses += 1
171.            elif swap == 'N':
172.                stayLosses += 1
173.
174.        # 计算交换和不交换的成功率
175.        totalSwaps = swapWins + swapLosses
176.        if totalSwaps != 0:  #避免除以0这种错误
177.            swapSuccess = round(swapWins / totalSwaps * 100, 1)
178.        else:
179.            swapSuccess = 0.0
180.
181.        totalStays = stayWins + stayLosses
182.        if (stayWins + stayLosses) != 0:  # 避免除数为0
183.            staySuccess = round(stayWins / totalStays * 100, 1)
184.        else:
185.            staySuccess = 0.0
186.
187.        print()
188.        print('Swapping:      ', end='')
189.        print('{} wins, {} losses, '.format(swapWins, swapLosses), end='')
190.        print('success rate {}%'.format(swapSuccess))
191.        print('Not swapping: ', end='')
192.        print('{} wins, {} losses, '.format(stayWins, stayLosses), end='')
193.        print('success rate {}%'.format(staySuccess))
194.        print()
195.        input('Press Enter repeat the experiment...')
```

探索程序

请尝试找出以下问题的答案。你需要尝试对代码进行一些修改,再次运行程序,查看修改后的效果。

1. 如果将第95行的 doorThatHasCar = random.randint(1, 3) 改为 doorThatHasCar = 1,则会发生什么?
2. 如果使用 print([FIRST_GOAT, SECOND_GOAT, THIRD_GOAT][showGoatDoor - 1]) 替换第119~124行的代码,则会发生什么?

项目 49

乘法表

在本项目中,我们将实现可用于生成从 0×0 到 12×12 的乘法表的程序。虽然本项目的程序简单,但是的确演示了一个有用的嵌套循环。

运行程序

运行 multiplicationtable.py,输出如下所示:

```
Multiplication Table, by Al Sweigart al@inventwithpython.com
  |  0   1   2   3   4   5   6   7   8   9  10  11  12
--+--------------------------------------------------------
 0|  0   0   0   0   0   0   0   0   0   0   0   0   0
 1|  0   1   2   3   4   5   6   7   8   9  10  11  12
 2|  0   2   4   6   8  10  12  14  16  18  20  22  24
 3|  0   3   6   9  12  15  18  21  24  27  30  33  36
 4|  0   4   8  12  16  20  24  28  32  36  40  44  48
 5|  0   5  10  15  20  25  30  35  40  45  50  55  60
 6|  0   6  12  18  24  30  36  42  48  54  60  66  72
 7|  0   7  14  21  28  35  42  49  56  63  70  77  84
 8|  0   8  16  24  32  40  48  56  64  72  80  88  96
 9|  0   9  18  27  36  45  54  63  72  81  90  99 108
10|  0  10  20  30  40  50  60  70  80  90 100 110 120
11|  0  11  22  33  44  55  66  77  88  99 110 121 132
12|  0  12  24  36  48  60  72  84  96 108 120 132 144
```

工作原理

如下面的代码所示,第 9 行用于输出表格的顶行。注意,该行在数字之间设置了足够大的距离以容纳最多 3 位的结果(为保持表格形式工整、美观)。由于这是一个 12×12 的乘法表,该间距可以容纳最大的 3 位数乘积 144。如果你希望创建一个更大的表,则可能需要增加列的间距。记住,标准终端窗口宽为 80 列、高为 24 行。因此,如果没有在窗口右侧换行,就无法创建更大的乘法表。

```
 1. """乘法表，作者：Al Sweigart al@inventwithpython.com
 2. 输出乘法表
 3. 标签：小，初学者，数学"""
 4.
 5. print('Multiplication Table, by Al Sweigart al@inventwithpython.com')
 6.
 7. # 输出水平数字标签
 8. print('  |  0   1   2   3   4   5   6   7   8   9  10  11  12')
 9. print('--+----------------------------------------------------')
10.
11. # 展示每一行的乘积
12. for number1 in range(0, 13):
13.
14.     # 输出垂直数字标签
15.     print(str(number1).rjust(2), end='')
16.
17.     # 输出分隔线
18.     print('|', end='')
19.
20.     for number2 in range(0, 13):
21.         # 在空白处输出乘积
22.         print(str(number1 * number2).rjust(3), end=' ')
23.
24.     print()  # 输出空行表示结束
```

探索程序

请尝试找出以下问题的答案。你需要尝试对代码进行一些修改，再次运行程序，查看修改后的效果。

1. 如果将第 12 行的 range(0, 13) 改为 range(0, 80)，会发生什么？
2. 如果将第 12 行的 range(0, 13) 改为 range(0, 100)，会发生什么？

项目 50

99瓶牛奶1

《99 瓶牛奶》是一首不明来历的民歌，以长度和重复性而闻名。歌词是："墙上有 99 瓶牛奶，99 瓶牛奶，拿下 1 瓶，传过去，墙上还有 98 瓶牛奶。"随着歌词的重复，瓶子的数量从 98 减少到 97，再从 97 减少到 96，直到归零——"墙上有 1 瓶牛奶，1 瓶牛奶，拿下来，传过去，墙上没有牛奶了！"

幸运的是，计算机在执行重复性任务方面表现出色。在本项目中，我们以编程方式复制了所有歌词。本项目的扩展版参见项目 51。

运行程序

运行 ninetyninebottles.py，输出如下所示：

```
Ninety-Nine Bottles, by Al Sweigart al@inventwithpython.com

(Press Ctrl-C to quit.)
99 bottles of milk on the wall,
99 bottles of milk,
Take one down, pass it around,
98 bottles of milk on the wall!

98 bottles of milk on the wall,
98 bottles of milk,
Take one down, pass it around,
97 bottles of milk on the wall!
--snip--
```

工作原理

上述歌曲歌词的重复部分使得很容易使用 `for` 循环（第 17~27 行）显示前 98 段歌词。然而，最后一段歌词有一些细微的差别，需要使用单独的代码（第 30~36 行）显示。这是因为最后一行 "No more bottles of milk on the wall!" 与循环中的重复行不同，而且 "bottle" 这个词是单数而不是复数。

```
 1. """99 瓶牛奶，作者：Al Sweigart al@inventwithpython.com
 2. 输出《99 瓶牛奶》这首歌的完整歌词！按 Ctrl-C 停止
 3. 标签：小，初学者，动画"""
 4.
```

```
 5. import sys, time
 6.
 7. print('Ninety-Nine Bottles, by Al Sweigart al@inventwithpython.com')
 8. print()
 9. print('(Press Ctrl-C to quit.)')
10.
11. time.sleep(2)
12.
13. bottles = 99   # 起始瓶数为 99
14. PAUSE = 2   # (!) 试着将 2 更改为 0, 以便能够立即看到完整的歌词
15.
16. try:
17.     while bottles > 1:   # 继续循环并显示歌词
18.         print(bottles, 'bottles of milk on the wall,')
19.         time.sleep(PAUSE)   #暂停数秒
20.         print(bottles, 'bottles of milk,')
21.         time.sleep(PAUSE)
22.         print('Take one down, pass it around,')
23.         time.sleep(PAUSE)
24.         bottles = bottles - 1   # 瓶数减 1
25.         print(bottles, 'bottles of milk on the wall!')
26.         time.sleep(PAUSE)
27.         print()   # 输出一行空行
28.
29.     # 显示最后一段
30.     print('1 bottle of milk on the wall,')
31.     time.sleep(PAUSE)
32.     print('1 bottle of milk,')
33.     time.sleep(PAUSE)
34.     print('Take it down, pass it around,')
35.     time.sleep(PAUSE)
36.     print('No more bottles of milk on the wall!')
37. except KeyboardInterrupt:
38.     sys.exit()   # 按下 Ctrl-C, 程序结束
```

输入源代码并运行几次后,请尝试对其进行更改。你也可以尝试以下操作。

- 为歌词重复的歌曲 *The Twelve Days of Christmas* 创建一个程序。
- 为其他类似的歌曲创建程序。你可以在维基百科中查找"Cumulative_song",找到这些歌曲清单。

探索程序

请尝试找出以下问题的答案。你需要尝试对代码进行一些修改,再次运行程序,查看修改后的效果。

1. 如果将第 24 行的 `bottles = bottles - 1` 改为 `bottles = bottles - 2`,会发生什么?
2. 如果将第 17 行的 `while bottles > 1:` 改为 `while bottles < 1:`,会发生什么?

项目 51

99瓶牛奶2

在本项目的程序中，对于《99瓶牛奶》这首歌曲，我们会通过删除1个字母、更改字母的大小写、调换2个字母或对1个字母进行重复，在每段歌词中引入"小瑕疵"。

随着歌曲的不断播放，这些"小瑕疵"累加起来，会形成一首听起来有点傻气的歌曲。在尝试运行本项目的程序之前，最好先运行一下项目50的程序。

运行程序

运行 ninetyninebottles2.py，输出如下所示：

```
niNety-nniinE BoOttels, by Al Sweigart al@inventwithpython.com
--snip--
99 bottles of milk on the wall,
99 bottles of milk,
Take one down, pass it around,
98 bottles of milk on the wall!

98 bottles of milk on the wall,
98 bottles of milk,
Take one d wn, pass it around,
97 bottles of milk on the wall!

97 bottles of milk on the wall,
97 bottels of milk,
Take one d wn, pass it around,
96 bottles of milk on the wall!
--snip--
75b otlte of mIl on teh wall,
75    ottels f miLk,
Take one d wn, pass it ar und,
74 bbOttles of milk on t e wall!
--snip--
1  otlE t of iml oo nteh lall,
1    o  Tle   FF FmMLIIkk,
Taake on  d wn,  pAasSs itt au nn d,
No more bottles of milk on the wall!
```

工作原理

Python 中的字符串是不可变的，也就是说，是不能更改的。如果字符串'Hello'存储在一个名为 greeting 的变量中，则代码 greeting = greeting + ' world!'实际上并没有改变'Hello'字符串，而是会创建一个新字符串'Hello world!'；以替换 greeting 中的'Hello'字符串。这里涉及的技术原理超出了本书的范围，但理解其中的区别很重要，这意味着不允许使用 greeting[0] = 'h'之类的代码，因为字符串是不可变的。然而，由于列表是可变的，我们可以创建一个单字符串列表（如第 58 行），更改列表中的字符，然后根据该列表创建一个字符串（见第 81 行）。这就是以下程序使用的改变或转换歌词字符串的方法。

```
 1. """99 瓶牛奶 2，作者：Al Sweigart al@inventwithpython.com
 2. 输出这首歌的歌词！这首歌每一段歌词变得越来越傻气。 按下 Ctrl-C，程序停止
 3. 标签：简短，动画，文字"""
 4.
 5. import random, sys, time
 6.
 7. # 创建常量
 8. # (!) 试着将这两个变量改为 0，这样就可以立即输出所有歌词了
 9. SPEED = 0.01  # 输出字母之间的停顿
10. LINE_PAUSE = 1.5  # 每行末尾的停顿
11.
12.
13. def slowPrint(text, pauseAmount=0.1):
14.     """一次一个地输出文本中的字符"""
15.     for character in text:
16.         # 设置 flush=True，以便立即输出文本
17.         print(character, flush=True, end='')  # end=''意味着没有换行符
18.         time.sleep(pauseAmount)  # 在每个字符之间停顿
19.     print()  # 输出一行空行
20.
21.
22. print('niNety-nniinE BoOttels, by Al Sweigart al@inventwithpython.com')
23. print()
24. print('(Press Ctrl-C to quit.)')
25.
26. time.sleep(2)
27.
28. bottles = 99   # 起始瓶数为 99
29.
30. # 此列表用于存储歌词字符串
31. lines = [' bottles of milk on the wall,',
32.          ' bottles of milk,',
33.          'Take one down, pass it around,',
34.          ' bottles of milk on the wall!']
35.
36. try:
37.     while bottles > 0:  # 继续循环并显示歌词
38.         slowPrint(str(bottles) + lines[0], SPEED)
39.         time.sleep(LINE_PAUSE)
40.         slowPrint(str(bottles) + lines[1], SPEED)
41.         time.sleep(LINE_PAUSE)
42.         slowPrint(lines[2], SPEED)
43.         time.sleep(LINE_PAUSE)
44.         bottles = bottles - 1  # 瓶数减 1
45.
46.         if bottles > 0:  # 输出当前一段歌词的最后一行
47.             slowPrint(str(bottles) + lines[3], SPEED)
```

```
48.         else:  #输出整首歌的最后一行
49.             slowPrint('No more bottles of milk on the wall!', SPEED)
50.
51.         time.sleep(LINE_PAUSE)
52.         print()    # 输出一行空行
53.
54.         # 随机选择一行歌词,并引入"小瑕疵"
55.         lineNum = random.randint(0, 3)
56.
57.         # 根据 line 字符串创建一个列表,以便我们可以编辑其内容。 Python 中的字符串是不可变的
58.         line = list(lines[lineNum])
59.
60.         effect = random.randint(0, 3)
61.         if effect == 0:  # 用空格替换字符
62.             charIndex = random.randint(0, len(line) - 1)
63.             line[charIndex] = ' '
64.         elif effect == 1:  # 更改字符的大小写
65.             charIndex = random.randint(0, len(line) - 1)
66.             if line[charIndex].isupper():
67.                 line[charIndex] = line[charIndex].lower()
68.             elif line[charIndex].islower():
69.                 line[charIndex] = line[charIndex].upper()
70.         elif effect == 2:  # 调换两个字符
71.             charIndex = random.randint(0, len(line) - 2)
72.             firstChar = line[charIndex]
73.             secondChar = line[charIndex + 1]
74.             line[charIndex] = secondChar
75.             line[charIndex + 1] = firstChar
76.         elif effect == 3:  # 重复一个字符
77.             charIndex = random.randint(0, len(line) - 2)
78.             line.insert(charIndex, line[charIndex])
79.
80.         # 将 line 列表转换回字符串,并将其放在 lines 中
81.         lines[lineNum] = ''.join(line)
82. except KeyboardInterrupt:
83.     sys.exit()  # 按下 Ctrl-C,程序结束
```

输入源代码并运行几次后,请试着对其进行更改。你可以根据标有!的注释对程序进行修改,也可以尝试以下操作。

- ❏ 交换两个相邻单词的顺序,其中"单词"是由空格分隔的文本。
- ❏ 在极少数情况下,让奶瓶数递增并迭代歌词。
- ❏ 更改整个单词的大小写。

探索程序

请尝试找出以下问题的答案。你需要尝试对代码进行一些修改,再次运行程序,查看修改后的效果。

1. 如果将第 44 行中的 `bottles = bottles - 1` 改为 `bottles = bottles - 2`,则会发生什么?
2. 如果将第 60 行的 `effect = random.randint(0, 3)` 改为 `effect = 0`,则会发生什么?
3. 如果删除或注释掉第 58 行的 `line = list(lines[lineNum])`,则会发生什么错误?

项目 52

数字系统计数器

我们习惯使用十进制数计数，这可能跟人有 10 根手指有关。十进制是以 10 为基数的计数系统，用 0~9 表示数字。不过，世界上也存在其他数字系统。计算机使用二进制，这是一种只有 0 和 1 两个数位的数字系统。程序员有时也使用十六进制，这是一种基数为 16 的数字系统，它使用数字 0~9 以及字母 A~F。

我们可以在任何数字系统中表示任何数字。本项目的程序以十进制、二进制和十六进制的方式来显示一系列数字。

运行程序

运行 numeralsystems.py，输出如下所示：

```
Numeral System Counters, by Al Sweigart al@inventwithpython.com

--snip--
Enter the starting number (e.g. 0) > 0
Enter how many numbers to display (e.g. 1000) > 20
DEC: 0     HEX: 0     BIN: 0
DEC: 1     HEX: 1     BIN: 1
DEC: 2     HEX: 2     BIN: 10
DEC: 3     HEX: 3     BIN: 11
DEC: 4     HEX: 4     BIN: 100
DEC: 5     HEX: 5     BIN: 101
DEC: 6     HEX: 6     BIN: 110
DEC: 7     HEX: 7     BIN: 111
DEC: 8     HEX: 8     BIN: 1000
DEC: 9     HEX: 9     BIN: 1001
DEC: 10    HEX: A     BIN: 1010
DEC: 11    HEX: B     BIN: 1011
DEC: 12    HEX: C     BIN: 1100
DEC: 13    HEX: D     BIN: 1101
DEC: 14    HEX: E     BIN: 1110
DEC: 15    HEX: F     BIN: 1111
DEC: 16    HEX: 10    BIN: 10000
DEC: 17    HEX: 11    BIN: 10001
DEC: 18    HEX: 12    BIN: 10010
DEC: 19    HEX: 13    BIN: 10011
```

工作原理

在 Python 中，你可以通过分别调用 bin() 和 hex() 函数获得数字对应的二进制数和十六

项目 52　数字系统计数器

进制数：

```
>>> bin(42)
'0b101010'
>>> hex(42)
'0x2a'
```

通过调用 `int()` 函数并提供转换基数，我们可以将这些字符串转换回十进制整数，如下所示：

```
>>> int('0b101010', 2)
42
>>> int('0x2a', 16)
42
```

记住，`bin()` 和 `hex()` 返回的二进制和十六进制形式的"数字"实际上是字符串值：`bin(42)` 返回字符串 `'0b101010'`，`hex(42)` 返回字符串 `'0x2a'`。在编程中，我们约定在二进制数前加上 0b 前缀，在十六进制数前加上 0x 前缀，这样就不会将二进制数 10000（十进制数 16）与十进制数 10000 混淆了。数字系统程序在显示数字之前会删除这些前缀。

```python
 1. """数字系统计数器，作者：Al Sweigart al@inventwithpython.com
 2. 用十进制、十六进制和二进制的方式来显示同一数字
 3. 标签：小,数学"""
 4.
 5.
 6. print('''Numeral System Counters, by Al Sweigart al@inventwithpython.com
 7.
 8. This program shows you equivalent numbers in decimal (base 10),
 9. hexadecimal (base 16), and binary (base 2) numeral systems.
10.
11. (Ctrl-C to quit.)
12. ''')
13.
14. while True:
15.     response = input('Enter the starting number (e.g. 0) > ')
16.     if response == '':
17.         response = '0'  # 默认从 0 开始
18.         break
19.     if response.isdecimal():
20.         break
21.     print('Please enter a number greater than or equal to 0.')
22. start = int(response)
23.
24. while True:
25.     response = input('Enter how many numbers to display (e.g. 1000) > ')
26.     if response == '':
27.         response = '1000'  # 默认显示 1000 个数
28.         break
29.     if response.isdecimal():
30.         break
31.     print('Please enter a number.')
32. amount = int(response)
33.
34. for number in range(start, start + amount):  # 主循环
35.     # 转换为十六进制/二进制形式并删除前缀
36.     hexNumber = hex(number)[2:].upper()
37.     binNumber = bin(number)[2:]
38.
39.     print('DEC:', number, '   HEX:', hexNumber, '   BIN:', binNumber)
```

输入源代码并运行几次后，请尝试对其进行更改。你也可以尝试以下操作。
- 使用 Python 的 `oct()` 函数为程序添加输出八进制（基数为 8 的数字系统）数的功能。
- 在互联网上搜索"number system conversion"（数字系统转换），了解如何实现 `bin()`、`oct()` 和 `hex()` 函数。

探索程序

请尝试找出以下问题的答案。你需要尝试对代码进行一些修改，再次运行程序，查看修改后的效果。

1. 如果将第 36 行的 `hex(number)[2:].upper()` 改为 `hex(number)[2:]`，会发生什么？
2. 如果将第 32 行的 `int(response)` 改为 `response`，会导致什么错误？

项目 53

元素周期表

元素周期表将所有已知的化学元素汇总到一个表中。本项目的程序展示了这个表，让玩家可以访问关于每一种元素的附加信息，比如它的原子序数、符号、熔点等。我从维基百科汇总了这些信息，并将其存储在 periodictable.csv 这个文件中。

运行程序

运行 periodictable.py，输出如下所示：

```
Periodic Table of Elements
By Al Sweigart al@inventwithpython.com
           Periodic Table of Elements
            1  2  3  4  5  6  7  8  9 10 11 12 13 14 15 16 17 18
         1  H                                                 He
         2  Li Be                            B  C  N  O  F  Ne
         3  Na Mg                            Al Si P  S  Cl Ar
         4  K  Ca Sc Ti V  Cr Mn Fe Co Ni Cu Zn Ga Ge As Se Br Kr
         5  Rb Sr Y  Zr Nb Mo Tc Ru Rh Pd Ag Cd In Sn Sb Te I  Xe
         6  Cs Ba La Hf Ta W  Re Os Ir Pt Au Hg Tl Pb Bi Po At Rn
         7  Fr Ra Ac Rf Db Sg Bh Hs Mt Ds Rg Cn Nh Fl Mc Lv Ts Og

            Ce Pr Nd Pm Sm Eu Gd Tb Dy Ho Er Tm Yb Lu
            Th Pa U  Np Pu Am Cm Bk Cf Es Fm Md No Lr
Enter a symbol or atomic number to examine, or QUIT to quit.
> 42
             Atomic Number: 42
                    Symbol: Mo
                   Element: Molybdenum
             Origin of name: Greek molýbdaina, 'piece of lead', from mólybdos, 'lead'
                     Group: 6
                    Period: 5
             Atomic weight: 95.95(1) u
                   Density: 10.22 g/cm^3
             Melting point: 2896 K
             Boiling point: 4912 K
     Specific heat capacity: 0.251 J/(g*K)
         Electronegativity: 2.16
```

```
Abundance  in earth's crust: 1.2 mg/kg
Press Enter to continue...
--snip--
```

工作原理

.csv（comma-separated values，逗号分隔值）文件是一种表示原始电子表格的文本文件。.csv 文件中的行均与元素周期表的行相对应，各列之间用逗号分隔。例如，periodictable.csv 中的前 3 行如下所示：

```
1. H,Hydrogen,"Greek elements hydro- and -gen, meaning 'water-forming--snip--
2. He,Helium,"Greek hḗlios, 'sun'",18,1,4.002602(2)[III][V],0.0001785--snip--
3. Li,Lithium,"Greek líthos, 'stone'",1,2,6.94[III][IV][V][VIII][VI],--snip--
```

利用 Python 的 csv 模块可以很容易地将数据从 .csv 文件导入字符串列表中，如第 11～14 行所示。第 28～54 行将该列表转换为字典，以便程序的其余部分可以轻松通过元素的名称或原子序数来取用信息。

```
 1. """元素周期表，作者：Al Sweigart al@inventwithpython.com
 2. 显示所有元素的原子信息
 3. 标签：简短,科学"""
 4.
 5. # 从维基百科网址找到元素周期表，复制并将其粘贴到 Excel 之类的电子表格程序中
 6. # 然后将该文件保存为 periodictable.csv
 7.
 8. import csv, sys, re
 9.
10. # 从 periodictable.csv 中读取所有数据
11. elementsFile = open('periodictable.csv', encoding='utf-8')
12. elementsCsvReader = csv.reader(elementsFile)
13. elements = list(elementsCsvReader)
14. elementsFile.close()
15.
16. ALL_COLUMNS = ['Atomic Number', 'Symbol', 'Element', 'Origin of name',
17.                'Group', 'Period', 'Atomic weight', 'Density',
18.                'Melting point', 'Boiling point',
19.                'Specific heat capacity', 'Electronegativity',
20.                'Abundance in earth\'s crust']
21.
22. # 为了调整文本，我们需要找到 ALL_COLUMNS 中最长的字符串
23. LONGEST_COLUMN = 0
24. for key in ALL_COLUMNS:
25.     if len(key) > LONGEST_COLUMN:
26.         LONGEST_COLUMN = len(key)
27.
28. # 将所有元素数据放到一个数据结构中
29. ELEMENTS = {}  # 存储所有元素数据的数据结构
30. for line in elements:
31.     element = {'Atomic Number':   line[0],
32.                'Symbol':          line[1],
33.                'Element':         line[2],
34.                'Origin of name':  line[3],
35.                'Group':           line[4],
36.                'Period':          line[5],
37.                'Atomic weight':   line[6] + ' u',    # 原子质量单位
38.                'Density':         line[7] + ' g/cm^3',  # 克/立方厘米
39.                'Melting point':   line[8] + ' K',    # 开尔文
```

```
40.                  'Boiling point':      line[9] + ' K',   # 开尔文
41.                  'Specific heat capacity':      line[10] + ' J/(g*K)',
42.                  'Electronegativity':           line[11],
43.                  'Abundance in earth\'s crust': line[12] + ' mg/kg'}
44.
45.     # 删除来自维百科的一些数据
46.     # 如硼的原子量
47.     # "10.81[III][IV][V][VI]" 应该变成 "10.81"
48.
49.     for key, value in element.items():
50.         # 删除[罗马数字]文本
51.         element[key] = re.sub(r'\[(I|V|X)+\]', '', value)
52.
53.     ELEMENTS[line[0]] = element   # 将原子序数映射到元素
54.     ELEMENTS[line[1]] = element   # 将符号映射到元素
55.
56. print('Periodic Table of Elements')
57. print('By Al Sweigart al@inventwithpython.com')
58. print()
59.
60. while True:   # 主循环
61.     # 显示表格并让玩家选择一个元素
62.     print('''          Periodic Table of Elements
63.      1  2  3  4  5  6  7  8  9  10 11 12 13 14 15 16 17 18
64.   1  H                                                    He
65.   2  Li Be                               B  C  N  O  F  Ne
66.   3  Na Mg                               Al Si P  S  Cl Ar
67.   4  K  Ca Sc Ti V  Cr Mn Fe Co Ni Cu Zn Ga Ge As Se Br Kr
68.   5  Rb Sr Y  Zr Nb Mo Tc Ru Rh Pd Ag Cd In Sn Sb Te I  Xe
69.   6  Cs Ba La Hf Ta W  Re Os Ir Pt Au Hg Tl Pb Bi Po At Rn
70.   7  Fr Ra Ac Rf Db Sg Bh Hs Mt Ds Rg Cn Nh Fl Mc Lv Ts Og
71.
72.            Ce Pr Nd Pm Sm Eu Gd Tb Dy Ho Er Tm Yb Lu
73.            Th Pa U  Np Pu Am Cm Bk Cf Es Fm Md No Lr''')
74.     print('Enter a symbol or atomic number to examine, or QUIT to quit.')
75.     response = input('> ').title()
76.
77.     if response == 'Quit':
78.         sys.exit()
79.
80.     # 显示所选元素的数据
81.     if response in ELEMENTS:
82.         for key in ALL_COLUMNS:
83.             keyJustified = key.rjust(LONGEST_COLUMN)
84.             print(keyJustified + ': ' + ELEMENTS[response][key])
85.         input('Press Enter to continue...')
```

探索程序

请尝试找出以下问题的答案。你需要尝试对代码进行一些修改，再次运行程序，查看修改后的效果。

1. 如果将第 77 行的 response == 'Quit' 改为 response == 'quit'，则会导致什么错误？
2. 如果删除或注释掉第 49 行和第 51 行的代码，则会发生什么？

项目 54

儿童隐语

儿童隐语（Pig Latin）是一款文字游戏，可将英语单词转换为对拉丁语的滑稽模仿。在这个游戏中，如果一个单词以辅音字母开头，则说话者删除该字母并将其放在词尾然后加上"ay"，例如，"pig"变成"igpay"，"latin"变成"atinlay"。如果单词以元音字母开头，则说话者在单词末尾添加"yay"即可，例如，"elephant"变成"elephantyay"，"umbrella"变成"umbrellayay"。

运行程序

运行 piglatin.py，输出如下所示：

```
Igpay Atinlay (Pig Latin)
By Al Sweigart al@inventwithpython.com

Enter your message:
> This is a very serious message.
Isthay isyay ayay eryvay erioussay essagemay.
(Copied pig latin to clipboard.)
```

工作原理

englishToPigLatin()函数接收英文字符串并返回其儿童隐语。main()函数只有在用户直接运行程序时才会被调用。你也可以自行编写 Python 程序，使用 import piglatin 语句导入 piglatin.py，然后调用 piglatin.englishToPigLatin()以使用 englishToPigLatin()函数。这种代码复用技术可使你节省重新编程所需的时间和精力。

```
 1. """儿童隐语，作者：Al Sweigart al@inventwithpython.com
 2. 将英语信息翻译成儿童隐语
 3. 标签：简短，文字"""
 4.
 5. try:
 6.     import pyperclip  # pyperclip 模块能够将文本复制到剪贴板
 7. except ImportError:
 8.     pass  # pyperclip 并不是必需模块，不安装也没什么影响
 9.
10. VOWELS = ('a', 'e', 'i', 'o', 'u', 'y')
11.
```

```
12.
13. def main():
14.     print('''Igpay Atinlay (Pig Latin)
15. By Al Sweigart al@inventwithpython.com
16.
17. Enter your message:''')
18.     pigLatin = englishToPigLatin(input('> '))
19.
20.     # 将所有单词重新连接成一个字符串中
21.     print(pigLatin)
22.
23.     try:
24.         pyperclip.copy(pigLatin)
25.         print('(Copied pig latin to clipboard.)')
26.     except NameError:
27.         pass  # 如果没有安装 pyperclip，那么不执行任何操作
28.
29.
30. def englishToPigLatin(message):
31.     pigLatin = ''  # 表示儿童隐语的字符串
32.     for word in message.split():
33.         # 将这个单词开头的非字母字符分开
34.         prefixNonLetters = ''
35.         while len(word) > 0 and not word[0].isalpha():
36.             prefixNonLetters += word[0]
37.             word = word[1:]
38.         if len(word) == 0:
39.             pigLatin = pigLatin + prefixNonLetters + ' '
40.             continue
41.
42.         # 将这个单词末尾的非字母字符分开
43.         suffixNonLetters = ''
44.         while not word[-1].isalpha():
45.             suffixNonLetters = word[-1] + suffixNonLetters
46.             word = word[:-1]
47.
48.         # 记住单词是全部大写还是只有首字母大写
49.         wasUpper = word.isupper()
50.         wasTitle = word.istitle()
51.
52.         word = word.lower()   # 把单词转换成小写形式，以便翻译
53.
54.         # 把单词开头的辅音字母分开
55.         prefixConsonants = ''
56.         while len(word) > 0 and not word[0] in VOWELS:
57.             prefixConsonants += word[0]
58.             word = word[1:]
59.
60.         # 在单词末尾加上隐语后缀
61.         if prefixConsonants != '':
62.             word += prefixConsonants + 'ay'
63.         else:
64.             word += 'yay'
65.
66.         # 将单词设置为全部大写或首字母大写
67.         if wasUpper:
68.             word = word.upper()
69.         if wasTitle:
70.             word = word.title()
71.
72.         # 将非字母添加回单词的开头或结尾
73.         pigLatin += prefixNonLetters + word + suffixNonLetters + ' '
```

```
74.        return pigLatin
75.
76.
77. if __name__ == '__main__':
78.     main()
```

探索程序

请尝试找出以下问题的答案。你需要尝试对代码进行一些修改，再次运行程序，查看修改后的效果。

1. 如果将第 32 行的 `message.split()` 改为 `message`，会发生什么？
2. 如果将第 10 行的 `('a', 'e', 'i', 'o', 'u', 'y')` 改为 `()`，会发生什么？
3. 如果将第 10 行的 `('a', 'e', 'i', 'o', 'u', 'y')` 改为 `('A', 'E', 'I', 'O', 'U', 'Y')`，会发生什么？

项目 55

强力球彩票

强力球彩票是一种令人兴奋的——输小钱——的方式。如果你购买一张 2 美元的彩票,就可以选择 6 个号码:从 1 到 69 开出的 5 个号码,以及从 1 到 26 开出的第 6 个"强力球"号码。号码的顺序无关紧要。如果彩票号码和你选中的 6 个号码相同,你将赢得 15.86 亿美元!不过一般来讲你不会赢的,因为你的胜算仅为 1/292201338。如果你花 200 美元买 100 张彩票,你的胜算仍然只有 1/2922013。你还是不会中头奖,而只会损失 100 倍的钱。

为了帮助你直观地了解自己中奖的可能性,本项目的程序模拟了多达 100 万张强力球彩票,然后将这些彩票号码与你选择的数字进行比较。现在,你无须花钱就可以享受没中彩票的"乐趣"。

有个有趣的事实:每组 6 个数字都和其他数字一样有可能中奖。因此,下次当你想购买彩票时,不妨选择数字 1、2、3、4、5 和 6,因为这些数字与更复杂的数字集合中大奖的可能性完全相同。

运行程序

运行 powerballlottery.py,输出如下所示:

```
Powerball Lottery, by Al Sweigart al@inventwithpython.com

Each powerball lottery ticket costs $2. The jackpot for this game
is $1.586 billion! It doesn't matter what the jackpot is, though,
because the odds are 1 in 292,201,338, so you won't win.

This simulation gives you the thrill of playing without wasting money.

Enter 5 different numbers from 1 to 69, with spaces between
each number. (For example: 5 17 23 42 50 51)
> 1 2 3 4 5
Enter the powerball number from 1 to 26.
> 6
How many times do you want to play? (Max: 1000000)
> 1000000
It costs $2000000 to play 1000000 times, but don't
worry. I'm sure you'll win it all back.
Press Enter to start...
The winning numbers are: 12 29 48 11 4  and 13 You lost.
The winning numbers are: 54 39 3  42 16 and 12 You lost.
```

```
The winning numbers are: 56  4   63 23 38 and 24 You lost.
--snip--
The winning numbers are: 46 29 10 62 17 and 21 You lost.
The winning numbers are: 5   20 18 65 30 and 10 You lost.
The winning numbers are: 54 30 58 10 1  and 18 You lost.
You have wasted $2000000
Thanks for playing!
```

工作原理

以下程序的输出看起来相当整齐，这是因为第 107 行的 `allWinningNums.ljust(21)` 使用足够多的空格填充数字以占据 21 列，无论中奖号码有多少位。这使得 `You lost.` 文本始终出现在屏幕上的相同位置，因此，即使程序快速输出一些行，该文本仍然是清晰、可读的。

```
 1. """强力球彩票, 作者: Al Sweigart al@inventwithpython.com
 2. 一个模拟买彩票的游戏, 让你可以体验未中彩票而又不浪费钱的兴奋感
 3. 标签: 简短, 诙谐, 模拟"""
 4.
 5. import random
 6.
 7. print('''Powerball Lottery, by Al Sweigart al@inventwithpython.com
 8.
 9. Each powerball lottery ticket costs $2. The jackpot for this game
10. is $1.586 billion! It doesn't matter what the jackpot is, though,
11. because the odds are 1 in 292,201,338, so you won't win.
12.
13. This simulation gives you the thrill of playing without wasting money.
14. ''')
15.
16. # 让玩家在 1 到 69 中选 5 个数输入
17. while True:
18.     print('Enter 5 different numbers from 1 to 69, with spaces between')
19.     print('each number. (For example: 5 17 23 42 50)')
20.     response = input('> ')
21.
22.     # 检查玩家是否输入了 5 个数
23.     numbers = response.split()
24.     if len(numbers) != 5:
25.         print('Please enter 5 numbers, separated by spaces.')
26.         continue
27.
28.     # 将字符串转换为整数
29.     try:
30.         for i in range(5):
31.             numbers[i] = int(numbers[i])
32.     except ValueError:
33.         print('Please enter numbers, like 27, 35, or 62.')
34.         continue
35.
36.     # 检查数字是否在 1 和 69 之间
37.     for i in range(5):
38.         if not (1 <= numbers[i] <= 69):
39.             print('The numbers must all be between 1 and 69.')
40.             continue
41.
42.     # 检查数字是否唯一
43.     # 为所有数字创建集合，删除重复数字
44.     if len(set(numbers)) != 5:
```

```
45.         print('You must enter 5 different numbers.')
46.         continue
47.
48.     break
49.
50. # 让玩家从 1 到 26 中选择"强力球"号码
51. while True:
52.     print('Enter the powerball number from 1 to 26.')
53.     response = input('> ')
54.
55.     # 将字符串转换为整数
56.     try:
57.         powerball = int(response)
58.     except ValueError:
59.         print('Please enter a number, like 3, 15, or 22.')
60.         continue
61.
62.     # 检查数字是否在 1 和 26 之间
63.     if not (1 <= powerball <= 26):
64.         print('The powerball number must be between 1 and 26.')
65.         continue
66.
67.     break
68.
69. # 输入你想玩游戏的次数
70. while True:
71.     print('How many times do you want to play? (Max: 1000000)')
72.     response = input('> ')
73.
74.     # 将字符串转换为整数
75.     try:
76.         numPlays = int(response)
77.     except ValueError:
78.         print('Please enter a number, like 3, 15, or 22000.')
79.         continue
80.
81.     # 检查数字是否在 1 和 1000000 之间
82.     if not (1 <= numPlays <= 1000000):
83.         print('You can play between 1 and 1000000 times.')
84.         continue
85.
86.     break
87.
88. # 运行程序进行模拟
89. price = '$' + str(2 * numPlays)
90. print('It costs', price, 'to play', numPlays, 'times, but don\'t')
91. print('worry. I\'m sure you\'ll win it all back.')
92. input('Press Enter to start...')
93.
94. possibleNumbers = list(range(1, 70))
95. for i in range(numPlays):
96.     # 生成彩票号码
97.     random.shuffle(possibleNumbers)
98.     winningNumbers = possibleNumbers[0:5]
99.     winningPowerball = random.randint(1, 26)
100.
101.    # 显示中奖号码
102.    print('The winning numbers are: ', end='')
103.    allWinningNums = ''
104.    for i in range(5):
105.        allWinningNums += str(winningNumbers[i]) + ' '
106.    allWinningNums += 'and ' + str(winningPowerball)
```

```
107.         print(allWinningNums.ljust(21), end='')
108.
109.         # 注意:集合是没有顺序的
110.         # set(numbers)和set(winningNumbers)中整数的顺序不重要
111.         if (set(numbers) == set(winningNumbers)
112.             and powerball == winningPowerball):
113.                 print()
114.                 print('You have won the Powerball Lottery! Congratulations,')
115.                 print('you would be a billionaire if this was real!')
116.                 break
117.         else:
118.             print(' You lost.')   # 句子前留一个空格
119.
120. print('You have wasted', price)
121. print('Thanks for playing!')
```

探索程序

请尝试找出以下问题的答案。你需要尝试对代码进行一些修改,再次运行程序,查看修改后的效果。

1. 如果将第 98 行的 `possibleNumbers[0:5]` 改为 `numbers`,以及将第 99 行的 `random.randint(1, 26)` 改为 `powerball`,会发生什么?
2. 如果删除或注释掉第 94 行的 `possibleNumbers = list(range(1, 70))`,会出现什么错误?

项目 56

素数

素数是只能被 1 和它自身整除的数。素数有多种实际应用，但没有算法可以预测它们，只能一个一个算出来。不过，我们确实知道有无数个素数有待发现。

本项目的程序通过蛮力计算找到素数，类似于项目 24。描述素数的另一种方式是：1 和该数本身是其全部的因数。

运行程序

运行 primenumbers.py，输出如下所示：

```
Prime Numbers, by Al Sweigart al@inventwithpython.com
--snip--
Enter a number to start searching for primes from:
(Try 0 or 1000000000000 (12 zeros) or another number.)
> 0
Press Ctrl-C at any time to quit. Press Enter to begin...
2, 3, 5, 7, 11, 13, 17, 19, 23, 29, 31, 37, 41, 43, 47, 53, 59, 61, 67, 71,
73, 79, 83, 89, 97, 101, 103, 107, 109, 113, 127, 131, 137, 139, 149, 151,
157, 163, 167, 173, 179, 181, 191, 193, 197, 199, 211, 223, 227, 229, 233,
239, 241, 251, 257, 263, 269, 271, 277, 281, 283, 293, 307, 311, 313, 317,
331, 337, 347, 349, 353, 359, 367, 373, 379, 383, 389, 397, 401, 409, 419,
421, 431, 433, 439, 443, 449, 457, 461, 463, 467, 479, 487, 491, 499, 503,
509, 521, 523, 541, 547, 557, 563, 569, 571, 577, 587, 593, 599, 601, 607,
613, 617, 619, 631, 641, 643, 647, --snip--
```

工作原理

isPrime() 函数用于接收一个整数，如果所接收的是一个素数，则返回 True，否则返回 False。如果你想理解这个程序，建议好好研究一下项目 24。isPrime() 函数本质上是查找给定数字的任何因数，只要找到一个，则返回 False。

程序中的算法可以快速找到素数，然而要想找到 googol（1 后跟 100 个零）那么大的素数，你需要使用高级算法，例如 Rabin-Miller 素数检验。拙作《Python 密码学编程》给出了这个算法的 Python 实现。

```
 1. """素数，作者：Al Sweigart al@inventwithpython.com
 2. 计算素数，素数是只能被1和它自身整除的数，在实际应用中多有涉及
 3. 标签：小，数学，动画"""
 4.
 5. import math, sys
 6.
 7. def main():
 8.     print('Prime Numbers, by Al Sweigart al@inventwithpython.com')
 9.     print('Prime numbers are numbers that are only evenly divisible by')
10.     print('one and themselves. They are used in a variety of practical')
11.     print('applications, but cannot be predicted. They must be')
12.     print('calculated one at a time.')
13.     print()
14.     while True:
15.         print('Enter a number to start searching for primes from:')
16.         print('(Try 0 or 1000000000000 (12 zeros) or another number.)')
17.         response = input('> ')
18.         if response.isdecimal():
19.             num = int(response)
20.             break
21.
22.     input('Press Ctrl-C at any time to quit. Press Enter to begin...')
23.
24.     while True:
25.         # 输出任意素数
26.         if isPrime(num):
27.             print(str(num) + ', ', end='', flush=True)
28.         num = num + 1  # 检查下一个数字
29.
30.
31. def isPrime(number):
32.     """如果数字是素数则返回True，否则返回False"""
33.     # 处理特殊情况
34.     if number < 2:
35.         return False
36.     elif number == 2:
37.         return True
38.
39.     #将number对"从2到number的平方根之间的所有数"取余
40.     for i in range(2, int(math.sqrt(number)) + 1):
41.         if number % i == 0:
42.             return False
43.     return True
44.
45.
46. # 程序运行入口（如果不是作为模块导入的话），运行游戏
47. if __name__ == '__main__':
48.     try:
49.         main()
50.     except KeyboardInterrupt:
51.         sys.exit()  # 按下Ctrl-C，程序结束
```

探索程序

请尝试找出以下问题的答案。你需要尝试对代码进行一些修改，再次运行程序，查看修改后的效果。

1. 如果将第 18 行的 response.isdecimal() 改为 response 并输入一个非数字的值作为搜索素数的起始数字，会出现什么错误？
2. 如果将第 34 行的 number < 2 改为 number > 2，则会发生什么？
3. 如果将第 41 行的 number % i == 0 改为 number % i != 0，则会发生什么？

项目 57 | 进度条

进度条是一种可视化元素,可以显示任务的完成进度。进度条通常用在文件下载或软件安装界面上。在本项目中,我们将创建 getProgressBar() 函数,用于根据传入的参数返回一个进度条字符串。这可以用于模拟下载文件。你可以在自己的项目中重复使用进度条代码。

运行程序

运行 progressbar.py,输出如下所示:

```
Progress Bar Simulation, by Al Sweigart
[████████              ] 24.6% 1007/4098
```

工作原理

进度条效果的实现要用到特定的技巧——在终端窗口中运行的程序可以执行的。正如转义字符'\n' 和 '\t'分别表示换行符和制表符一样,转义字符'\b'表示退格字符。如果我们"输出"退格字符,文本光标将向左移动并清除之前输出的字符。这仅适用于文本光标所在的当前行。如果运行代码 print('Hello\b\b\b\b\bHowdy'),Python 将输出文本 Hello,然后将文本光标向后移动 5 个空格,然后输出文本 Howdy。Howdy 文本将覆盖 Hello。

基于上述原理,我们可以创建一个单行动画进度条:输出一个进度条,然后将文本光标移到开头,接着输出更新的进度条。这种技术可以用来生成任何文本动画,尽管它会被限制在终端窗口中占据 1 行,但不需要使用 bext 之类的模块。

创建程序后,你可以通过运行 import progressbar 并输出从 progressbar.getProgressBar() 返回的字符串,从而在其他 Python 程序中显示进度条。

```
1. """进度条,作者:Al Sweigart al@inventwithpython.com
2. 一个进度条动画示例,可以在其他程序中使用
3. 标签:小,模块"""
4.
5. import random, time
```

```
 6.
 7. BAR = chr(9608)  # 字符 9608 表示的是■
 8.
 9. def main():
10.     # 模拟下载
11.     print('Progress Bar Simulation, by Al Sweigart')
12.     bytesDownloaded = 0
13.     downloadSize = 4096
14.     while bytesDownloaded < downloadSize:
15.         # 下载随机数量的"字节"
16.         bytesDownloaded += random.randint(0, 100)
17.
18.         # 获取进度条字符串
19.         barStr = getProgressBar(bytesDownloaded, downloadSize)
20.
21.         # 不要在末尾输出空行,
22.         # 立即将输出的字符串刷新到屏幕上
23.         print(barStr, end='', flush=True)
24.
25.         time.sleep(0.2)  # 暂停一下
26.
27.         # 输出退格符,将文本光标移动到行首
28.         print('\b' * len(barStr), end='', flush=True)
29.
30.
31. def getProgressBar(progress, total, barWidth=40):
32.     """返回一个字符串,用于表示一个进度条,
33.     该进度条包含 barWidth 个块,并已从总进度中取得进度"""
34.
35.     progressBar = ''   # 进度条将是一个字符串值
36.     progressBar += '['  # 创建进度条的左端
37.
38.     # 确保进度块的数量在 0 和总量之间
39.     if progress > total:
40.         progress = total
41.     if progress < 0:
42.         progress = 0
43.
44.     # 计算要显示的"进度块"的数量
45.     numberOfBars = int((progress / total) * barWidth)
46.
47.     progressBar += BAR * numberOfBars  # 添加进度条
48.     progressBar += ' ' * (barWidth - numberOfBars)  # 添加空白
49.     progressBar += ']'  # 创建进度条的右端
50.
51.     # 计算完成百分比
52.     percentComplete = round(progress / total * 100, 1)
53.     progressBar += ' ' + str(percentComplete) + '%'  # 增加百分比
54.
55.     # 把所有数字加起来
56.     progressBar += ' ' + str(progress) + '/' + str(total)
57.
58.     return progressBar   # 返回进度条字符串
59.
60.
61. # 程序运行入口(如果不是作为模块导入的话)
62. if __name__ == '__main__':
63.     main()
```

输入源代码并运行几次后,请尝试对其进行更改。你也可以尝试以下操作。

❑ 创建一个单行的旋转器的动画,让|、/、-和\字符交替显示,以产生旋转效果。

- 创建一个程序，显示从左向右移动的滚动字幕。
- 创建一个单行动画，以 4 个等号为一个单元来回移动显示，类似于电视节目 *Knight Rider* 中汽车机器人的红色扫描灯或电视节目 *Battlestar Galactica* 中的 Cylog 机器人脸。

探索程序

请尝试找出以下问题的答案。你需要尝试对代码进行一些修改，再次运行程序，查看修改后的效果。

1. 如果删除或注释掉第 28 行的 print('\b' * len(barStr), end='', flush=True)，会发生什么？
2. 如果交换第 47 行和第 48 行代码的顺序，会发生什么？
3. 如果将第 52 行的 round(progress / total * 100, 1)改为 round(progress / total * 100)，会发生什么？

项目 58

彩虹

在本项目中，我们将编写一个简单的程序，用于显示一条五颜六色的、在屏幕上来回移动的彩虹。上述效果的实现基于这样一个事实，即当出现新的文本行时，现有文本会向上滚动，使其看起来像是在移动。这个程序适合初学者，它类似于项目 15 的程序。

运行程序

运行 rainbow1.py，输出结果是锯齿形的（实际是彩色的），如下所示：

工作原理

这个程序会连续输出相同的彩虹图案。发生变化的是输出在彩虹左边的空格字符的数量。

增加空格的数量会产生彩虹向右移动的效果，减少空格的数量则会产生彩虹向左移动的效果。indent 变量用于跟踪空格的数量。如果将 indentIncreasing 变量设置为 True，那么 indent 应该增加，直到达到 60——此时它被改为 False。其余代码用于减少空格数，一旦达到 0，indentIncreasing 会再次变回 True，重复彩虹的锯齿形显示效果。

```
 1. """彩虹，作者：Al Sweigart al@inventwithpython.com
 2. 显示一个简单的彩虹动画
 3. 标签：小，艺术，bext 模块，初学者，动画"""
 4.
 5. import time, sys
 6.
 7. try:
 8.     import bext
 9. except ImportError:
10.     print('This program requires the bext module, which you')
11.     print('can install by following the instructions at')
12.     print('https://pypi.org/project/Bext/')
13.     sys.exit()
14.
15. print('Rainbow, by Al Sweigart al@inventwithpython.com')
16. print('Press Ctrl-C to stop.')
17. time.sleep(3)
18.
19. indent = 0  # 要缩进多少个空格
20. indentIncreasing = True  # 缩进的空格是否增多
21.
22. try:
23.     while True:  # 主循环
24.         print(' ' * indent, end='')
25.         bext.fg('red')
26.         print('##', end='')
27.         bext.fg('yellow')
28.         print('##', end='')
29.         bext.fg('green')
30.         print('##', end='')
31.         bext.fg('blue')
32.         print('##', end='')
33.         bext.fg('cyan')
34.         print('##', end='')
35.         bext.fg('purple')
36.         print('##')
37.
38.         if indentIncreasing:
39.             # 增加缩进的空格数
40.             indent = indent + 1
41.             if indent == 60:  # (!) 把 60 改为 10 或 30
42.                 # 改变方向
43.                 indentIncreasing = False
44.         else:
45.             # 减少缩进的空格数
46.             indent = indent - 1
47.             if indent == 0:
48.                 # 改变方向
49.                 indentIncreasing = True
50.
51.         time.sleep(0.02)  # 添加一个轻微的停顿
52. except KeyboardInterrupt:
53.     sys.exit()  # 按下 Ctrl-C，程序结束
```

探索程序

请尝试找出以下问题的答案。你需要尝试对代码进行一些修改，再次运行程序，查看修改后的效果。

1. 如果将第 43 行的 False 改为 True，则会发生什么？
2. 如果将所有 bext.fg() 调用的参数改为 'random'，则会发生什么？

项目 59

石头剪刀布

本项目介绍的这个版本的游戏也称为 Rochambeau 或 jan-ken-pon，玩家与计算机进行游戏，可以选择石头、布或剪刀。石头赢剪刀，剪刀赢布，布赢石头。本项目的程序添加短暂的停顿以制造悬念。

这个游戏的衍生版本参见项目 60。

运行程序

运行 rockpaperscissors.py，输出如下所示：

```
Rock, Paper, Scissors, by Al Sweigart al@inventwithpython.com
- Rock beats scissors.
- Paper beats rocks.
- Scissors beats paper.

0 Wins, 0 Losses, 0 Ties
Enter your move: (R)ock (P)aper (S)cissors or (Q)uit
> r
ROCK versus...
1...
2...
3...
SCISSORS
You win!
1 Wins, 0 Losses, 0 Ties
Enter your move: (R)ock (P)aper (S)cissors or (Q)uit
--snip--
```

工作原理

石头剪刀布的游戏逻辑相当简单，我们在这里使用 `if-elif` 语句来实现它。为了增加一些悬念，第 44～50 行在显示对手（计算机）的动作之前倒计时，在计数之间有短暂的停顿。这段时间能让玩家期待游戏结果而兴奋。如果没有停顿，一旦玩家输入了动作，结果就会立刻出现，会让人感觉有些乏味。这也表明不需要很多代码就可以改善玩家的用户体验。

项目 59　石头剪刀布

```
1.  """石头剪刀布, 作者: Al Sweigart al@inventwithpython.com
2.  关于运气的经典游戏
3.  标签: 简短, 游戏"""
4.
5.  import random, time, sys
6.
7.  print('''Rock, Paper, Scissors, by Al Sweigart al@inventwithpython.com
8.  - Rock beats scissors.
9.  - Paper beats rocks.
10. - Scissors beats paper.
11. ''')
12.
13. # 这些变量记录了赢、输和平局的次数
14. wins = 0
15. losses = 0
16. ties = 0
17.
18. while True:  # 主循环
19.     while True:  # 一直不断询问, 直到玩家输入 R、P、S 或 Q
20.         print('{} Wins, {} Losses, {} Ties'.format(wins, losses, ties))
21.         print('Enter your move: (R)ock (P)aper (S)cissors or (Q)uit')
22.         playerMove = input('> ').upper()
23.         if playerMove == 'Q':
24.             print('Thanks for playing!')
25.             sys.exit()
26.
27.         if playerMove == 'R' or playerMove == 'P' or playerMove == 'S':
28.             break
29.         else:
30.             print('Type one of R, P, S, or Q.')
31.
32.     # 显示玩家的选择
33.     if playerMove == 'R':
34.         print('ROCK versus...')
35.         playerMove = 'ROCK'
36.     elif playerMove == 'P':
37.         print('PAPER versus...')
38.         playerMove = 'PAPER'
39.     elif playerMove == 'S':
40.         print('SCISSORS versus...')
41.         playerMove = 'SCISSORS'
42.
43.     # 数到 3, 然后短暂地停顿一下
44.     time.sleep(0.5)
45.     print('1...')
46.     time.sleep(0.25)
47.     print('2...')
48.     time.sleep(0.25)
49.     print('3...')
50.     time.sleep(0.25)
51.
52.     # 显示计算机的选择
53.     randomNumber = random.randint(1, 3)
54.     if randomNumber == 1:
55.         computerMove = 'ROCK'
56.     elif randomNumber == 2:
57.         computerMove = 'PAPER'
58.     elif randomNumber == 3:
59.         computerMove = 'SCISSORS'
60.     print(computerMove)
61.     time.sleep(0.5)
62.
```

```
63.    # 显示并记录游戏结果
64.    if playerMove == computerMove:
65.        print('It\'s a tie!')
66.        ties = ties + 1
67.    elif playerMove == 'ROCK' and computerMove == 'SCISSORS':
68.        print('You win!')
69.        wins = wins + 1
70.    elif playerMove == 'PAPER' and computerMove == 'ROCK':
71.        print('You win!')
72.        wins = wins + 1
73.    elif playerMove == 'SCISSORS' and computerMove == 'PAPER':
74.        print('You win!')
75.        wins = wins + 1
76.    elif playerMove == 'ROCK' and computerMove == 'PAPER':
77.        print('You lose!')
78.        losses = losses + 1
79.    elif playerMove == 'PAPER' and computerMove == 'SCISSORS':
80.        print('You lose!')
81.        losses = losses + 1
82.    elif playerMove == 'SCISSORS' and computerMove == 'ROCK':
83.        print('You lose!')
84.        losses = losses + 1
```

输入源代码并运行几次后，请尝试对其进行更改。你也可以尝试以下操作。

❏ 将"蜥蜴"和"斯波克"动作添加到游戏中。蜥蜴能击败斯波克和布，但会被石头和剪刀击败。斯波克可以击败剪刀和石头，但会被蜥蜴和布击败。

❏ 允许玩家每次胜利赢得1分，每次失败失去1分。每次获胜后，玩家还可以冒"积分翻倍或被清零"的风险，在后续回合中有可能会赢得2、4、8、16以及更多的积分。

探索程序

请尝试找出以下问题的答案。你需要尝试对代码进行一些修改，再次运行程序，查看修改后的效果。

1. 如果将第53行的 `random.randint(1, 3)` 改为 `random.randint(1, 300)`，则会出现什么错误？
2. 如果将第64行的 `playerMove == computerMove` 改为 `True`，则会发生什么？

项目 60　石头剪刀布（无敌版）

本项目中这个版本的石头剪刀布游戏与项目 59 非常相似，不同之处在于玩家总是会赢。选择计算机动作的代码被设置为总是选择会被击败的动作。你可以将此游戏介绍给自己的朋友——他们一开始获胜时可能会很兴奋，看看他们需要多长时间才能意识到游戏中的"大获全胜"是游戏"故意为之"的。

运行程序

运行 rockpaperscissorsalwayswin.py，输出如下所示：

```
Rock, Paper, Scissors, by Al Sweigart al@inventwithpython.com
- Rock beats scissors.
- Paper beats rocks.
- Scissors beats paper.
0 Wins, 0 Losses, 0 Ties
Enter your move: (R)ock (P)aper (S)cissors or (Q)uit
> p
PAPER versus...
1...
2...
3...
ROCK
You win!
1 Wins, 0 Losses, 0 Ties
Enter your move: (R)ock (P)aper (S)cissors or (Q)uit
> s
SCISSORS versus...
1...
2...
3...
PAPER
You win!
2 Wins, 0 Losses, 0 Ties
--snip--
SCISSORS versus...
1...
2...
3...
PAPER
You win!
```

```
413 Wins, 0 Losses, 0 Ties
Enter your move: (R)ock (P)aper (S)cissors or (Q)uit
--snip--
```

工作原理

你可能会注意到，本项目的程序比项目 59 的程序要短。这是合理的，因为现在不必为计算机随机生成动作并计算游戏结果，所以可以从原始代码中删除相当多的代码。现在也没有使用变量跟踪输和平局的次数，因为无论如何输和平局的次数都是零。

```
 1. """石头剪刀布（无敌版），作者：Al Sweigart al@inventwithpython.com
 2.
 3. 关于运气的经典游戏，不过你总是会赢
 4. 标签：小，游戏，诙谐"""
 5.
 6. import time, sys
 7.
 8. print('''Rock, Paper, Scissors, by Al Sweigart al@inventwithpython.com
 9. - Rock beats scissors.
10. - Paper beats rocks.
11. - Scissors beats paper.
12. ''')
13.
14. # 这些变量记录了赢的次数
15. wins = 0
16.
17. while True:    # 主循环
18.     while True:    # 一直继续询问，直到玩家输入 R、P、S 或 Q
19.         print('{} Wins, 0 Losses, 0 Ties'.format(wins))
20.         print('Enter your move: (R)ock (P)aper (S)cissors or (Q)uit')
21.         playerMove = input('> ').upper()
22.         if playerMove == 'Q':
23.             print('Thanks for playing!')
24.             sys.exit()
25.
26.         if playerMove == 'R' or playerMove == 'P' or playerMove == 'S':
27.             break
28.         else:
29.             print('Type one of R, P, S, or Q.')
30.
31.     # 显示玩家的选择
32.     if playerMove == 'R':
33.         print('ROCK versus...')
34.     elif playerMove == 'P':
35.         print('PAPER versus...')
36.     elif playerMove == 'S':
37.         print('SCISSORS versus...')
38.
39.     # 数到3，然后短暂地停顿一下
40.     time.sleep(0.5)
41.     print('1...')
42.     time.sleep(0.25)
43.     print('2...')
44.     time.sleep(0.25)
45.     print('3...')
46.     time.sleep(0.25)
47.
48.     # 显示计算机的选择
```

```
49.     if playerMove == 'R':
50.         print('SCISSORS')
51.     elif playerMove == 'P':
52.         print('ROCK')
53.     elif playerMove == 'S':
54.         print('PAPER')
55.
56.     time.sleep(0.5)
57.
58.     print('You win!')
59.     wins = wins + 1
```

输入源代码并运行几次后，请尝试对其进行更改。你也可以尝试以下操作。

❏ 将"蜥蜴"和"斯波克"动作添加到游戏中。蜥蜴能击败斯波克和布，但会被石头和剪刀击败。斯波克可以击败剪刀和石头，但会被蜥蜴和布击败。

❏ 允许玩家每次胜利赢得 1 分，每次失败失去 1 分。每次获胜后，玩家还可以冒"积分翻倍或被清零"的风险，在后续回合中有可能会赢得 2、4、8、16 以及更多的积分。

探索程序

请尝试找出以下问题的答案。你需要尝试对代码进行一些修改，再次运行程序，查看修改后的效果。

1. 如果删除或注释掉第 31～55 行代码，则会发生什么？
2. 如果将第 21 行的 input('> ').upper() 改为 input('> ')，则会发生什么？

项目 61

ROT13密码

ROT13 密码是最简单的加密算法之一，代表"回转 13 位"。ROT13 密码将字母 A～Z 表示为数字 0～25，这样加密的字母与字面字母相差 13 个位置，例如 A 变为 N，B 变为 O，以此类推。加密过程与解密过程完全相同，因此通过编程实现起来非常简单。不过，加密过程很容易被破解。有鉴于此，通常你会发现 ROT13 密码用于隐藏不太敏感的信息，例如剧透信息或其他一些细枝末节的答案，以免被无意中阅读到。如需了解更多有关 ROT13 密码的信息，请访问维基百科。如果希望更广泛地了解密码和密码破解，请阅读拙作《Python 密码学编程》(人民邮电出版社)。

运行程序

运行 rot13cipher.py，输出如下所示：

```
ROT13 Cipher, by Al Sweigart al@inventwithpython.com

Enter a message to encrypt/decrypt (or QUIT):
> Meet me by the rose bushes tonight.
The translated message is:
Zrrg zr ol gur ebfr ohfurf gbavtug.

(Copied to clipboard.)
Enter a message to encrypt/decrypt (or QUIT):
--snip--
```

工作原理

本项目的代码与项目 6 的代码有很多共同之处，只不过本项目的代码更简单，因为它总是使用密钥 13。由于执行加密和解密是同一份代码（第 27～39 行），因此不需要询问玩家究竟是想加密还是解密。

本项目代码与项目 6 的代码的另一个区别在于，本项目的代码维持原始消息的大小写形式，没有自动将消息转换为大写形式。例如，Hello 加密为 Uryyb，而 HELLO 加密为 URYYB。

```
1. """ROT13 密码，作者：Al Sweigart al@inventwithpython.com
2. 用于加密和解密文本的最简单的移位密码
3. 标签：小，密码学"""
4.
```

```python
5. try:
6.     import pyperclip  # pyperclip 模块能够将文本复制到剪贴板
7. except ImportError:
8.     pass  # pyperclip 并不是必需模块, 即使不安装, 也没什么影响
9.
10. # 创建常量
11. UPPER_LETTERS = 'ABCDEFGHIJKLMNOPQRSTUVWXYZ'
12. LOWER_LETTERS = 'abcdefghijklmnopqrstuvwxyz'
13.
14. print('ROT13 Cipher, by Al Sweigart al@inventwithpython.com')
15. print()
16.
17. while True:  # 主循环
18.     print('Enter a message to encrypt/decrypt (or QUIT):')
19.     message = input('> ')
20.
21.     if message.upper() == 'QUIT':
22.         break  # 结束主循环
23.
24.     # 用第 13 个字母替换当前消息中的字母
25.     translated = ''
26.     for character in message:
27.         if character.isupper():
28.             # 拼接大写翻译字符
29.             transCharIndex = (UPPER_LETTERS.find(character) + 13) % 26
30.             translated += UPPER_LETTERS[transCharIndex]
31.         elif character.islower():
32.             # 拼接小写翻译字符
33.             transCharIndex = (LOWER_LETTERS.find(character) + 13) % 26
34.             translated += LOWER_LETTERS[transCharIndex]
35.         else:
36.             # 拼接未翻译的字符
37.             translated += character
38.
39.     # 显示翻译结果
40.     print('The translated message is:')
41.     print(translated)
42.     print()
43.
44.     try:
45.         # 将翻译结果复制到剪贴板
46.         pyperclip.copy(translated)
47.         print('(Copied to clipboard.)')
48.     except:
49.         pass
```

探索程序

请尝试找出以下问题的答案。你需要尝试对代码进行一些修改, 再次运行程序, 查看修改后的效果。

1. 如果将第 27 行的 `character.isupper()` 改为 `character.islower()`, 则会发生什么?

2. 如果将第 41 行的 `print(translated)` 改为 `print(message)`, 则会发生什么?

项目 62

旋转立方体

本项目的特色是用三角函数实现立方体旋转动画。你可以修改 `rotatePoint()` 和 `line()` 函数的代码,以便将其用于自己的动画程序中。

我们用于绘制立方体的块文本字符看起来不像细直线,这种绘图称为线框模型,因为它仅渲染对象表面的边缘。立方体和球体(由三角形组成的粗糙球体)的线框模型如下所示:

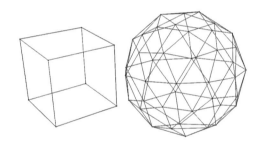

运行程序

运行 rotatingcube.py,输出结果为线框立方体,如下所示:

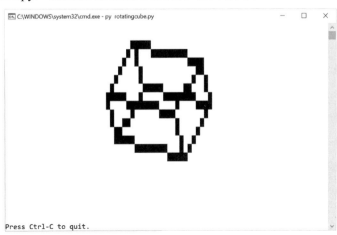

工作原理

本项目的代码主要包含两个部分：`line()`函数和`rotatePoint()`函数。立方体有8个顶点，每个角有1个顶点。程序将这些角作为(x, y, z)元组存储在`CUBE_CORNERS`列表中。这些点还定义了立方体边缘线的连接。如果所有点以相同的量向同一方向旋转，就会产生立方体在旋转的视觉效果。

```
 1. """旋转立方体，作者：Al Sweigart al@inventwithpython.com
 2. 旋转立方体动画
 3. 标签：大，艺术，数学"""
 4.
 5. # 这个程序必须在终端窗口/命令提示符窗口中运行
 6.
 7. import math, time, sys, os
 8.
 9. # 创建常量
10. PAUSE_AMOUNT = 0.1  # 暂停时间为0.1
11. WIDTH, HEIGHT = 80, 24
12. SCALEX = (WIDTH - 4) // 8
13. SCALEY = (HEIGHT - 4) // 8
14. # 文本单元格的高度是其宽度的2倍，所以设置纵向缩放
15. SCALEY *= 2
16. TRANSLATEX = (WIDTH - 4) // 2
17. TRANSLATEY = (HEIGHT - 4) // 2
18.
19. # (!) 试着把9608改为'#'或'*'或其他字符
20. LINE_CHAR = chr(9608)  # 字符9608表示的是'█'
21.
22. # (!) 试着将其中两个值设置为0，使立方体只沿着一个轴旋转
23. X_ROTATE_SPEED = 0.03
24. Y_ROTATE_SPEED = 0.08
25. Z_ROTATE_SPEED = 0.13
26.
27. # 本程序将X、Y、Z坐标存储在列表中，X坐标位于索引0，Y坐标位于索引1，Z坐标位于索引2
28. # 当访问这些列表中的坐标时，这些常量使代码更具可读性
29. X = 0
30. Y = 1
31. Z = 2
32.
33.
34. def line(x1, y1, x2, y2):
35.     """返回给定点之间直线上的点列表
36.
37.     使用Bresenham线性算法"""
38.     points = []  # 存储直线上的点
39.     # "陡峭"是指直线的倾斜角大于45度或小于-45度
40.
41.     # 检查起点和终点相邻的特殊情况
42.     # 该函数无法正确处理这些情况
43.     # 返回硬编码列表
44.     if (x1 == x2 and y1 == y2 + 1) or (y1 == y2 and x1 == x2 + 1):
45.         return [(x1, y1), (x2, y2)]
46.
47.     isSteep = abs(y2 - y1) > abs(x2 - x1)
48.     if isSteep:
49.         # 这个算法只处理非陡峭的直线
50.         # 所以让我们把斜率改为非陡度线，再改回来
51.         x1, y1 = y1, x1  # 交换x1和y1
52.         x2, y2 = y2, x2  # 交换x2和y2
```

```
53.        isReversed = x1 > x2   # 如果直线从右向左移动，则为 True
54.
55.        if isReversed:  # 让直线上的点从右向左移动
56.            x1, x2 = x2, x1    # 交换 x1 和 x2
57.            y1, y2 = y2, y1    # 交换 y1 和 y2
58.
59.            deltax = x2 - x1
60.            deltay = abs(y2 - y1)
61.            extray = int(deltax / 2)
62.            currenty = y2
63.            if y1 < y2:
64.                ydirection = 1
65.            else:
66.                ydirection = -1
67.            # 计算这条直线上每个 x 对应的 y 值
68.            for currentx in range(x2, x1 - 1, -1):
69.                if isSteep:
70.                    points.append((currenty, currentx))
71.                else:
72.                    points.append((currentx, currenty))
73.                extray -= deltay
74.                if extray <= 0:   # 若 extray<=0，只改变 y 值
75.                    currenty -= ydirection
76.                    extray += deltax
77.        else:  # 让直线上的点从左向右移动
78.            deltax = x2 - x1
79.            deltay = abs(y2 - y1)
80.            extray = int(deltax / 2)
81.            currenty = y1
82.            if y1 < y2:
83.                ydirection = 1
84.            else:
85.                ydirection = -1
86.            # 计算这条直线上每个 x 对应的 y 值
87.            for currentx in range(x1, x2 + 1):
88.                if isSteep:
89.                    points.append((currenty, currentx))
90.                else:
91.                    points.append((currentx, currenty))
92.                extray -= deltay
93.                if extray < 0:   # 若 extray<0，只改变 y 值
94.                    currenty += ydirection
95.                    extray += deltax
96.        return points
97.
98.
99.    def rotatePoint(x, y, z, ax, ay, az):
100.        """返回由 x、y、z 参数旋转得到的(x, y, z)元组
101.
102.        旋转发生在(0,0,0)周围，角度为 ax、ay、az（弧度）
103.        各轴方向：
104.              -y
105.               |
106.              +-- +x
107.              /
108.            +z
109.        """
110.
111.        # 绕 x 轴旋转
112.        rotatedX = x
113.        rotatedY = (y * math.cos(ax)) - (z * math.sin(ax))
114.        rotatedZ = (y * math.sin(ax)) + (z * math.cos(ax))
```

```
115.            x, y, z = rotatedX, rotatedY, rotatedZ
116.
117.            # 绕 y 轴旋转
118.            rotatedX = (z * math.sin(ay)) + (x * math.cos(ay))
119.            rotatedY = y
120.            rotatedZ = (z * math.cos(ay)) - (x * math.sin(ay))
121.            x, y, z = rotatedX, rotatedY, rotatedZ
122.
123.            # 绕 z 轴旋转
124.            rotatedX = (x * math.cos(az)) - (y * math.sin(az))
125.            rotatedY = (x * math.sin(az)) + (y * math.cos(az))
126.            rotatedZ = z
127.
128.            return (rotatedX, rotatedY, rotatedZ)
129.
130.
131.    def adjustPoint(point):
132.        """调整三维空间中（x,y,z）点到适合的二维空间中（x,y）点在屏幕上显示
133.        按 SCALEX 和 SCALEY 缩放二维点的大小，
134.        然后按 TRANSLATEX 和 TRANSLATEY 移动点"""
135.        return (int(point[X] * SCALEX + TRANSLATEX),
136.                int(point[Y] * SCALEY + TRANSLATEY))
137.
138.
139.    """CUBE_CORNERS 用于存储立方体角点的（x,y,z）坐标
140.    图中标记了立方体中每个角的索引
141.        0---1
142.       /|  /|
143.      2---3 |
144.      | 4-|-5
145.      |/  |/
146.      6---7"""
147.    CUBE_CORNERS = [[-1, -1, -1],  # 点 0
148.                    [ 1, -1, -1],  # 点 1
149.                    [-1, -1,  1],  # 点 2
150.                    [ 1, -1,  1],  # 点 3
151.                    [-1,  1, -1],  # 点 4
152.                    [ 1,  1, -1],  # 点 5
153.                    [-1,  1,  1],  # 点 6
154.                    [ 1,  1,  1]]  # 点 7
155.
156.    # rotatedCorners 将 CUBE_CORNERS 旋转 rx、ry 和 rz 后的（x,y,z）坐标存储起来
157.    rotatedCorners = [None, None, None, None, None, None, None, None]
158.    # 每个轴的旋转量
159.    xRotation = 0.0
160.    yRotation = 0.0
161.    zRotation = 0.0
162.
163.    try:
164.        while True:  # 主循环
165.            # 沿着不同的轴旋转立方体
166.            xRotation += X_ROTATE_SPEED
167.            yRotation += Y_ROTATE_SPEED
168.            zRotation += Z_ROTATE_SPEED
169.            for i in range(len(CUBE_CORNERS)):
170.                x = CUBE_CORNERS[i][X]
171.                y = CUBE_CORNERS[i][Y]
172.                z = CUBE_CORNERS[i][Z]
173.                rotatedCorners[i] = rotatePoint(x, y, z, xRotation,
174.                    yRotation, zRotation)
175.
176.            # 得到立方体中的点
```

```
177.        cubePoints = []
178.        for fromCornerIndex, toCornerIndex in ((0, 1) (1, 3), (3, 2),
179.        (2, 0), (0, 4), (1, 5), (2, 6), (3, 7), (4, 5), (5, 7), (7, 6), (6, 4)):
180.            fromX, fromY = adjustPoint(rotatedCorners[fromCornerIndex])
181.            toX, toY = adjustPoint(rotatedCorners[toCornerIndex])
182.            pointsOnLine = line(fromX, fromY, toX, toY)
183.            cubePoints.extend(pointsOnLine)
184.
185.        # 消除重复点
186.        cubePoints = tuple(frozenset(cubePoints))
187.
188.        # 在屏幕上显示立方体
189.        for y in range(HEIGHT):
190.            for x in range(WIDTH):
191.                if (x, y) in cubePoints:
192.                    # 显示完整的块
193.                    print(LINE_CHAR, end='', flush=False)
194.                else:
195.                    # 显示空的空间
196.                    print(' ', end='', flush=False)
197.            print(flush=False)
198.        print('Press Ctrl-C to quit.', end='', flush=True)
199.
200.        time.sleep(PAUSE_AMOUNT)  # 暂停一下
201.
202.        # 清屏
203.        if sys.platform == 'win32':
204.            os.system('cls')    # 在 Windows 中使用 cls 命令
205.        else:
206.            os.system('clear')  # 在 macOS 和 Linux 中使用 clear 命令
207.
208. except KeyboardInterrupt:
209.     print('Rotating Cube, by Al Sweigart al@inventwithpython.com')
210.     sys.exit()  # 按下 Ctrl-C, 程序结束
```

输入源代码并运行几次后,请试着对其进行更改。你可以根据标有!的注释对程序进行修改,也可以尝试以下操作。

- 修改 CUBE_CORNERS 和第 184 行的元组以创建不同的线框模型,例如金字塔形和平面六边形。
- 将 CUBE_CORNERS 的坐标增加 1.5,使立方体围绕屏幕中心旋转,而不是围绕自己的中心旋转。

探索程序

请尝试找出以下问题的答案。你需要尝试对代码进行一些修改,再次运行程序,查看修改后的效果。

1. 如果删除或注释掉第 203~206 行的代码,会发生什么?
2. 如果将第 178 行的元组改为 ((0, 1), (1, 3), (3, 2), (2, 0), (0,4), (4, 5), (5, 1)),会发生什么?

项目 63

乌尔皇室游戏

乌尔皇室游戏有着4000余年历史，起源于美索不达米亚文明。考古学家在挖掘乌尔皇室墓地的过程中，发现了这款游戏的游戏板。游戏规则是根据游戏板和一块巴比伦泥板重建的，与飞行棋（Parcheesi）的规则相类似。玩家要兼具运气和技巧才能获胜。

乌尔皇室公墓中发现的5个游戏板之一

游戏开始时，两名玩家各有7枚棋子，先将7枚棋子都移动到目标位置的玩家是赢家。玩家轮流掷4个骰子。这些骰子是棱锥状的四面体。每个骰子有两个标记点，故骰子出现标记或非标记的机会均等。本项目的游戏不使用骰子，而是使用以正面作为标记点的硬币代替。对于出现的每个标记点，玩家可以将棋子移动1格。这意味着棋子每次可以走0~4格，尽管最有可能走2格。

棋子沿着下方所示的路径行进。1个格子中只能有1枚棋子。如果玩家的棋子在共享中间的共有路径上落在对手的棋子上，则将对手的棋子送回"家"。如果棋子落到中间的花格上（程序会提醒重新走棋），那么已经落在该格子内的对手棋子是安全的，不会被送回"家"。如果棋子落在其他4个花格中的任何一个上，玩家就可以再次掷骰子，再走一次棋。该游戏使用字母X和O表示棋子。

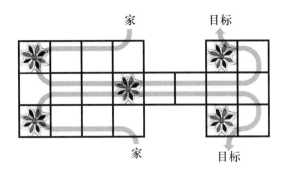

每位玩家的棋子从"家"到"目标"的行进路径

运行程序

运行 royalgameofur.py，输出如下所示：

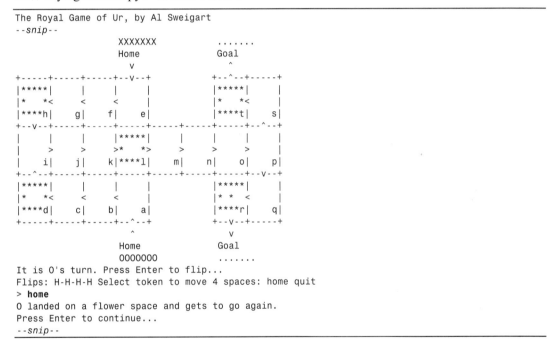

工作原理

类似项目 43 播棋，本项目游戏板上的空格被标以字母 a~t。掷骰子后，玩家可以在棋盘上选择自己的一个棋子进行移动，也可以从棋盘下的"家"中拿一枚棋子移动到棋盘上。以下程序使用字典表示棋盘，键为 a~t，X 和 O 值代表棋子（" "代表空格）。

此外，该字典具有键 x_home、o_home、x_goal 和 o_goal，这些键的值是 7 个字符的字符串，表示"家"和"目标"的"满座"程度。这些字符串中的 X 或 O 字符代表"家"或"目标"的棋子，而 . 代表一个空位。`displayBoard()` 函数会在屏幕上显示这 7 个字符的字符串。

```python
 1. """乌尔皇室游戏，作者：Al Sweigart al@inventwithpython.com
 2. 一种来自美索不达米亚的有约 5000 年历史的棋盘游戏。两名玩家在冲向球门时互相撞倒对方
 3. 标签：大，棋盘，游戏，双人游戏
 4. """
 5.
 6. import random, sys
 7.
 8. X_PLAYER = 'X'
 9. O_PLAYER = 'O'
10. EMPTY = ' '
11.
12. # 创建棋格的常量
13. X_HOME = 'x_home'
14. O_HOME = 'o_home'
15. X_GOAL = 'x_goal'
16. O_GOAL = 'o_goal'
17.
18. # 棋格按从左到右、从上到下的顺序排列
19. ALL_SPACES = 'hgfetsijklmnopdcbarq'
20. X_TRACK  = 'HefghijklmnopstG'   # H 代表家，G 代表目标
21. O_TRACK  = 'HabcdijklmnopqrG'
22.
23. FLOWER_SPACES = ('h', 't', 'l', 'd', 'r')
24.
25. BOARD_TEMPLATE = """
26.                {}                    {}
27.               Home                  Goal
28.                v                     ^
29.     +-----+-----+-----+--v--+     +--^--+-----+
30.     |*****|     |     |     |     |*****|     |
31.     |* {} *<  {}  <  {}  <  {}  |     |* {} *<  {}  |
32.     |****h|    g|    f|    e|     |****t|    s|
33.     +--v--+-----+-----+-----+-----+-----+--^--+
34.     |     |     |     |*****|     |     |     |     |
35.     |  {} >  {} >  {} >* {} *>  {} >  {} >  {} >  {} |
36.     |    i|    j|    k|****l|    m|    n|    o|    p|
37.     +--^--+-----+-----+-----+-----+-----+--v--+
38.     |*****|     |     |     |     |*****|     |
39.     |* {} *<  {}  <  {}  <  {}  |     |* {} *<  {}  |
40.     |****d|    c|    b|    a|     |****r|    q|
41.     +-----+-----+-----+--^--+     +--v--+-----+
42.                ^                     v
43.               Home                  Goal
44.                {}                    {}
45. """
46.
47.
48. def main():
49.     print('''The Royal Game of Ur, by Al Sweigart
50.
51. This is a 5,000 year old game. Two players must move their tokens
52. from their home to their goal. On your turn you flip four coins and can
53. move one token a number of spaces equal to the heads you got.
54.
55. Ur is a racing game; the first player to move all seven of their tokens
56. to their goal wins. To do this, tokens must travel from their home to
57. their goal:
58.
59.               X Home        X Goal
60.                 v             ^
```

```
61. +---+---+---+-v-+       +-^-+---+
62. |v<<<<<<<<<<<<  |       |  ^|<< |
63. |v |   |   |   |        |   | ^ |
64. +v--+---+---+---+---+---+---+-^-+
65. |>>>>>>>>>>>>>>>>>>>>>>>>>>>^  |
66. |>>>>>>>>>>>>>>>>>>>>>>>>>>>v  |
67. +^--+---+---+---+---+---+-v-+
68. |^  |   |   |   |       |   | v |
69. |^<<<<<<<<<<<<  |       | v<<<< |
70. +---+---+---+-^-+       +-v-+---+
71.               ^               v
72.             O Home         O Goal
73.
74. If you land on an opponent's token in the middle track, it gets sent
75. back home. The **flower** spaces let you take another turn. Tokens in
76. the middle flower space are safe and cannot be landed on.''')
77.     input('Press Enter to begin...')
78.
79.     gameBoard = getNewBoard()
80.     turn = O_PLAYER
81.     while True:    # 主循环
82.         # 为当前回合创建变量
83.         if turn == X_PLAYER:
84.             opponent = O_PLAYER
85.             home = X_HOME
86.             track = X_TRACK
87.             goal = X_GOAL
88.             opponentHome = O_HOME
89.         elif turn == O_PLAYER:
90.             opponent = X_PLAYER
91.             home = O_HOME
92.             track = O_TRACK
93.             goal = O_GOAL
94.             opponentHome = X_HOME
95.
96.         displayBoard(gameBoard)
97.
98.         input('It is ' + turn + '\'s turn. Press Enter to flip...')
99.
100.        flipTally = 0
101.        print('Flips: ', end='')
102.        for i in range(4):    # 掷4枚硬币
103.            result = random.randint(0, 1)
104.            if result == 0:
105.                print('T', end='')    # 硬币背面
106.            else:
107.                print('H', end='')    # 硬币正面
108.            if i != 3:
109.                print('-', end='')    # 输出分隔符
110.            flipTally += result
111.        print('  ', end='')
112.
113.        if flipTally == 0:
114.            input('You lose a turn. Press Enter to continue...')
115.            turn = opponent    # 轮到另一个玩家掷骰子
116.            continue
117.
118.        # 询问玩家的下一步走子
119.        validMoves = getValidMoves(gameBoard, turn, flipTally)
120.
121.        if validMoves == []:
122.            print('There are no possible moves, so you lose a turn.')
```

```python
123.            input('Press Enter to continue...')
124.            turn = opponent  # 轮到另一个玩家掷骰子
125.            continue
126.
127.        while True:
128.            print('Select move', flipTally, 'spaces: ', end='')
129.            print(' '.join(validMoves) + ' quit')
130.            move = input('> ').lower()
131.
132.            if move == 'quit':
133.                print('Thanks for playing!')
134.                sys.exit()
135.            if move in validMoves:
136.                break  # 如果玩家的下一步走子有效，则循环结束
137.
138.            print('That is not a valid move.')
139.
140.        # 在棋盘上执行下一步走子
141.        if move == 'home':
142.            # 如果玩家从"家"里移出，则"家"里的棋子数减1
143.            gameBoard[home] -= 1
144.            nextTrackSpaceIndex = flipTally
145.        else:
146.            gameBoard[move] = EMPTY  # 将出发地的棋格设置为空
147.            nextTrackSpaceIndex = track.index(move) + flipTally
148.
149.        movingOntoGoal = nextTrackSpaceIndex == len(track) - 1
150.        if movingOntoGoal:
151.            gameBoard[goal] += 1
152.            # 判断玩家是否赢了
153.            if gameBoard[goal] == 7:
154.                displayBoard(gameBoard)
155.                print(turn, 'has won the game!')
156.                print('Thanks for playing!')
157.                sys.exit()
158.        else:
159.            nextBoardSpace = track[nextTrackSpaceIndex]
160.            # 判断对手那里是否有棋子
161.            if gameBoard[nextBoardSpace] == opponent:
162.                gameBoard[opponentHome] += 1
163.
164.            # 将目的地的棋格标记上玩家的棋子
165.            gameBoard[nextBoardSpace] = turn
166.
167.            # 判断玩家是否落在花格上，以及是否可以再走一次棋
168.            if nextBoardSpace in FLOWER_SPACES:
169.                print(turn, 'landed on a flower space and goes again.')
170.                input('Press Enter to continue...')
171.            else:
172.                turn = opponent  # 转到另一个玩家掷骰子
173.
174.
175. def getNewBoard():
176.     """
177.     返回表示棋盘状态的字典。键是棋格标签的字符串，值是 X_PLAYER、O_PLAYER 或 EMPTY
178.     游戏中还有计数器，以提示玩家出发地和目的地有多少标记
179.     """
180.     board = {X_HOME: 7, X_GOAL: 0, O_HOME: 7, O_GOAL: 0}
181.     # 开始时将每个棋格设置为空
182.     for spaceLabel in ALL_SPACES:
183.         board[spaceLabel] = EMPTY
184.     return board
```

```
185.
186.
187. def displayBoard(board):
188.     """在屏幕上显示棋盘"""
189.     # 通过输出许多空行来清空屏幕，让原来的棋盘不再可见
190.     print('\n' * 60)
191.
192.     xHomeTokens = ('X' * board[X_HOME]).ljust(7, '.')
193.     xGoalTokens = ('X' * board[X_GOAL]).ljust(7, '.')
194.     oHomeTokens = ('O' * board[O_HOME]).ljust(7, '.')
195.     oGoalTokens = ('O' * board[O_GOAL]).ljust(7, '.')
196.
197.     # 按照从左到右、从上到下的顺序，添加用来填充 BOARD_TEMPLATE 的字符串
198.     spaces = []
199.     spaces.append(xHomeTokens)
200.     spaces.append(xGoalTokens)
201.     for spaceLabel in ALL_SPACES:
202.         spaces.append(board[spaceLabel])
203.     spaces.append(oHomeTokens)
204.     spaces.append(oGoalTokens)
205.
206.     print(BOARD_TEMPLATE.format(*spaces))
207.
208.
209. def getValidMoves(board, player, flipTally):
210.     validMoves = []  # 包含带有可移动棋子的棋格
211.     if player == X_PLAYER:
212.         opponent = O_PLAYER
213.         track = X_TRACK
214.         home = X_HOME
215.     elif player == O_PLAYER:
216.         opponent = X_PLAYER
217.         track = O_TRACK
218.         home = O_HOME
219.
220.     # 判断玩家是否可以从"家"里移动棋子
221.     if board[home] > 0 and board[track[flipTally]] == EMPTY:
222.         validMoves.append('home')
223.
224.     # 判断哪些棋格中有玩家可以移动的棋子
225.     for trackSpaceIndex, space in enumerate(track):
226.         if space == 'H' or space == 'G' or board[space] != player:
227.             continue
228.         nextTrackSpaceIndex = trackSpaceIndex + flipTally
229.         if nextTrackSpaceIndex >= len(track):
230.             # 玩家必须朝着目的地移动正确的步数，否则无法移动
231.             continue
232.         else:
233.             nextBoardSpaceKey = track[nextTrackSpaceIndex]
234.             if nextBoardSpaceKey == 'G':
235.                 # 这颗棋子可以被移出棋盘
236.                 validMoves.append(space)
237.                 continue
238.             if board[nextBoardSpaceKey] in (EMPTY, opponent):
239.                 # 如果下一个棋格是受保护的中间棋格，则玩家只能在它为空的情况下移动到那里
240.                 if nextBoardSpaceKey == 'l' and board['l'] == opponent:
241.                     continue  # 跳过此步，棋格受到保护
242.                 validMoves.append(space)
243.
244.     return validMoves
```

```
245.
246.
247. if __name__ == '__main__':
248.     main()
```

探索程序

请尝试找出以下问题的答案。你需要尝试对代码进行一些修改，再次运行程序，查看修改后的效果。

1. 如果将第 149 行的 nextTrackSpaceIndex == len(track) - 1 改为 nextTrackSpaceIndex == 1，则会发生什么？
2. 如果将第 103 行的 result = random.randint(0, 1) 改为 result = 1，则会发生什么？
3. 如果将第 180 行的 board = {X_HOME: 7, X_GOAL: 0, O_HOME: 7, O_GOAL: 0} 改为 board = {}，则会导致什么错误？

项目 64

7段显示模块

7段显示模块是一种 LCD 组件，用于在袖珍计算器、微波炉以及其他小型电子设备中显示数字。通过 LCD 中 7 个线段的不同组合，7 段显示模块可以用来表示数字 0~9，如下所示：

```
 _       _  _     _  _  _  _  _
| |  |  _| _| |_| |_ |_   | |_| |_|
|_|  | |_  _|   |  _| |_| | |_|  _|
```

7 段显示模块的优点是其他程序可以将其作为模块导入。在项目 14 和项目 19 中，我们导入了 sevseg.py 文件，以便调用其中的 getSevSegStr() 函数。有关 7 段显示模块及其衍生版本的更多信息可参见维基百科。

运行程序

即使 sevseg.py 是一个模块文件，它也可以直接运行，能输出数字样例，如下所示：

```
This module is meant to be imported rather than run.
For example, this code:
    import sevseg
    myNumber = sevseg.getSevSegStr(42, 3)
    print(myNumber)

Will print 42, zero-padded to three digits:
         _   _
|  |  |_| _|
|  |    | |_
```

工作原理

getSevSegStr() 函数用于创建一个包含 3 个字符串的列表。3 个字符串分别代表数字的顶行、中行和底行，第 23~71 行有一个长长的 if-elif 语句列表，用于将每个数字（以及小数点和减号）的行连接到这 3 个字符串上。这 3 个字符串在第 80 行用换行符连接在一起，以便函数返回一个适合传递给 print() 的多行字符串。

```
1. """7段显示模块，作者：Al Sweigart al@inventwithpython.com
2. 一个用于倒计时和数字时钟程序的 7 段显示模块
3. 标签：简短,模块"""
4. """A labeled seven-segment display, with each segment labeled A to G:
```

```
 5.         __A__
 6.        |     |      Each digit in a seven-segment display:
 7.        F     B
 8.        |__G__|         __     __    __    __    __    __    __    __
 9.        |     |     |  |    |    __|   __|  |__|  |__   |__     |  |__|  |__|
10.        E     C
11.        |__D__|"""     |__|    |   __|   __|     |   __|  |__|    |  |__|    _|
12.
13.
14.     def getSevSegStr(number, minWidth=0):
15.         """返回一个由数字组成的 7 段字符串。 如果返回的字符串的宽度小于最小宽度,
16.         则返回的字符串将用 0 填充"""
17.
18.         # 将数字转换为字符串,如果它是整数或浮点数
19.         number = str(number).zfill(minWidth)
20.
21.         rows = ['', '', '']
22.         for i, numeral in enumerate(number):
23.             if numeral == '.':          # 呈现小数点
24.                 rows[0] += ' '
25.                 rows[1] += ' '
26.                 rows[2] += '.'
27.                 continue  # 跳过数字之间的空格
28.             elif numeral == '-':        # 呈现负号
29.                 rows[0] += '    '
30.                 rows[1] += ' __ '
31.                 rows[2] += '    '
32.             elif numeral == '0':        # 呈现 0
33.                 rows[0] += ' __ '
34.                 rows[1] += '|  |'
35.                 rows[2] += '|__|'
36.             elif numeral == '1':        # 呈现 1
37.                 rows[0] += '    '
38.                 rows[1] += '   |'
39.                 rows[2] += '   |'
40.             elif numeral == '2':        # 呈现 2
41.                 rows[0] += ' __ '
42.                 rows[1] += ' __|'
43.                 rows[2] += '|__ '
44.             elif numeral == '3':        # 呈现 3
45.                 rows[0] += ' __ '
46.                 rows[1] += ' __|'
47.                 rows[2] += ' __|'
48.             elif numeral == '4':        # 呈现 4
49.                 rows[0] += '    '
50.                 rows[1] += '|__|'
51.                 rows[2] += '   |'
52.             elif numeral == '5':        # 呈现 5
53.                 rows[0] += ' __ '
54.                 rows[1] += '|__ '
55.                 rows[2] += ' __|'
56.             elif numeral == '6':        # 呈现 6
57.                 rows[0] += ' __ '
58.                 rows[1] += '|__ '
59.                 rows[2] += '|__|'
60.             elif numeral == '7':        # 呈现 7
61.                 rows[0] += ' __ '
62.                 rows[1] += '   |'
63.                 rows[2] += '   |'
64.             elif numeral == '8':        # 呈现 8
65.                 rows[0] += ' __ '
66.                 rows[1] += '|__|'
67.                 rows[2] += '|__|'
```

```
68.        elif numeral == '9':    # 呈现9
69.            rows[0] += ' __ '
70.            rows[1] += '|__|'
71.            rows[2] += '  __|'
72.
73.        # 如果这不是最后一个数字
74.        # 则添加一个空格（空格用于数字之间）
75.        if i != len(number) - 1 and number[i + 1] != '.':
76.            rows[0] += ' '
77.            rows[1] += ' '
78.            rows[2] += ' '
79.
80.    return '\n'.join(rows)
81.
82.
83. # 如果这个程序不是作为模块导入，则显示数字 00 到 99
84. if __name__ == '__main__':
85.     print('This module is meant to be imported rather than run.')
86.     print('For example, this code:')
87.     print('    import sevseg')
88.     print('    myNumber = sevseg.getSevSegStr(42, 3)')
89.     print('    print(myNumber)')
90.     print()
91.     print('...will print 42, zero-padded to three digits:')
92.     print(' __      __ ')
93.     print('|  |  | |__   __|')
94.     print('|__|  | | __   __|')
```

输入源代码并运行几次后，请试着对其进行更改。你也可以尝试以下操作。

❏ 为数字创建新字体，例如使用 chr(9608) 返回的块字符串表示数字（占 5 行）。

❏ 查阅有关 7 段显示模块的维基百科文章，了解如何显示字母，并让 sevseg.py 支持显示字母。

❏ 请通过互联网了解什么是 16 段显示模块，并创建一个 sixteenseg.py 模块，以生成这种样式的数字。

探索程序

请尝试找出以下问题的答案。你需要尝试对代码进行一些修改，再次运行程序，查看修改后的效果。

1. 如果将第 76 行、第 77 行和第 78 行的单空格字符串改为空串，则会发生什么？
2. 如果将第 14 行的 minWidth=0 默认参数改为 minWidth=8，则会发生什么？

项目 65

"闪灵地毯"

1980 年，由斯坦利·库布里克（Stanley Kubrick）执导的恐怖悬疑片《闪灵》在美国上映，电影中的故事发生在瞭望酒店。酒店地毯的六边形设计成为这部著名电影的标志性部分。地毯上遍布交错且环环相扣的六边形，其令人眩晕的效果与这部电影的恐怖格调甚是吻合。与项目 35 类似，运行本项目中的程序，可以在计算机屏幕上输出重复的图案。

注意，这个程序会使用原始字符串，在前引号之前加上小写的 r，这样字符串中的反斜线不会被解释为转义字符。

运行程序

运行 shiningcarpet.py，输出如下所示:

```
_ \ \_/ __ \ \ \_/ __ \ \ \_/ __ \ \ \_/ __ \ \ \_/ __ \
 \ \_/ __ \ \ \_/ __ \ \ \_/ __ \ \ \_/ __ \ \ \_/ __ \
\ \ \_/ __ \ \ \_/ __ \ \ \_/ __ \ \ \_/ __ \ \ \_/ __
/ / / \_ / / / \_ / / / \_ / / / \_ / / / \_ / / / \_
_/ / / \_ / / / \_ / / / \_ / / / \_ / / / \_ / / / \
_/ / / \_ / / / \_ / / / \_ / / / \_ / / / \_ / / /
\ \ \_/ __ \ \ \_/ __ \ \ \_/ __ \ \ \_/ __ \ \ \_/ __
 \ \_/ __ \ \ \_/ __ \ \ \_/ __ \ \ \_/ __ \ \ \_/ __
\ \ \_/ __ \ \ \_/ __ \ \ \_/ __ \ \ \_/ __ \ \ \_/ __
/ / / \_ / / / \_ / / / \_ / / / \_ / / / \_ / / / \_
_/ / / \_ / / / \_ / / / \_ / / / \_ / / / \_ / / / \
_/ / / \_ / / / \_ / / / \_ / / / \_ / / / \_ / / /
\ \ \_/ __ \ \ \_/ __ \ \ \_/ __ \ \ \_/ __ \ \ \_/ __
 \ \_/ __ \ \ \_/ __ \ \ \_/ __ \ \ \_/ __ \ \ \_/ __
\ \ \_/ __ \ \ \_/ __ \ \ \_/ __ \ \ \_/ __ \ \ \_/ __
/ / / \_ / / / \_ / / / \_ / / / \_ / / / \_ / / / \_
_/ / / \_ / / / \_ / / / \_ / / / \_ / / / \_ / / / \
_/ / / \_ / / / \_ / / / \_ / / / \_ / / / \_ / / /
```

工作原理

创建以下这样的程序（或类似于项目 35 的程序）不是直接从编码开始，而是要先在文本编辑器中绘制镶嵌图案。绘制出图案后，我们就可以将其缩减为最小平铺单元，具体如下：

```
_ \ \ \_/ __
 \ \ \___/ _
 \ \ \_____/
 / / / ___ \_
_/ / / / _ \__
__/ / / / \___
```

将该文本图案复制并粘贴到源代码中后，我们可以围绕它编写程序的其余部分。软件不是只要坐下来从头到尾编写代码就能一蹴而就的，专业软件开发者都会经历多次修改、实验和调试。程序最终可能只有 9 行核心代码，但程序小并不意味着投入很少的努力就能完成。

```
 1. """闪灵地毯，作者：Al Sweigart al@inventwithpython.com
 2. 展示了《闪灵》中地毯图案的镶嵌效果
 3. 标签：小，初学者，艺术"""
 4.
 5. # 创建常量
 6. X_REPEAT = 6  # 水平镶嵌的次数
 7. Y_REPEAT = 4  # 垂直镶嵌的次数
 8.
 9. for i in range(Y_REPEAT):
10.     print(r'_ \ \ \_/ __' * X_REPEAT)
11.     print(r' \ \ \___/ _' * X_REPEAT)
12.     print(r'\ \ \_____/ ' * X_REPEAT)
13.     print(r'/ / / ___ \_' * X_REPEAT)
14.     print(r'_/ / / _ \__' * X_REPEAT)
15.     print(r'__/ / / \___' * X_REPEAT)
```

探索程序

请试着使这个程序创建出如下图案：

项目 65 "闪灵地毯"

```
   \_   \_   \_   \_   \_   \_   \_   \_   \_
 _/ \_/ \_/ \_/ \_/ \_/ \_/ \_/ \_/ \_/ \_/ \
 \   /   \   /   \   /   \   /   \   /   \
  \_/ __\_/ __\_/ __\_/ __\_/ __\_/ __\_/ __\
 _/ \_/ \_/ \_/ \_/ \_/ \_/ \_/ \_/ \_/ \_/ \
 \   /   \   /   \   /   \   /   \   /   \
  \_/ \_/ \_/ \_/ \_/ \_/ \_/ \_/ \_/ \_/ \_/

 / __ \ ^ / __ \ ^ / __ \ ^ / __ \ ^ / __ \ ^ / __ \ ^
 /    \ vvv /    \ vvv /    \ vvv /    \ vvv /    \ vvv /    \ vvv
 |()  ()|   |()  ()|   |()  ()|   |()  ()|   |()  ()|   |()  ()|
  \ ^ / __ \ ^ / __ \ ^ / __ \ ^ / __ \ ^ / __ \ ^ / __ \ ^ /
   \ vvv /    \ vvv /    \ vvv /    \ vvv /    \ vvv /    \ vvv /
 )|   |()  ()|   |()  ()|   |()  ()|   |()  ()|   |()  ()|   |() (
```

项目 66

简单替换密码

在本项目中,我们将实现把一个字母替换成另一个字母的程序。字母 A 有 26 种可能的替换,B 有 25 种可能的替换,C 有 24 种可能的替换,以此类推,那么可能的密钥总数为 26×25×24×23×…×1! 即便是超级计算机,也会因为有太多的密钥而无法暴力破解,所以项目 7 中使用的密码破解方法对简单替换密码无效。然而,狡猾的攻击者可以利用已知的弱点来破解密码。如果希望了解有关密码和密码破解的更多信息,请阅读拙作《Python 密码学编程》(人民邮电出版社)。

运行程序

运行 simplesubcipher.py,输出如下所示:

```
Simple Substitution Cipher, by Al Sweigart
A simple substitution cipher has a one-to-one translation for each
symbol in the plaintext and each symbol in the ciphertext.
Do you want to (e)ncrypt or (d)ecrypt?
> e
Please specify the key to use.
Or enter RANDOM to have one generated for you.
> random
The key is WNOMTRCEHDXBFVSLKAGZIPYJQU. KEEP THIS SECRET!
Enter the message to encrypt.
> Meet me by the rose bushes tonight.
The encrypted message is:
Fttz ft nq zet asgt nigetg zsvhcez.
Full encrypted text copied to clipboard.

Simple Substitution Cipher, by Al Sweigart
A simple substitution cipher has a one-to-one translation for each
symbol in the plaintext and each symbol in the ciphertext.
Do you want to (e)ncrypt or (d)ecrypt?
> d
Please specify the key to use.
> WNOMTRCEHDXBFVSLKAGZIPYJQU
Enter the message to decrypt.
> Fttz ft nq zet asgt nigetg zsvhcez.
The decrypted message is:
Meet me by the rose bushes tonight.
Full decrypted text copied to clipboard.
```

工作原理

字母表中的字母如何使用以 WNOM 开头的密钥进行加密呢？如下所示，构成密钥的 26 个字母分别对应于其下方的字母。如要解密，将底部的字母替换为其上方的相应字母即可。

```
A B C D E F G H I J K L M N O P Q R S T U V W X Y Z
↕ ↕ ↕ ↕ ↕ ↕ ↕ ↕ ↕ ↕ ↕ ↕ ↕ ↕ ↕ ↕ ↕ ↕ ↕ ↕ ↕ ↕ ↕ ↕ ↕ ↕
W N O M T R C E H D X B F V S L K A G Z I P Y J Q U
```

使用此密钥，字母 A 加密为 W（W 解密为 A），字母 B 加密为 N，以此类推。LETTERS 和 key 变量分别被赋予 charsA 和 charsB（如果解密则相反）。charsA 中的任何消息字符都会被替换为 charsB 中的相应字符，从而生成最终转换后的消息字符。

```python
 1. """简单替换密码，作者：Al Sweigart al@inventwithpython.com
 2. 简单替换密码会对明文中的每个符号和密文中的每个符号进行一对一的转换
 3. 标签：简短，密码学，数学"""
 4.
 5. import random
 6.
 7. try:
 8.     import pyperclip  # pyperclip 模块能够将文本复制到剪贴板
 9. except ImportError:
10.     pass  # 即便没有安装 pyperclip，也没什么大不了
11.
12. # 每一个可能被加密/解密的符号
13. LETTERS = 'ABCDEFGHIJKLMNOPQRSTUVWXYZ'
14.
15.
16. def main():
17.     print('''Simple Substitution Cipher, by Al Sweigart
18. A simple substitution cipher has a one-to-one translation for each
19. symbol in the plaintext and each symbol in the ciphertext.''')
20.
21.     # 让用户指定是加密还是解密
22.     while True:  # 继续询问，直到用户输入 e 或 d
23.         print('Do you want to (e)ncrypt or (d)ecrypt?')
24.         response = input('> ').lower()
25.         if response.startswith('e'):
26.             myMode = 'encrypt'
27.             break
28.         elif response.startswith('d'):
29.             myMode = 'decrypt'
30.             break
31.         print('Please enter the letter e or d.')
32.
33.     # 让用户指定要使用的密钥
34.     while True:  # 继续询问，直到用户输入有效的密钥
35.         print('Please specify the key to use.')
36.         if myMode == 'encrypt':
37.             print('Or enter RANDOM to have one generated for you.')
38.         response = input('> ').upper()
39.         if response == 'RANDOM':
40.             myKey = generateRandomKey()
41.             print('The key is {}. KEEP THIS SECRET!'.format(myKey))
42.             break
43.         else:
44.             if checkKey(response):
45.                 myKey = response
46.                 break
```

```
47.
48.     # 让用户指定要加密/解密的消息
49.     print('Enter the message to {}.'.format(myMode))
50.     myMessage = input('> ')
51.
52.     # 执行加密/解密
53.     if myMode == 'encrypt':
54.         translated = encryptMessage(myMessage, myKey)
55.     elif myMode == 'decrypt':
56.         translated = decryptMessage(myMessage, myKey)
57.
58.     # 展示结果
59.     print('The %sed message is:' % (myMode))
60.     print(translated)
61.
62.     try:
63.         pyperclip.copy(translated)
64.         print('Full %sed text copied to clipboard.' % (myMode))
65.     except:
66.         pass  # 如果没有安装pyperclip，则什么也不做
67.
68.
69. def checkKey(key):
70.     """如果密钥有效则返回True，否则返回False"""
71.     keyList = list(key)
72.     lettersList = list(LETTERS)
73.     keyList.sort()
74.     lettersList.sort()
75.     if keyList != lettersList:
76.         print('There is an error in the key or symbol set.')
77.         return False
78.     return True
79.
80.
81. def encryptMessage(message, key):
82.     """使用密钥加密消息"""
83.     return translateMessage(message, key, 'encrypt')
84.
85.
86. def decryptMessage(message, key):
87.     """使用密钥解密消息"""
88.     return translateMessage(message, key, 'decrypt')
89.
90.
91. def translateMessage(message, key, mode):
92.     """使用密钥加密或解密消息"""
93.     translated = ''
94.     charsA = LETTERS
95.     charsB = key
96.     if mode == 'decrypt':
97.         # 对于解密，我们可以使用与加密相同的代码
98.         # 我们只需要交换key和LETTERS字符串的位置
99.         charsA, charsB = charsB, charsA
100.
101.     # 循环遍历消息中的每个符号
102.     for symbol in message:
103.         if symbol.upper() in charsA:
104.             # 加密/解密符号
105.             symIndex = charsA.find(symbol.upper())
106.             if symbol.isupper():
107.                 translated += charsB[symIndex].upper()
108.             else:
```

```
109.                translated += charsB[symIndex].lower()
110.            else:
111.                # 该符号不在 LETTERS 中，则添加该符号
112.                translated += symbol
113.
114.    return translated
115.
116.
117. def generateRandomKey():
118.     """生成并返回一个随机加密密钥"""
119.     key = list(LETTERS)         # 根据 LETTERS 字符串生成列表
120.     random.shuffle(key)         # 随机打乱列表
121.     return ''.join(key)         # 从列表中获取一个字符串
122.
123.
124. # 程序运行入口（如果不是作为模块导入的话）
125. if __name__ == '__main__':
126.     main()
```

探索程序

请尝试找出以下问题的答案。你需要尝试对代码进行一些修改，再次运行程序，查看修改后的效果。

1. 如果删除或注释掉第 120 行的 `random.shuffle(key)` 并输入 RANDOM 作为密钥，则会发生什么？
2. 如果将第 13 行的 `LETTERS` 字符串扩展为 `'ABCDEFGHIJKLMNOPQRSTUVWXYZ 1234567890'`，则会发生什么？

项目 67

正弦消息

在本项目中，我们通过编写程序，用不断向上滚动的波浪形图案显示用户选择的消息。这种效果是用正弦三角函数 math.sin() 实现的，该函数可实现正弦波功能。即使你不理解其中的数学原理也不要紧，本项目的程序很简短，而且很容易复制使用。

运行程序

运行 sinemessage.py，输出如下所示：

```
Sine Message, by Al Sweigart al@inventwithpython.com
(Press Ctrl-C to quit.)

What message do you want to display? (Max 39 chars.)
> I <3 Programming!
                              I <3 Programming!
                           I <3 Programming!
                         I <3 Programming!
                           I <3 Programming!
                             I <3 Programming!
                               I <3 Programming!
                                I <3 Programming!
                               I <3 Programming!
                             I <3 Programming!
                           I <3 Programming!
                         I <3 Programming!
                       I <3 Programming!
                     I <3 Programming!
                   I <3 Programming!
                 I <3 Programming!
             I <3 Programming!
         I <3 Programming!
    I <3 Programming!
 I <3 Programming!
I <3 Programming!
 I <3 Programming!
   I <3 Programming!
      I <3 Programming!
--snip--
```

工作原理

`math` 模块中的 `math.sin()` 函数接收一个参数（我们称之为 `x`），并返回另一个数字，即 `x` 的正弦值。一些数学应用程序会用到正弦函数，在本项目的程序中，正弦函数只是用于创建整洁的波浪形效果。我们将一个名为 `step` 的变量传递给 `math.sin()`。该变量的值从 0 开始，其值在主程序循环的每次迭代中增加 0.25。

我们将使用 `math.sin()` 的返回值来计算应该在用户消息的两侧输出多少个填充空格。由于 `math.sin()` 返回一个-1.0~1.0 的浮点数，但我们想要的最小填充量是零，而不是负值，因此以下程序第 28 行将 `math.sin()` 的返回值加 1，使有效范围为 0.0~2.0。我们当然需要零或者两个以上的空格，因此第 28 行将这个数字乘上名为 `multiplier` 的变量以增加填充量。该乘积值表示的是输出用户消息前在其左侧补充的空格数。

程序结果是我们在运行程序时看到的波浪消息动画。本项目程序的代码如下：

```
 1. """正弦消息，作者：Al Sweigart al@inventwithpython.com
 2. 创建一个正弦波效果的信息
 3. 标签：小，艺术"""
 4.
 5. import math, shutil, sys, time
 6.
 7. # 获取终端窗口的大小
 8. WIDTH, HEIGHT = shutil.get_terminal_size()
 9. # 如果不自动添加换行符，我们无法输出到 Windows 上的最后一列，所以将宽度减 1
10. WIDTH -= 1
11.
12. print('Sine Message, by Al Sweigart al@inventwithpython.com')
13. print('(Press Ctrl-C to quit.)')
14. print()
15. print('What message do you want to display? (Max', WIDTH // 2, 'chars.)')
16. while True:
17.     message = input('> ')
18.     if 1 <= len(message) <= (WIDTH // 2):
19.         break
20.     print('Message must be 1 to', WIDTH // 2, 'characters long.')
21.
22. step = 0.0  # 这个"步长"决定了正弦波的陡峭程度
23. # sin 的值在-1.0 和 1.0 之间，所以我们需要用 multiplier 来改变它
24. multiplier = (WIDTH - len(message)) / 2
25. try:
26.     while True:  # 主循环
27.         sinOfStep = math.sin(step)
28.         padding = ' ' * int((sinOfStep + 1) * multiplier)
29.         print(padding + message)
30.         time.sleep(0.1)
31.         step += 0.25  # (!) 尝试将 0.25 更改为 0.1 或 0.5
32. except KeyboardInterrupt:
33.     sys.exit()  # 按下 Ctrl-C，程序结束
```

输入源代码并运行几次后，请试着对其进行更改。你可以根据标有!的注释对程序进行修改。

探索程序

请尝试找出以下问题的答案。你需要尝试对代码进行一些修改，再次运行程序，查看修改后的效果。

1. 如果将第 27 行的 math.sin(step) 改为 math.cos(step)，会发生什么？
2. 如果将第 27 行的 math.sin(step) 改为 math.sin(0)，会发生什么？

项目 68

滑动拼图

滑动拼图这个经典的游戏基于一个 4×4 的板子，其上有 15 个带编号的图块和一个空白空间，游戏目标是滑动图块直到编号数字按正确顺序从左到右、从上到下排列。图块只能滑动，不能直接取下来重新排列。这种拼图游戏的有些衍生版本会用打乱顺序的图像。玩家最后会拼出一幅完整的图像。

运行程序

运行 slidingtilepuzzle.py，输出如下所示：

```
Sliding Tile Puzzle, by Al Sweigart al@inventwithpython.com
    Use the WASD keys to move the tiles
    back into their original order:
            1  2  3  4
            5  6  7  8
            9 10 11 12
           13 14 15
Press Enter to begin...
+------+------+------+------+
|      |      |      |      |
|  5   |  10  |      |  11  |
|      |      |      |      |
+------+------+------+------+
|      |      |      |      |
|  6   |  3   |  7   |  2   |
|      |      |      |      |
+------+------+------+------+
|      |      |      |      |
|  14  |  1   |  15  |  8   |
|      |      |      |      |
+------+------+------+------+
|      |      |      |      |
|  9   |  13  |  4   |  12  |
|      |      |      |      |
+------+------+------+------+
                    (W)
Enter WASD (or QUIT): (A) ( ) (D)
> w
```

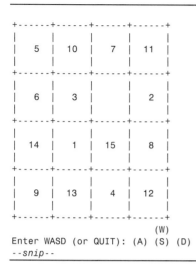

```
                     (W)
Enter WASD (or QUIT): (A) (S) (D)
--snip--
```

工作原理

表示滑动拼图游戏板的数据结构是一个二维列表。每个内层列表代表4×4板子的一列，包含用于对图块进行编号的字符串（或者表示空白空间的BLANK字符串）。getNewBoard()函数用于返回此二维列表，其中所有图块位于起始位置，右下角为空白空间。Python可以用a, b = b, a 这样的语句交换两个变量的值。以下程序在第100~107行使用该语句规则交换空白空间和相邻的图块，模拟将编号图块滑入空白空间的操作。getNewPuzzle()函数通过随机执行200次这样的交换来生成新的拼图。代码如下：

```
 1. """滑动拼图，作者：Al Sweigart al@inventwithpython.com
 2. 让有编号的图块按正确顺序排列
 3. 标签：大，游戏，谜题"""
 4.
 5. import random, sys
 6.
 7. BLANK = '  '  # 注意：这个字符串是两个空格，而不是一个
 8.
 9.
10. def main():
11.     print('''Sliding Tile Puzzle, by Al Sweigart al@inventwithpython.com
12.
13.     Use the WASD keys to move the tiles
14.     back into their original order:
15.           1  2  3  4
16.           5  6  7  8
17.           9 10 11 12
18.          13 14 15     ''')
19.     input('Press Enter to begin...')
20.
21.     gameBoard = getNewPuzzle()
22.
23.     while True:
24.         displayBoard(gameBoard)
25.         playerMove = askForPlayerMove(gameBoard)
```

```
26.        makeMove(gameBoard, playerMove)
27.
28.        if gameBoard == getNewBoard():
29.            print('You won!')
30.            sys.exit()
31.
32.
33. def getNewBoard():
34.     """返回一个代表新拼图的二维列表"""
35.     return [['1 ', '5 ', '9 ', '13'], ['2 ', '6 ', '10', '14'],
36.             ['3 ', '7 ', '11', '15'], ['4 ', '8 ', '12', BLANK]]
37.
38.
39. def displayBoard(board):
40.     """在屏幕上显示给定的游戏板"""
41.     labels = [board[0][0], board[1][0], board[2][0], board[3][0],
42.               board[0][1], board[1][1], board[2][1], board[3][1],
43.               board[0][2], board[1][2], board[2][2], board[3][2],
44.               board[0][3], board[1][3], board[2][3], board[3][3]]
45.     boardToDraw = """
46. +------+------+------+------+
47. |      |      |      |      |
48. |  {}  |  {}  |  {}  |  {}  |
49. |      |      |      |      |
50. +------+------+------+------+
51. |      |      |      |      |
52. |  {}  |  {}  |  {}  |  {}  |
53. |      |      |      |      |
54. +------+------+------+------+
55. |      |      |      |      |
56. |  {}  |  {}  |  {}  |  {}  |
57. |      |      |      |      |
58. +------+------+------+------+
59. |      |      |      |      |
60. |  {}  |  {}  |  {}  |  {}  |
61. |      |      |      |      |
62. +------+------+------+------+
63. """.format(*labels)
64.     print(boardToDraw)
65.
66.
67. def findBlankSpace(board):
68.     """返回空白空间位置的(x, y)元组"""
69.     for x in range(4):
70.         for y in range(4):
71.             if board[x][y] == '  ':
72.                 return (x, y)
73.
74.
75. def askForPlayerMove(board):
76.     """让玩家选择要滑动的图块"""
77.     blankx, blanky = findBlankSpace(board)
78.
79.     w = 'W' if blanky != 3 else ' '
80.     a = 'A' if blankx != 3 else ' '
81.     s = 'S' if blanky != 0 else ' '
82.     d = 'D' if blankx != 0 else ' '
83.
84.     while True:
85.         print('                        ({})'.format(w))
86.         print('Enter WASD (or QUIT): ({}) ({}) ({})'.format(a, s, d))
87.
```

```
 88.         response = input('> ').upper()
 89.         if response == 'QUIT':
 90.             sys.exit()
 91.         if response in (w + a + s + d).replace(' ', ''):
 92.             return response
 93.
 94.
 95. def makeMove(board, move):
 96.     """在给定的游戏板上执行给定的走法"""
 97.     # 注意：这个函数会假设下一步走法是有效的
 98.     bx, by = findBlankSpace(board)
 99.
100.     if move == 'W':
101.         board[bx][by], board[bx][by + 1] = board[bx][by + 1], board[bx][by]
102.     elif move == 'A':
103.         board[bx][by], board[bx + 1][by] = board[bx + 1][by], board[bx][by]
104.     elif move == 'S':
105.         board[bx][by], board[bx][by - 1] = board[bx][by - 1], board[bx][by]
106.     elif move == 'D':
107.         board[bx][by], board[bx - 1][by] = board[bx - 1][by], board[bx][by]
108.
109.
110. def makeRandomMove(board):
111.     """向随机方向移动拼图"""
112.     blankx, blanky = findBlankSpace(board)
113.     validMoves = []
114.     if blanky != 3:
115.         validMoves.append('W')
116.     if blankx != 3:
117.         validMoves.append('A')
118.     if blanky != 0:
119.         validMoves.append('S')
120.     if blankx != 0:
121.         validMoves.append('D')
122.
123.     makeMove(board, random.choice(validMoves))
124.
125.
126. def getNewPuzzle(moves=200):
127.     """随机移动已拼完的拼图，以获得一个新拼图"""
128.     board = getNewBoard()
129.
130.     for i in range(moves):
131.         makeRandomMove(board)
132.     return board
133.
134.
135. # 程序运行入口（如果不是作为模块导入的话）
136. if __name__ == '__main__':
137.     main()
```

输入源代码并运行几次后，请试着对其进行更改。你也可以尝试以下操作。

- 创建一个更难的 5×5 的滑动拼图。
- 创建一种"自动解决"模式，使之可以保存当前的图块排列状态，然后尝试不超过 40 次的随机移动。如果在 40 次以内拼完，则停止；否则，会加载保存的状态，并让玩家尝试另外 40 次随机移动，直到拼完。

探索程序

请尝试找出以下问题的答案。你需要尝试对代码进行一些修改，再次运行程序，查看修改后的效果。

1. 如果将第 21 行的 getNewPuzzle() 改为 getNewPuzzle(1)，会发生什么？
2. 如果将第 21 行的 getNewPuzzle() 改为 getNewPuzzle(0)，会发生什么？
3. 如果删除或注释掉第 30 行的 sys.exit()，会发生什么？

项目 69

蜗牛赛跑

蜗牛赛跑的"快节奏"刺激着实让人抓狂,不过,如果用可爱的ASCII艺术画来展示,或许会让你觉得这种"快节奏"刺激没那么让人"如坐针毡"。每只蜗牛(由@字符代表壳,v代表两个眼柄)缓慢但坚定地向终点线移动。最多可以有8只蜗牛参赛,每只蜗牛都有可定制的名字,它们的身后会留下一条黏液痕迹。本项目很适合初学者动手实践。

运行程序

运行 snailrace.py,输出如下所示:

```
Snail Race, by Al Sweigart al@inventwithpython.com

@v <-- snail

How many snails will race? Max: 8
> 3
Enter snail #1's name:
> Alice
Enter snail #2's name:
> Bob
Enter snail #3's name:
> Carol
START                           FINISH
|                               |
      Alice
......@v
      Bob
.....@v
        Carol
........@v
--snip--
```

工作原理

以下程序会用到两个数据结构(存储在两个变量中)。其中,`snailNames` 是一个字符串列表,用于存储蜗牛的名字。`snailProgress` 是一个字典,其键表示蜗牛的名字,其值则表示蜗牛移动了多少步。第77～80行代码用于读取这两个变量中的数据，以在计算机屏幕的合适位置绘制蜗牛。

```
1.  """蜗牛赛跑，作者：Al Sweigart al@inventwithpython.com
2.  "快节奏"的蜗牛赛跑行动
3.  标签：简短，艺术，初学者，游戏，多人游戏"""
4.
5.  import random, time, sys
6.
7.  # 创建常量
8.  MAX_NUM_SNAILS = 8
9.  MAX_NAME_LENGTH = 20
10. FINISH_LINE = 40  # (!) 请试着修改这个数字
11.
12. print('''Snail Race, by Al Sweigart al@inventwithpython.com
13.
14.     @v <-- snail
15.
16. ''')
17.
18. # 指定参赛蜗牛的数量
19. while True:  # 继续询问，直到玩家输入一个数字
20.     print('How many snails will race? Max:', MAX_NUM_SNAILS)
21.     response = input('> ')
22.     if response.isdecimal():
23.         numSnailsRacing = int(response)
24.         if 1 < numSnailsRacing <= MAX_NUM_SNAILS:
25.             break
26.     print('Enter a number between 2 and', MAX_NUM_SNAILS)
27.
28. # 输入每只蜗牛的名字
29. snailNames = []  # 用于存放蜗牛名字的列表
30. for i in range(1, numSnailsRacing + 1):
31.     while True:  # 继续询问，直到玩家输入有效的名字
32.         print('Enter snail #' + str(i) + "'s name:")
33.         name = input('> ')
34.         if len(name) == 0:
35.             print('Please enter a name.')
36.         elif name in snailNames:
37.             print('Choose a name that has not already been used.')
38.         else:
39.             break  # 输入的名字是有效的
40.     snailNames.append(name)
41.
42. # 在起跑线上显示每只蜗牛
43. print('\n' * 40)
44. print('START' + (' ' * (FINISH_LINE - len('START')) + 'FINISH'))
45. print('|' + (' ' * (FINISH_LINE - len('|')) + '|'))
46. snailProgress = {}
47. for snailName in snailNames:
48.     print(snailName[:MAX_NAME_LENGTH])
49.     print('@v')
50.     snailProgress[snailName] = 0
51.
52. time.sleep(1.5)  # 比赛开始前的倒计时
53.
54. while True:  # 主循环
55.     # 随机挑选蜗牛使其向前移动
56.     for i in range(random.randint(1, numSnailsRacing // 2)):
57.         randomSnailName = random.choice(snailNames)
58.         snailProgress[randomSnailName] += 1
```

```
59.
60.         # 检查蜗牛是否已经到达终点线
61.         if snailProgress[randomSnailName] == FINISH_LINE:
62.             print(randomSnailName, 'has won!')
63.             sys.exit()
64.
65.     # (!) 实验：给蜗牛冠以你的名字，并添加一个小窍门，让它前进得更快
66.
67.     time.sleep(0.5)   # (!) 实验：尝试将 0.5 改为其他值
68.
69.     # (!) 实验：如果把这行代码注释掉，会发生什么？
70.     print('\n' * 40)
71.
72.     # 显示起跑线和终点线
73.     print('START' + (' ' * (FINISH_LINE - len('START')) + 'FINISH'))
74.     print('|' + (' ' * (FINISH_LINE - 1) + '|'))
75.
76.     # 显示蜗牛(附有名字标签)
77.     for snailName in snailNames:
78.         spaces = snailProgress[snailName]
79.         print((' ' * spaces) + snailName[:MAX_NAME_LENGTH])
80.         print(('.' * snailProgress[snailName]) + '@v')
```

输入源代码并运行几次后，请试着对其进行更改。你可以根据标有!的注释对程序进行修改，也可以尝试以下操作。

- 添加随机的"速度提升"功能，将蜗牛向前快速移动 4 步而不是 1 步。
- 添加蜗牛在比赛中可以随机进入"睡眠模式"的功能。这种模式会使它们停下来少移动几步，并在其旁边加上"zzz"。
- 添加对终点的限制，以防蜗牛同时到达终点线。

探索程序

请尝试找出以下问题的答案。你需要尝试对代码进行一些修改，再次运行程序，查看修改后的效果。

1. 如果将第 79 行的 `snailName[:MAX_NAME_LENGTH]` 改为 `snailNames[0]`，则会发生什么？
2. 如果将第 49 行的 `print('@v')` 改为 `print('v@')`，则会发生什么？

| 项目 70

虚拟算盘

早在电子计算器发明很多年之前，算盘这种计算工具就已广泛应用于世界上的许多国家和地区。下方的虚拟算盘，每一列代表 1 个数位，其上的算珠代表该数位的数字。例如，如果最右列有 2 颗算珠被移到上方，同时次右列有 3 颗算珠被移到上方，就代表数字 32。在本项目中，我们将实现用于模拟算盘珠算的程序。（我知道使用计算机来模拟计算机出现之前的计算工具的运算颇具讽刺意味。）

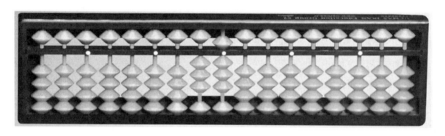

算盘的每一列代表不同的数位。最右边的一列代表个位，其左边的一列代表十位，再左边的一列代表百位，以此类推。计算机键盘上部的 Q、W、E、R、T、Y、U、I、O 和 P 键可以用于增加对应数位的数字，而 A、S、D、F、G、H、J、K、L 和；键则可以用于减少对应数位的数字。虚拟算盘上的算珠可以滑动，用以反映当前的数字。你也可以在程序中直接输入数字。

水平分隔线下方的 4 颗算珠是 "地珠"，向分隔线方向拨起一个算珠则该数位加 1。水平分隔线上方的算珠是 "天珠"，向分隔线方向拨下 1 颗算珠则该数位加 5。所以，在十位列中向下拨动 1 颗天珠，同时向上拨动 3 颗地珠，则表示数字 80。

运行程序

运行 soroban.py，输出如下所示：

```
Soroban - The Japanese Abacus
By Al Sweigart al@inventwithpython.com

 +===============================+
 I O O O O O O O O O O I
 I | | | | | | | | | | I
 I | | | | | | | | | | I
```

```
+==============================+
I   |   |   |   |   |   |   |   |   |   I
I   |   |   |   |   |   |   |   |   |   I
I   O   O   O   O   O   O   O   O   O   I
I   O   O   O   O   O   O   O   O   O   I
I   O   O   O   O   O   O   O   O   O   I
I   O   O   O   O   O   O   O   O   O   I
+==O==O==O==O==O==O==O==O==O==O==+
  +q  w  e  r  t  y  u  i  o  p
  -a  s  d  f  g  h  j  k  l  ;
(Enter a number, "quit", or a stream of up/down letters.)
> pppiiiii
+==============================+
I   O   O   O   O   O   O   |   O   O   I
I   |   |   |   |   |   |   |   |   |   I
I   |   |   |   |   |   |   O   |   |   I
+==============================+
I   |   |   |   |   |   |   |   |   O   I
I   |   |   |   |   |   |   |   |   O   I
I   O   O   O   O   O   O   O   O   O   I
I   O   O   O   O   O   O   O   |   I
I   O   O   O   O   O   O   O   |   I
I   O   O   O   O   O   O   O   O   I
+==O==O==O==O==O==O==O==5==O==3==+
  +q  w  e  r  t  y  u  I  o  p
  -a  s  d  f  g  h  j  k  l  ;
(Enter a number, "quit", or a stream of up/down letters.)
--snip--
```

工作原理

在本项目的程序中，我们用 `displayAbacus()` 函数接收一个参数 `number`，以确定应该在虚拟算盘上呈现算珠的位置。虚拟算盘总是有 90 个可能的位置要么为 "O" 珠，要么为 "|" 表示的列，如第 125~137 行的多行字符串中的花括号所示。第 137 行的 10 个花括号用于表示 `number` 数字本身。

我们需要创建一个字符串列表来填充（其实是替换）这些花括号，填充顺序是从左到右、从上到下。`displayAbacus()` 中的代码根据 `hasBead` 列表中的 `True` 值显示 "O" 算珠，`False` 值显示 "|"。该列表中的前 10 个值用于顶部 "天" 行。如果该列的数字为 0、1、2、3 或 4，将在该行中放置一个算珠，因为除非该列的数字为 0~4，否则天珠不会出现在该行中。接下来，我们为剩余的行继续向 `hasBead` 中添加布尔值。

第 116~121 行使用 `hasBead` 创建一个包含实际'O'和'|'字符串的 `abacusChar` 列表，该列表在与第 124 行的 `numberList` 结合后，形成一个 `chars` 列表，`chars` 列表中的元素被用于替换珠算多行字符串 ASCII 艺术画的花括号。

```
1. """虚拟算盘，作者：Al Sweigart al@inventwithpython.com
2. 算盘的运算模拟
3. 标签：大，艺术，数学，模拟"""
4.
5. NUMBER_OF_DIGITS = 10
6.
7.
8. def main():
```

```python
  9.  print('Soroban - The Japanese Abacus')
 10.  print('By Al Sweigart al@inventwithpython.com')
 11.  print()
 12.
 13.  abacusNumber = 0   # 这表示算盘上的数字
 14.
 15.  while True:   # 主循环
 16.      displayAbacus(abacusNumber)
 17.      displayControls()
 18.
 19.      commands = input('> ')
 20.      if commands == 'quit':
 21.          # 退出当前程序
 22.          break
 23.      elif commands.isdecimal():
 24.          # 设置算盘表示的数字
 25.          abacusNumber = int(commands)
 26.      else:
 27.          # 处理递增/递减命令
 28.          for letter in commands:
 29.              if letter == 'q':
 30.                  abacusNumber += 1000000000
 31.              elif letter == 'a':
 32.                  abacusNumber -= 1000000000
 33.              elif letter == 'w':
 34.                  abacusNumber += 100000000
 35.              elif letter == 's':
 36.                  abacusNumber -= 100000000
 37.              elif letter == 'e':
 38.                  abacusNumber += 10000000
 39.              elif letter == 'd':
 40.                  abacusNumber -= 10000000
 41.              elif letter == 'r':
 42.                  abacusNumber += 1000000
 43.              elif letter == 'f':
 44.                  abacusNumber -= 1000000
 45.              elif letter == 't':
 46.                  abacusNumber += 100000
 47.              elif letter == 'g':
 48.                  abacusNumber -= 100000
 49.              elif letter == 'y':
 50.                  abacusNumber += 10000
 51.              elif letter == 'h':
 52.                  abacusNumber -= 10000
 53.              elif letter == 'u':
 54.                  abacusNumber += 1000
 55.              elif letter == 'j':
 56.                  abacusNumber -= 1000
 57.              elif letter == 'i':
 58.                  abacusNumber += 100
 59.              elif letter == 'k':
 60.                  abacusNumber -= 100
 61.              elif letter == 'o':
 62.                  abacusNumber += 10
 63.              elif letter == 'l':
 64.                  abacusNumber -= 10
 65.              elif letter == 'p':
 66.                  abacusNumber += 1
 67.              elif letter == ';':
 68.                  abacusNumber -= 1
 69.
 70.      # 算盘不能显示负数
```

```
71.         if abacusNumber < 0:
72.             abacusNumber = 0  # 将所有负数改为 0
73.         # 算盘不能显示大于 9999999999 的数字
74.         if abacusNumber > 9999999999:
75.             abacusNumber = 9999999999
76.
77.
78.  def displayAbacus(number):
79.      numberList = list(str(number).zfill(NUMBER_OF_DIGITS))
80.
81.      hasBead = []  # : 存储了每颗算珠是否在该位置的布尔值
82.
83.      # 顶部的天珠表示数字 0、1、2、3 和 4
84.      for i in range(NUMBER_OF_DIGITS):
85.          hasBead.append(numberList[i] in '01234')
86.
87.      # 底部的地珠表示数字 5、6、7、8 和 9
88.      for i in range(NUMBER_OF_DIGITS):
89.          hasBead.append(numberList[i] in '56789')
90.
91.      # 第 1 行（最上面）的地珠表示除 0 以外的所有数字
92.      for i in range(NUMBER_OF_DIGITS):
93.          hasBead.append(numberList[i] in '12346789')
94.
95.      # 第 2 行地珠表示数字 2、3、4、7、8 和 9
96.      for i in range(NUMBER_OF_DIGITS):
97.          hasBead.append(numberList[i] in '234789')
98.
99.      # 第 3 行地珠表示数字 0、3、4、5、8 和 9
100.     for i in range(NUMBER_OF_DIGITS):
101.         hasBead.append(numberList[i] in '034589')
102.
103.     # 第 4 行地珠表示数字 0、1、2、4、5、6 和 9
104.     for i in range(NUMBER_OF_DIGITS):
105.         hasBead.append(numberList[i] in '014569')
106.
107.     # 第 5 行地珠表示数字 0、1、2、5、6 和 7
108.     for i in range(NUMBER_OF_DIGITS):
109.         hasBead.append(numberList[i] in '012567')
110.
111.     # 第 6 行地珠表示数字 0、1、2、3、5、6、7 和 8
112.     for i in range(NUMBER_OF_DIGITS):
113.         hasBead.append(numberList[i] in '01235678')
114.
115.     # 将 True 或 False 值转换为 O 或 | 字符
116.     abacusChar = []
117.     for i, beadPresent in enumerate(hasBead):
118.         if beadPresent:
119.             abacusChar.append('O')
120.         else:
121.             abacusChar.append('|')
122.
123.     # 用 O 或 | 字符绘制算盘
124.     chars = abacusChar + numberList
125.     print("""
126. +================================+
127. I {} {} {} {} {} {} {} {} {} {} I
128. I  |  |  |  |  |  |  |  |  |  | I
129. I {} {} {} {} {} {} {} {} {} {} I
130. +================================+
131. I {} {} {} {} {} {} {} {} {} {} I
132. I {} {} {} {} {} {} {} {} {} {} I
```

```
133. I  {}  {}  {}  {}  {}  {}  {}  {}  {}  I
134. I  {}  {}  {}  {}  {}  {}  {}  {}  {}  I
135. I  {}  {}  {}  {}  {}  {}  {}  {}  {}  I
136. I  {}  {}  {}  {}  {}  {}  {}  {}  {}  I
137. +=={}=={}=={}=={}=={}=={}=={}=={}=={}==+""".format(*chars))
138.
139.
140. def displayControls():
141.     print('  +q w e r t y u i o p')
142.     print('  -a s d f g h j k l ;')
143.     print('(Enter a number, "quit", or a stream of up/down letters.)')
144.
145.
146. if __name__ == '__main__':
147.     main()
```

探索程序

请尝试找出以下问题的答案。你需要尝试对代码进行一些修改，再次运行程序，查看修改后的效果。

1. 如果将第 13 行的 abacusNumber = 0 改为 abacusNumber = 9999，则会发生什么？
2. 如果将第 119 行的 abacusChar.append('O') 改为 abacusChar.append('@')，则会发生什么？

项目 71

声音模拟

在本项目中，我们用第三方 playsound 模块模拟 4 种不同的声音（分别对应键盘上的 A、S、D 和 F 键）。一旦成功重复游戏提供的模式，这串字母会变得越来越长。通过短期记忆的方式，你能记住多少种声音？

如果查看代码，你会发现 playsound.playsound() 函数被传入了要播放的声音文件名。

运行程序前，请将声音文件与 soundmimic.py 放在同一文件夹中。如需了解有关 playsound 模块的更多信息，请访问 Python 官方网站。使用 macOS 的用户，请务必下载并安装 pyobjc 模块，才能让 playsound 模块正常工作。

运行程序

运行 soundmimic.py，输出如下所示：

```
Sound Mimic, by Al Sweigart al@inventwithpython.com
Try to memorize a pattern of A S D F letters (each with its own sound)
as it gets longer and longer.
Press Enter to begin...
<screen clears>
Pattern: S
<screen clears>
Enter the pattern:
> s
Correct!
<screen clears>
Pattern: S F
<screen clears>
Enter the pattern:
> sf
Correct!
<screen clears>
Pattern: S F F
<screen clears>
Enter the pattern:
> sff
Correct!
<screen clears>
Pattern: S F F D
--snip--
```

注意，如遇到"指定设备未打开，或不被 MCI 所识别"之类的错误，请将 `playsound` 模块第 55 行代码 `command = ' '.join(command).encode('utf-16')` 改为 `command = ' '.join(command)`，方可运行。

工作原理

在本项目中，我们导入了 `playsound` 模块，以便播放声音文件。`playsound` 模块中有一个 `playsound()` 函数，我们可以将 .wav 或 .mp3 文件的名称传递给该函数进行播放。在游戏的每一轮中，程序将随机选择的字母（A、S、D 或 F）附加到模式列表中，并播放该列表中的声音。随着模式列表变长，播放器必须记住的声音文件的模式也会变长。

```python
 1. """声音模拟，作者：Al Sweigart al@inventwithpython.com
 2. 有声模式匹配游戏。请试着记住越来越多的字母
 3. 受到西蒙电子游戏的启发
 4. 标签：简短，初学者，游戏"""
 5.
 6. import random, sys, time
 7.
 8. # 从下列 URL 下载声音文件（或使用你自己的）
 9. # https://inventwithpython.com/soundA.wav
10. # https://inventwithpython.com/soundS.wav
11. # https://inventwithpython.com/soundD.wav
12. # https://inventwithpython.com/soundF.wav
13.
14. try:
15.     import playsound
16. except ImportError:
17.     print('The playsound module needs to be installed to run this')
18.     print('program. On Windows, open a Command Prompt and run:')
19.     print('pip install playsound')
20.     print('On macOS and Linux, open a Terminal and run:')
21.     print('pip3 install playsound')
22.     sys.exit()
23.
24. print('''Sound Mimic, by Al Sweigart al@inventwithpython.com
25. Try to memorize a pattern of A S D F letters (each with its own sound)
26. as it gets longer and longer.''')
27.
28. input('Press Enter to begin...')
29.
30. pattern = ''
31. while True:
32.     print('\n' * 60)  # 输出空行来清空屏幕
33.
34.     # 在模式中添加一个随机字母
35.     pattern = pattern + random.choice('ASDF')
36.
37.     # 显示模式（并播放它们的声音）
38.     print('Pattern: ', end='')
39.     for letter in pattern:
40.         print(letter, end=' ', flush=True)
41.         playsound.playsound('sound' + letter + '.wav')
42.
43.     time.sleep(1)  # 在结尾添加一个短暂的停顿
44.     print('\n' * 60)  # 通过输出空行来清空屏幕
45.
```

```
46.    # 让玩家输入模式
47.    print('Enter the pattern:')
48.    response = input('> ').upper()
49.
50.    if response != pattern:
51.        print('Incorrect!')
52.        print('The pattern was', pattern)
53.    else:
54.        print('Correct!')
55.
56.    for letter in pattern:
57.        playsound.playsound('sound' + letter + '.wav')
58.
59.    if response != pattern:
60.        print('You scored', len(pattern) - 1, 'points.')
61.        print('Thanks for playing!')
62.        break
63.
64.    time.sleep(1)
```

探索程序

请尝试找出以下问题的答案。你需要尝试对代码进行一些修改，再次运行程序，查看修改后的效果。

1. 如果删除或注释掉第 44 行的 `print('\n' * 60)`，会发生什么？
2. 如果将第 59 行的 `response != pattern` 改为 `False`，会发生什么？

| 项目 72

"海绵宝宝的嘲弄"

你可能见过"海绵宝宝的嘲弄"表情包：一张海绵宝宝的图片，上面有一个标题，文字在大小写字母之间切换，以强调嘲弄的语气，比如，uSiNg SpOnGeBoB MeMeS dOeS NoT mAkE YoU wItTy（意思是使用海绵宝宝表情包不会让你变得机智）。出于随机性原因，文本并不严格交替大小写。

本项目的程序使用 upper() 和 lower() 字符串方法将你的消息转换为"海绵宝宝的嘲弄"表情包。其他程序也可以使用 import spongecase 将其作为模块导入，并调用 spongecase.englishToSpongecase() 函数。

运行程序

运行 spongecase.py，输出如下所示：

```
sPoNgEcAsE, bY aL sWeIGaRt Al@iNvEnTwItHpYtHoN.cOm

eNtEr YoUr MeSsAgE:
> Using SpongeBob memes does not make you witty.

uSiNg SpOnGeBoB MeMeS dOeS NoT mAkE YoU wItTy.
(cOpIed SpOnGeTexT to ClIpbOaRd.)
```

工作原理

在本项目中，我们在第 34 行用 for 循环来迭代消息字符串中的每个字符。useUpper 变量包含一个布尔值，用于指示将字符转为大写（如果为 True）或小写（如果为 False）形式。第 45 行和第 46 行在 90%的迭代中切换 useUpper 中的布尔值（将其取反）。这意味着结果几乎总是在大写和小写字母之间切换。

```
 1. """海绵宝宝的嘲弄，作者: Al Sweigart al@inventwithpython.com
 2. 将英文信息翻译为"海绵宝宝的嘲弄"
 3. 标签：小，初学者，文本"""
 4.
 5. import random
 6.
 7. try:
 8.     import pyperclip  # pyperclip 模块能够将文本复制到剪贴板
 9. except ImportError:
```

```
10.     pass  # pyperclip 模块并不是必需的，不安装的话也没有什么影响
11.
12.
13. def main():
14.     """以文本形式运行程序"""
15.     print('''sPoNgEtExT, bY aL sWeIGaRt Al@iNvEnTwItHpYtHoN.cOm
16.
17. eNtEr YoUr MeSsAgE:''')
18.     spongecase = englishToSpongecase(input('> '))
19.     print()
20.     print(spongecase)
21.
22.     try:
23.         pyperclip.copy(spongecase)
24.         print('(cOpIed SpOnGeCasE to ClIpbOaRd.)')
25.     except:
26.         pass  # 如果没有安装 pyperclip，则不用执行任何操作
27.
28.
29. def englishToSpongecase(message):
30.     """返回给定字符串的"海绵宝宝的嘲弄"文本格式"""
31.     spongecase = ''
32.     useUpper = False
33.
34.     for character in message:
35.         if not character.isalpha():
36.             spongecase += character
37.             continue
38.
39.         if useUpper:
40.             spongecase += character.upper()
41.         else:
42.             spongecase += character.lower()
43.
44.         # 转换大小写的概率为 90%
45.         if random.randint(1, 100) <= 90:
46.             useUpper = not useUpper  # 转换大小写
47.     return spongecase
48.
49.
50. # 程序运行入口（如果不是作为模块导入的话）
51. if __name__ == '__main__':
52.     main()
```

探索程序

请尝试找出以下问题的答案。你需要尝试对代码进行一些修改，再次运行程序，查看修改后的效果。

1. 如果将第 45 行的 `random.randint(1, 100)` 改为 `random.randint(80, 100)`，则会发生什么？
2. 如果删除或注释掉第 46 行的 `useUpper = not useUpper`，则会发生什么？

项目 73

数独

数独是非常流行的益智游戏，常见于报纸或 App。数独的盘面是一个 9×9 的网格，玩家必须在其中将数字 1~9 各填 1 次，并且只能在每行、每列和 3×3 的子网格中各填 1 次。游戏从一些已经用数字（称为给定值）填充的空格开始。格式标准的数独谜题仅存在一个有效的答案。

运行程序

运行 sudoku.py，输出如下所示：

```
Sudoku Puzzle, by Al Sweigart al@inventwithpython.com
--snip--
  A B C   D E F   G H I
1 . . .  | . . . | . . .
2 . 7 9  | . 5 . | 1 8 .
3 8 . .  | . . . | . . 7
  ------+-------+------
4 . . 7  | 3 . 6 | 8 . .
5 4 5 .  | 7 . 8 | . 9 6
6 . . 3  | 5 . 2 | 7 . .
  ------+-------+------
7 7 . .  | . . . | . . 5
8 . 1 6  | . 3 . | 4 2 .
9 . . .  | . . . | . . .
Enter a move, or RESET, NEW, UNDO, ORIGINAL, or QUIT:
(For example, a move looks like "B4 9".)
--snip--
```

工作原理

SudokuGrid 类的对象是表示数独网格的数据结构，我们可以通过调用它提供的方法来修改或检索有关网格的信息。例如，makeMove() 方法用于在网格上填数字，resetGrid() 方法用于将网格恢复到原始状态，如果答案数字都已填写到网格中，则 isSolved() 方法返回 True。

从第 134 行开始是程序的主要部分，使用一个 SudokuGrid 对象及其方法创建游戏。你也可以将该类复制并粘贴到自己创建的其他数独程序中，重用其功能。

```python
 1. """数独,作者: Al Sweigart al@inventwithpython.com
 2. 经典的 9x9 数字填空游戏
 3. 标签:大,游戏,面向对象,谜题"""
 4.
 5. import copy, random, sys
 6.
 7. # 这个游戏需要一个存储谜题的 sudokupuzzle.txt 文件
 8. # 以下是此文件中的内容示例
 9. # ..3.2.6..9..3.5...1..18.64....81.29..7.......8..67.82....26.95..8..2.3..9..5.1.3..
10. # 2...8.3...6..7..84.3.5..2.9...1.54.8.........4.27.6...3.1..7.4.72.4..6...4.1...3
11. # ......9.7...42.18....7.5.261..9.4...5....4....5.7..992.1.8....34.59...5.7......
12. # .3..5..4...8.1.5..46....12.7.5.2.8....6.3....4.1.9.3.25.....98..1.2.6...8..6..2.
13.
14. # 创建常量
15. EMPTY_SPACE = '.'
16. GRID_LENGTH = 9
17. BOX_LENGTH = 3
18. FULL_GRID_SIZE = GRID_LENGTH * GRID_LENGTH
19.
20.
21. class SudokuGrid:
22.     def __init__(self, originalSetup):
23.         # originalSetup 是用于谜题设置的 81 个字符的字符串,
24.         # 带有数字和句点(用于表示空格)
25.         self.originalSetup = originalSetup
26.
27.         # 数独网格的状态由字典表示,该字典具有(x,y) 键和数字值(作为字符串)
28.         self.grid = {}
29.         self.resetGrid()    # 将网格状态设置为原始设置
30.         self.moves = []     # 跟踪每次移动,以便实现恢复功能
31.
32.     def resetGrid(self):
33.         """将 self.grid 跟踪的网格状态重置为
34.         self.originalSetup 中的状态"""
35.         for x in range(1, GRID_LENGTH + 1):
36.             for y in range(1, GRID_LENGTH + 1):
37.                 self.grid[(x, y)] = EMPTY_SPACE
38.
39.         assert len(self.originalSetup) == FULL_GRID_SIZE
40.         i = 0  # i 为 0~80
41.         y = 0  # y 为 0~8
42.         while i < FULL_GRID_SIZE:
43.             for x in range(GRID_LENGTH):
44.                 self.grid[(x, y)] = self.originalSetup[i]
45.                 i += 1
46.             y += 1
47.
48.     def makeMove(self, column, row, number):
49.         """将数字放在网格上的列(从 A 到 I 的字母)和行
50.         (从 1 到 9 的整数)处"""
51.         x = 'ABCDEFGHI'.find(column)    # 将其转换为整数
52.         y = int(row) - 1
53.
54.         # 检查移动是否在"给定"号码上进行
55.         if self.originalSetup[y * GRID_LENGTH + x] != EMPTY_SPACE:
56.             return False
57.
58.         self.grid[(x, y)] = number    # 将此数字放在网格上
59.
60.         # 我们需要单独存储字典对象的副本
61.         self.moves.append(copy.copy(self.grid))
62.         return True
63.
64.     def undo(self):
```

```
65.          """将当前网格状态设置为 self.moves 列表中的前一个状态"""
66.          if self.moves == []:
67.              return  # self.moves 中没有状态, 所以不执行任何操作
68.
69.          self.moves.pop()  # 删除当前状态
70.
71.          if self.moves == []:
72.              self.resetGrid()
73.          else:
74.              # 将网格状态恢复到上一步
75.              self.grid = copy.copy(self.moves[-1])
76.
77.      def display(self):
78.          """在屏幕上显示网格的当前状态"""
79.          print('  A B C  D E F  G H I')  # 显示列标签
80.          for y in range(GRID_LENGTH):
81.              for x in range(GRID_LENGTH):
82.                  if x == 0:
83.                      # 显示行标签
84.                      print(str(y + 1) + ' ', end='')
85.
86.                  print(self.grid[(x, y)] + ' ', end='')
87.                  if x == 2 or x == 5:
88.                      # 显示一条垂直线
89.                      print('| ', end='')
90.              print()  # 输出一行空行
91.
92.              if y == 2 or y == 5:
93.                  # 显示一条水平线
94.                  print('   ------+------+------')
95.
96.      def _isCompleteSetOfNumbers(self, numbers):
97.          """如果包含数字 1~9, 则返回 True"""
98.          return sorted(numbers) == list('123456789')
99.
100.     def isSolved(self):
101.         """如果当前网格处于已填写的状态, 则返回 True"""
102.         # 判断每一行的数字是否正确
103.         for row in range(GRID_LENGTH):
104.             rowNumbers = []
105.             for x in range(GRID_LENGTH):
106.                 number = self.grid[(x, row)]
107.                 rowNumbers.append(number)
108.             if not self._isCompleteSetOfNumbers(rowNumbers):
109.                 return False
110.
111.         # 判断每一列的数字是否正确
112.         for column in range(GRID_LENGTH):
113.             columnNumbers = []
114.             for y in range(GRID_LENGTH):
115.                 number = self.grid[(column, y)]
116.                 columnNumbers.append(number)
117.             if not self._isCompleteSetOfNumbers(columnNumbers):
118.                 return False
119.
120.         # 判断每个网格中的数字是否正确
121.         for boxx in (0, 3, 6):
122.             for boxy in (0, 3, 6):
123.                 boxNumbers = []
124.                 for x in range(BOX_LENGTH):
125.                     for y in range(BOX_LENGTH):
126.                         number = self.grid[(boxx + x, boxy + y)]
127.                         boxNumbers.append(number)
128.                     if not self._isCompleteSetOfNumbers(boxNumbers):
```

```
129.              return False
130.
131.          return True
132.
133.
134.  print('''Sudoku Puzzle, by Al Sweigart al@inventwithpython.com
135.
136.  Sudoku is a number placement logic puzzle game. A Sudoku grid is a 9x9
137.  grid of numbers. Try to place numbers in the grid such that every row,
138.  column, and 3x3 box has the numbers 1 through 9 once and only once.
139.
140.  For example, here is a starting Sudoku grid and its solved form:
141.
142.      5 3 . | . 7 . | . . .          5 3 4 | 6 7 8 | 9 1 2
143.      6 . . | 1 9 5 | . . .          6 7 2 | 1 9 5 | 3 4 8
144.      . 9 8 | . . . | . 6 .          1 9 8 | 3 4 2 | 5 6 7
145.      ------+-------+------          ------+-------+------
146.      8 . . | . 6 . | . . 3          8 5 9 | 7 6 1 | 4 2 3
147.      4 . . | 8 . 3 | . . 1  -->     4 2 6 | 8 5 3 | 7 9 1
148.      7 . . | . 2 . | . . 6          7 1 3 | 9 2 4 | 8 5 6
149.      ------+-------+------          ------+-------+------
150.      . 6 . | . . . | 2 8 .          9 6 1 | 5 3 7 | 2 8 4
151.      . . . | 4 1 9 | . . 5          2 8 7 | 4 1 9 | 6 3 5
152.      . . . | . 8 . | . 7 9          3 4 5 | 2 8 6 | 1 7 9
153.  ''')
154.  input('Press Enter to begin...')
155.
156.
157.  # 加载 sudokupuzzles.txt 文件
158.  with open('sudokupuzzles.txt') as puzzleFile:
159.      puzzles = puzzleFile.readlines()
160.
161.  # 删除每个谜题末尾的换行符
162.  for i, puzzle in enumerate(puzzles):
163.      puzzles[i] = puzzle.strip()
164.
165.  grid = SudokuGrid(random.choice(puzzles))
166.
167.  while True:  # 主循环
168.      grid.display()
169.
170.      # 检查谜题是否已解决
171.      if grid.isSolved():
172.          print('Congratulations! You solved the puzzle!')
173.          print('Thanks for playing!')
174.          sys.exit()
175.
176.      # 获取玩家的下一步动作
177.      while True:  # 继续询问，直到玩家输入有效的下一步动作
178.          print()  # 输出换行符
179.          print('Enter a move, or RESET, NEW, UNDO, ORIGINAL, or QUIT:')
180.          print('(For example, a move looks like "B4 9".)')
181.
182.          action = input('> ').upper().strip()
183.
184.          if len(action) > 0 and action[0] in ('R', 'N', 'U', 'O', 'Q'):
185.              # 玩家输入了一个有效的动作
186.              break
187.
188.          if len(action.split()) == 2:
189.              space, number = action.split()
190.              if len(space) != 2:
191.                  continue
192.
```

```
193.            column, row = space
194.            if column not in list('ABCDEFGHI'):
195.                print('There is no column', column)
196.                continue
197.            if not row.isdecimal() or not (1 <= int(row) <= 9):
198.                print('There is no row', row)
199.                continue
200.            if not (1 <= int(number) <= 9):
201.                print('Select a number from 1 to 9, not ', number)
202.                continue
203.            break  # 玩家输入了一个有效的动作
204.
205.    print()  # 输出空行
206.
207.    if action.startswith('R'):
208.        # 重置网格
209.        grid.resetGrid()
210.        continue
211.
212.    if action.startswith('N'):
213.        # 得到一个新的谜题
214.        grid = SudokuGrid(random.choice(puzzles))
215.        continue
216.
217.    if action.startswith('U'):
218.        # 撤销上一步动作
219.        grid.undo()
220.        continue
221.
222.    if action.startswith('O'):
223.        # 查看原始数字
224.        originalGrid = SudokuGrid(grid.originalSetup)
225.        print('The original grid looked like this:')
226.        originalGrid.display()
227.        input('Press Enter to continue...')
228.
229.    if action.startswith('Q'):
230.        # 退出游戏
231.        print('Thanks for playing!')
232.        sys.exit()
233.
234.    # 处理玩家输入的下一步动作
235.    if grid.makeMove(column, row, number) == False:
236.        print('You cannot overwrite the original grid\'s numbers.')
237.        print('Enter ORIGINAL to view the original grid.')
238.        input('Press Enter to continue...')
```

探索程序

请尝试找出以下问题的答案。你需要尝试对代码进行一些修改，再次运行程序，查看修改后的效果。

1. 如果删除或重命名 sudokupuzzles.txt 文件并运行程序，则会发生什么错误？
2. 如果将第 84 行的 str(y + 1) 改为 str(y)，则会发生什么？
3. 如果将第 92 行的 if y == 2 or y == 5: 改为 if y == 1 or y == 6:，则会发生什么？

| 项目 74

语音合成

在本项目中,我们将演示第三方模块 pyttsx3 的用法。你输入的任何消息都将由操作系统的文本转语音功能播放出来。尽管语音合成是计算机科学领域极其复杂的分支之一,但由于 pyttsx3 模块提供了简单的语音合成接口,因此使得本项目的程序适合初学者。一旦学会如何使用该模块,你就可以将语音合成功能添加到自己的程序中。

运行程序

运行 texttospeechtalker.py,输出如下所示:

```
Text To Speech Talker, by Al Sweigart al@inventwithpython.com
Text-to-speech using the pyttsx3 module, which in turn uses
the NSSpeechSynthesizer (on macOS), SAPI5 (on Windows), or
eSpeak (on Linux) speech engines.

Enter the text to speak, or QUIT to quit.
> Hello. My name is Guido van Robot.
<computer speaks text out loud>
> quit
Thanks for playing!
```

工作原理

本项目中的程序很短,因为 pyttsx3 模块帮我们实现了语音合成代码。要使用该模块,请按照本书前言中的引导进行安装。完成安装后,我们可以使用 import pyttsx3 导入 Python 脚本并调用 pyttsx3.init() 函数。该函数将返回一个语音合成引擎的引擎对象。该对象有一个 say() 方法,我们可以将一串文本传递给该方法,然后调用 runAndWait() 方法让计算机播放该文本对应的语音。

```
 1. """语音合成,作者:Al Sweigart al@inventwithpython.com
 2. 本项目会用到 pyttsx3 模块的文本转语音功能
 3.
 4. 标签: 小,初学者"""
 5.
 6. import sys
```

```
  7.
  8. try:
  9.     import pyttsx3
 10. except ImportError:
 11.     print('The pyttsx3 module needs to be installed to run this')
 12.     print('program. On Windows, open a Command Prompt and run:')
 13.     print('pip install pyttsx3')
 14.     print('On macOS and Linux, open a Terminal and run:')
 15.     print('pip3 install pyttsx3')
 16.     sys.exit()
 17.
 18. tts = pyttsx3.init()    # 初始化 TTS 引擎
 19.
 20. print('Text To Speech Talker, by Al Sweigart al@inventwithpython.com')
 21. print('Text-to-speech using the pyttsx3 module, which in turn uses')
 22. print('the NSSpeechSynthesizer (on macOS), SAPI5 (on Windows), or')
 23. print('eSpeak (on Linux) speech engines.')
 24. print()
 25. print('Enter the text to speak, or QUIT to quit.')
 26. while True:
 27.     text = input('> ')
 28.
 29.     if text.upper() == 'QUIT':
 30.         print('Thanks for playing!')
 31.         sys.exit()
 32.
 33.     tts.say(text)    # 为 TTS 引擎传入一些文本
 34.     tts.runAndWait()    # 让 TTS 引擎播放语音
```

探索程序

这只是一个简单的程序，没有很多可对其进行定制的选项。你不妨考虑一下自己的哪些其他程序可以从语音合成中受益。

项目 75

3张牌蒙特

3张牌蒙特是一种常见的骗局,一些人特别容易上当。3张扑克牌面朝下放在一个纸板箱上,其中1张是红心皇后牌。庄家迅速重新排列扑克牌,然后让对方挑出红心皇后牌。庄家可以使用各种各样的诡计隐藏牌或其他作弊手段,让受骗者永远都不能赢。通常,庄家在人群中有"托儿",他们与庄家暗中合作,假装赢了赌局(让受骗者认为他们也能赢)或故意输得很惨(让受骗者认为他们可以做得更好)。

在本项目中,我们用程序显示3张牌,然后快速呈现一系列交换操作。最后,程序会清空屏幕,玩家必须选择1张牌。你能保持对红心皇后牌的跟踪吗?为了提供更逼真的3张牌蒙特游戏体验,你可以启用作弊功能,确保玩家总是赢不了,即使他们选择了正确的牌。

运行程序

运行 threecardmonte.py,输出结果如下:

```
Three-Card Monte, by Al Sweigart al@inventwithpython.com
Find the red lady (the Queen of Hearts)! Keep an eye on how
the cards move.
Here are the cards:
 ___   ___   ___
|J  | |Q  | |8  |
| ♦ | | ♥ | | ♣ |
|__J| |__Q| |__8|
Press Enter when you are ready to begin...
swapping left and middle...
swapping right and middle...
swapping middle and left...
swapping right and left...
swapping left and middle...
--snip--
<screen clears>
Which card has the Queen of Hearts? (LEFT MIDDLE RIGHT)
> middle

 ___   ___   ___
|Q  | |8  | |J  |
| ♥ | | ♣ | | ♦ |
|__Q| |__8| |__J|
You lost!
Thanks for playing, sucker!
```

工作原理

在本项目的程序中,我们用(rank,suit)元组来表示 1 张扑克牌。其中,rank 表示牌面的字符串,例如'2''10''Q'或'K'。suit 则是由红心、梅花、黑桃或方块符号组成的字符串。鉴于无法使用键盘输入这些符号,程序在第 14~17 行调用 chr()函数来生成它们,例如,元组('9','♦') 即代表方块 9。

程序并不是直接输出这些元组,第 25~39 行的 displayCards()函数对这些元组进行了解释,并在屏幕上渲染出 ASCII 艺术画,如同项目 4 的做法。该函数的 cards 参数是一个扑克牌元组列表,可实现将多张牌显示在同一行。

```
 1. """3张牌蒙特,作者:Al Sweigart al@inventwithpython.com
 2. 交换卡牌后找到红心皇后牌
 3. (在现实生活中,骗子会用手抚摸红心皇后牌,所以你总是输)
 4. 标签: 简短, 卡牌, 游戏"""
 5.
 6. import random, time
 7.
 8. # 设置常量
 9. NUM_SWAPS = 16  # (!) 尝试将 16 更改为 30 或 100
10. DELAY = 0.8  # (!) 尝试将 0.8 更改为 2.0 或 0.0
11.
12. # 卡牌套装字符
13. HEARTS   = chr(9829)  # 字符 9829 表示的是♥
14. DIAMONDS = chr(9830)  # 字符 9830 表示的是♦
15. SPADES   = chr(9824)  # 字符 9824 表示的是♠
16. CLUBS    = chr(9827)  # 字符 9827 表示的是♣
17.
18. # 3 张卡牌列表的索引
19. LEFT = 0
20. MIDDLE = 1
21. RIGHT = 2
22.
23.
24. def displayCards(cards):
25.     """显示"cards"中的卡牌,这是(等级,花色)的列表元组"""
26.     rows = ['', '', '', '', '']  # 存储要显示的文本
27.
28.     for i, card in enumerate(cards):
29.         rank, suit = card  # 卡牌是元组数据结构
30.         rows[0] += ' ___  '  # 输出卡牌的顶行
31.         rows[1] += '|{} | '.format(rank.ljust(2))
32.         rows[2] += '| {} | '.format(suit)
33.         rows[3] += '|_{}| '.format(rank.rjust(2, '_'))
34.
35.     # 在屏幕上输出每一行
36.     for i in range(5):
37.         print(rows[i])
38.
39.
40. def getRandomCard():
41.     """返回一张不是红心皇后牌的随机卡牌"""
42.     while True:  # 不断生成卡牌,直到获得非红心皇后牌
43.         rank = random.choice(list('23456789JQKA') + ['10'])
44.         suit = random.choice([HEARTS, DIAMONDS, SPADES, CLUBS])
45.
46.         # 只要不是红心皇后牌,就返回这张卡牌
47.         if rank != 'Q' and suit != HEARTS:
```

```
 48.         return (rank, suit)
 49.
 50.
 51. print('Three-Card Monte, by Al Sweigart al@inventwithpython.com')
 52. print()
 53. print('Find the red lady (the Queen of Hearts)! Keep an eye on how')
 54. print('the cards move.')
 55. print()
 56.
 57. # 显示原始排列
 58. cards = [('Q', HEARTS), getRandomCard(), getRandomCard()]
 59. random.shuffle(cards)    # 将红心皇后牌放在随机位置
 60. print('Here are the cards:')
 61. displayCards(cards)
 62. input('Press Enter when you are ready to begin...')
 63.
 64. # 输出洗牌后的结果
 65. for i in range(NUM_SWAPS):
 66.     swap = random.choice(['l-m', 'm-r', 'l-r', 'm-l', 'r-m', 'r-l'])
 67.
 68.     if swap == 'l-m':
 69.         print('swapping left and middle...')
 70.         cards[LEFT], cards[MIDDLE] = cards[MIDDLE], cards[LEFT]
 71.     elif swap == 'm-r':
 72.         print('swapping middle and right...')
 73.         cards[MIDDLE], cards[RIGHT] = cards[RIGHT], cards[MIDDLE]
 74.     elif swap == 'l-r':
 75.         print('swapping left and right...')
 76.         cards[LEFT], cards[RIGHT] = cards[RIGHT], cards[LEFT]
 77.     elif swap == 'm-l':
 78.         print('swapping middle and left...')
 79.         cards[MIDDLE], cards[LEFT] = cards[LEFT], cards[MIDDLE]
 80.     elif swap == 'r-m':
 81.         print('swapping right and middle...')
 82.         cards[RIGHT], cards[MIDDLE] = cards[MIDDLE], cards[RIGHT]
 83.     elif swap == 'r-l':
 84.         print('swapping right and left...')
 85.         cards[RIGHT], cards[LEFT] = cards[LEFT], cards[RIGHT]
 86.
 87.     time.sleep(DELAY)
 88.
 89. # 输出空行，以隐藏洗牌过程消息
 90. print('\n' * 60)
 91.
 92. # 要求玩家找到红心皇后牌
 93. while True:    # 继续询问，直到玩家输入 LEFT、MIDDLE 或 RIGHT
 94.     print('Which card has the Queen of Hearts? (LEFT MIDDLE RIGHT)')
 95.     guess = input('> ').upper()
 96.
 97.     # 获取玩家输入的卡牌的索引
 98.     if guess in ['LEFT', 'MIDDLE', 'RIGHT']:
 99.         if guess == 'LEFT':
100.             guessIndex = 0
101.         elif guess == 'MIDDLE':
102.             guessIndex = 1
103.         elif guess == 'RIGHT':
104.             guessIndex = 2
105.         break
106.
107. # (!) 取消注释这段代码，让玩家总是输
108. # if cards[guessIndex] == ('Q', HEARTS):
109. #     # 玩家赢了，故移动红心皇后牌
```

```
110. #    possibleNewIndexes = [0, 1, 2]
111. #    possibleNewIndexes.remove(guessIndex)    # 删除红心皇后牌的索引
112. #    newInd = random.choice(possibleNewIndexes)    # 选择一个新索引
113. #    # 将红心皇后牌放在新索引处
114. #    cards[guessIndex], cards[newInd] = cards[newInd], cards[guessIndex]
115.
116. displayCards(cards)    # 显示所有卡牌
117.
118. # 判断玩家是否赢了
119. if cards[guessIndex] == ('Q', HEARTS):
120.     print('You won!')
121.     print('Thanks for playing!')
122. else:
123.     print('You lost!')
124.     print('Thanks for playing, sucker!')
```

输入源代码并运行几次后，请尝试对其进行更改。你可以根据标有!的注释对程序进行修改，也可以尝试以下操作。

- 使用项目 57 中的退格输出技术短暂显示每条交换消息，在输出下一条消息之前通过输出一系列字符"\b"将之前的消息清除。
- 创建 4 张牌蒙特游戏程序，以增加难度。

探索程序

请尝试找出以下问题的答案。你需要尝试对代码进行一些修改，再次运行程序，查看修改后的效果。

1. 如果将第 58 行的 [('Q', HEARTS), getRandomCard(), getRandomCard()] 改为 [('Q', HEARTS), ('Q', HEARTS), ('Q', HEARTS)]，则会发生什么？
2. 如果将第 43 行的 list('23456789JQKA') 改为 list('ABCDEFGHIJK')，则会发生什么？
3. 如果删除或注释掉第 87 行的 time.sleep(DELAY)，则会发生什么？

项目 76

井字棋

井字棋是一款经典的在 3×3 网格上玩的纸笔游戏。玩家轮流向格子中填 X 或 O 标记,争取首先完成"三点一线"。通常井字棋游戏会以平局结束,不过如果对手疏忽大意,你就可能会智胜对手。

运行程序

运行 tictactoe.py,输出如下所示:

```
Welcome to Tic-Tac-Toe!
     | |    1 2 3
    -+-+-
     | |    4 5 6
    -+-+-
     | |    7 8 9
What is X's move? (1-9)
> 1
    X| |    1 2 3
    -+-+-
     | |    4 5 6
    -+-+-
     | |    7 8 9
What is O's move? (1-9)
--snip--
    X|O|X   1 2 3
    -+-+-
    X|O|O   4 5 6
    -+-+-
    O|X|X   7 8 9
The game is a tie!
Thanks for playing!
```

工作原理

为了在程序中表示井字棋棋盘,我们使用一个带有键 1~9 的字典来代表棋盘上的每一格。

编了号的空格的排列方式与手机键盘的相同。字典中的值包括玩家标记的字符"X"或"O"以及代表空格的" "。

```
 1. """井字棋, 作者: Al Sweigart al@inventwithpython.com
 2. 经典的棋盘游戏
 3. 标签: 简短, 棋盘, 游戏, 双人游戏"""
 4.
 5. ALL_SPACES = ['1', '2', '3', '4', '5', '6', '7', '8', '9']
 6. X, O, BLANK = 'X', 'O', ' '  # 字符串值, 常量
 7.
 8.
 9. def main():
10.     print('Welcome to Tic-Tac-Toe!')
11.     gameBoard = getBlankBoard()  # 创建井字棋棋盘字典
12.     currentPlayer, nextPlayer = X, O  # X 在前, O 在后
13.
14.     while True:  # 主循环
15.         # 在屏幕上显示棋盘
16.         print(getBoardStr(gameBoard))
17.
18.         # 不断询问玩家, 直到他们输入一个 1 和 9 之间的数字
19.         move = None
20.         while not isValidSpace(gameBoard, move):
21.             print('What is {}\'s move? (1-9)'.format(currentPlayer))
22.             move = input('> ')
23.         updateBoard(gameBoard, move, currentPlayer)  # 根据玩家的操作更新棋盘
24.
25.         # 判断游戏是否结束
26.         if isWinner(gameBoard, currentPlayer):  # 判断谁是获胜者
27.             print(getBoardStr(gameBoard))
28.             print(currentPlayer + ' has won the game!')
29.             break
30.         elif isBoardFull(gameBoard):  # 判断是否平局
31.             print(getBoardStr(gameBoard))
32.             print('The game is a tie!')
33.             break
34.         # 轮到另一个玩家向格子中填标记
35.         currentPlayer, nextPlayer = nextPlayer, currentPlayer
36.     print('Thanks for playing!')
37.
38.
39. def getBlankBoard():
40.     """创建一个新的空的井字棋棋盘"""
41.     # 棋格数字映射:          1|2|3
42.     #                       -+-+-
43.     #                       4|5|6
44.     #                       -+-+-
45.     #                       7|8|9
46.     # 键为 1 和 9 之间的数字, 值为 X、O 或 BLANK
47.     board = {}
48.     for space in ALL_SPACES:
49.         board[space] = BLANK  # 所有格子最初均无标记
50.     return board
51.
52.
53. def getBoardStr(board):
54.     """返回表示棋盘的文本"""
55.     return '''
56. {}|{}|{}  1 2 3
57. -+-+-
58. {}|{}|{}  4 5 6
```

```
59.             -+-+-
60.             {}|{}|{}    7 8 9'''.format(board['1'], board['2'], board['3'],
61.                                         board['4'], board['5'], board['6'],
62.                                         board['7'], board['8'], board['9'])
63.
64.
65. def isValidSpace(board, space):
66.     """如果棋盘上的格子编号是1～9且格子为空,则返回True"""
67.     return space in ALL_SPACES and board[space] == BLANK
68.
69.
70. def isWinner(board, player):
71.     """如果玩家获胜,则返回True"""
72.     # 为了提升可读性,此处使用简短变量名
73.     b, p = board, player
74.     # 检查3行、3列和2条对角线上的3个标记
75.     return ((b['1'] == b['2'] == b['3'] == p) or  # 第1行
76.             (b['4'] == b['5'] == b['6'] == p) or  # 第2行
77.             (b['7'] == b['8'] == b['9'] == p) or  # 第3行
78.             (b['1'] == b['4'] == b['7'] == p) or  # 第1列
79.             (b['2'] == b['5'] == b['8'] == p) or  # 第2列
80.             (b['3'] == b['6'] == b['9'] == p) or  # 第3列
81.             (b['3'] == b['5'] == b['7'] == p) or  # 对角线
82.             (b['1'] == b['5'] == b['9'] == p))    # 对角线
83.
84.
85. def isBoardFull(board):
86.     """如果棋盘上的每个格子均已填上标记,则返回True"""
87.     for space in ALL_SPACES:
88.         if board[space] == BLANK:
89.             return False  # 如果有任一格子未填上标记,则返回False
90.     return True  # 格子均已填上标记,所以返回True
91.
92.
93. def updateBoard(board, space, mark):
94.     """设置棋盘上要标记的格子"""
95.     board[space] = mark
96.
97.
98. if __name__ == '__main__':
99.     main()  # 程序运行入口(如果不是作为模块导入的话)
```

探索程序

请尝试找出以下问题的答案。你需要尝试对代码进行一些修改,再次运行程序,查看修改后的效果。

1. 如果将第6行的 X, O, BLANK = 'X', 'O', ' ' 改为 X, O, BLANK = 'X', 'X', ' ',则会发生什么?
2. 如果将第95行的 board[space] = mark 改为 board[space] = X,则会发生什么?
3. 如果将第49行的 board[space] = BLANK 改为 board[space] = X,则会发生什么?

项目 77

汉诺塔

汉诺塔是一款堆叠移动益智游戏。游戏中有 3 根柱子，上面堆叠着各种大小的盘子。游戏的目标是将一根柱子上的所有盘子移动到另一根柱子上。不过要注意，玩家每次只能移动 1 个盘子，并且任何时候都不能将较大的盘子放在较小的盘子之上。努力找出其中蕴含的模式规则有助于解决这个难题，你能找到它吗？（提示：先试着将 TOTAL_DISKS 变量设为 3 或 4，以解决较为简单的问题。）

运行程序

运行 towerofhanoi.py，输出如下所示：

```
The Tower of Hanoi, by Al Sweigart al@inventwithpython.com

Move the tower of disks, one disk at a time, to another tower. Larger
disks cannot rest on top of a smaller disk.

More info at https://en.wikipedia.com/wiki/Tower_of_Hanoi
```

```
Enter the letters of "from" and "to" towers, or QUIT.
(e.g. AB to moves a disk from tower A to tower B.)
> ab
```

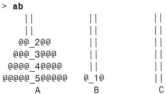

```
Enter the letters of "from" and "to" towers, or QUIT.
(e.g. AB to moves a disk from tower A to tower B.)
--snip--
```

工作原理

本项目中的程序使用一个整数列表数据结构表示汉诺塔。列表中的每个整数表示盘子的大小。列表中的第一个整数表示底部的盘子，最后一个整数表示顶部的盘子。例如，[5, 4, 2] 代表以下汉诺塔：

```
    ||
    ||
   @@_2@@
  @@@@_4@@@@
 @@@@@_5@@@@@
```

Python 列表的 `append()` 和 `pop()` 方法分别用于从列表的尾部添加和删除元素。正如 `someList[0]` 和 `someList[1]` 允许我们访问列表中正数第一个和第二个值那样，Python 允许我们使用负索引访问列表末端的值，例如 `someList[-1]` 和 `someList[-2]` 可以用于访问列表中最后一个和倒数第二个值。这对于查找当前位于塔顶的盘子很有帮助。

```
1.  """汉诺塔，作者: Al Sweigart al@inventwithpython.com
2.  一个堆叠移动的益智游戏
3.  标签：简短，游戏，谜题"""
4.
5.  import copy
6.  import sys
7.
8.  TOTAL_DISKS = 5  # 盘子越多，难度越大
9.
10. # 从 A 柱子上的盘子开始移动
11. COMPLETE_TOWER = list(range(TOTAL_DISKS, 0, -1))
12.
13.
14. def main():
15.     print("""The Tower of Hanoi, by Al Sweigart al@inventwithpython.com
16.
17. Move the tower of disks, one disk at a time, to another tower. Larger
18. disks cannot rest on top of a smaller disk.
19. """
20.     )
21.
22.     # 创建柱子。列表的末尾是柱子顶端
23.     towers = {'A': copy.copy(COMPLETE_TOWER), 'B': [], 'C': []}
24.
25.     while True:  # 运行一个回合
26.         # 显示柱子和盘子
27.         displayTowers(towers)
28.
29.         # 询问玩家接下来如何移动
30.         fromTower, toTower = askForPlayerMove(towers)
31.
32.         # 将顶部的盘子从 fromTower 移动到 toTower
33.         disk = towers[fromTower].pop()
34.         towers[toTower].append(disk)
35.
36.         # 判断玩家是否通关了
37.         if COMPLETE_TOWER in (towers['B'], towers['C']):
38.             displayTowers(towers)  # 最后一次显示柱子
39.             print('You have solved the puzzle! Well done!')
40.             sys.exit()
```

```
 41.
 42.
 43. def askForPlayerMove(towers):
 44.     """询问玩家接下来如何移动。 返回(fromTower, toTower)"""
 45.
 46.     while True:  # 不断询问玩家,直到他们输入有效的移动
 47.         print('Enter the letters of "from" and "to" towers, or QUIT.')
 48.         print('(e.g. AB to moves a disk from tower A to tower B.)')
 49.         response = input('> ').upper().strip()
 50.
 51.         if response == 'QUIT':
 52.             print('Thanks for playing!')
 53.             sys.exit()
 54.
 55.         # 确保玩家输入了有效的柱子字母
 56.         if response not in ('AB', 'AC', 'BA', 'BC', 'CA', 'CB'):
 57.             print('Enter one of AB, AC, BA, BC, CA, or CB.')
 58.             continue  # 再次询问玩家接下来如何移动
 59.
 60.         # 语法糖 —— 使用更具描述性的变量名称
 61.         fromTower, toTower = response[0], response[1]
 62.
 63.         if len(towers[fromTower]) == 0:
 64.             # "from" 柱子上不能没有盘子
 65.             print('You selected a tower with no disks.')
 66.             continue  # 再次询问玩家接下来如何移动
 67.         elif len(towers[toTower]) == 0:
 68.             # 任何盘子都可以移动到空的"to"柱子上
 69.             return fromTower, toTower
 70.         elif towers[toTower][-1] < towers[fromTower][-1]:
 71.             print('Can\'t put larger disks on top of smaller ones.')
 72.             continue  # 再次询问玩家接下来如何移动
 73.         else:
 74.             # 这是一个有效的移动,因此返回选定的柱子
 75.             return fromTower, toTower
 76.
 77.
 78. def displayTowers(towers):
 79.     """显示当前状态"""
 80.
 81.     # 展示3根柱子
 82.     for level in range(TOTAL_DISKS, -1, -1):
 83.         for tower in (towers['A'], towers['B'], towers['C']):
 84.             if level >= len(tower):
 85.                 displayDisk(0)  # 显示没有盘子的柱子
 86.             else:
 87.                 displayDisk(tower[level])  # 显示盘子
 88.         print()
 89.
 90.     # 显示柱子标签 A、B 和 C
 91.     emptySpace = ' ' * (TOTAL_DISKS)
 92.     print('{0} A{0}{0} B{0}{0} C\n'.format(emptySpace))
 93.
 94.
 95. def displayDisk(width):
 96.     """显示给定宽度的盘子。 宽度为 0 表示没有盘子"""
 97.     emptySpace = ' ' * (TOTAL_DISKS - width)
 98.
 99.     if width == 0:
100.         # 显示没有盘子的柱子
101.         print(emptySpace + '||' + emptySpace, end='')
102.     else:
```

```
103.            # 显示盘子
104.            disk = '@' * width
105.            numLabel = str(width).rjust(2, '_')
106.            print(emptySpace + disk + numLabel + disk + emptySpace, end='')
107.
108.
109.    # 程序运行入口(如果不是作为模块导入的话)
110.    if __name__ == '__main__':
111.        main()
```

探索程序

请尝试找出以下问题的答案。你需要尝试对代码进行一些修改，再次运行程序，查看修改后的效果。

1. 如果删除或注释掉第70行、第71行和第72行的代码，则会发生什么？
2. 如果将第97行的 emptySpace = ' ' * (TOTAL_DISKS - width) 改为 emptySpace = ' '，则会发生什么？
3. 如果将第99行的 width == 0 改为 width != 0，则会发生什么？

| 项目 78

脑筋急转弯

把一块黄色石头扔进蓝色池塘,它会变成什么?英国有7月4日吗?医生如何做到30天不睡觉?不管你认为这些问题的答案是什么,你可能都已经错了。在本项目中,有54个精心设计的问题,它们的答案简单、明显但问题具有误导性。找到真正的答案需要"动脑子"。

复制本书中的代码会减少你体验游戏的乐趣,因为你会提前知晓答案。

运行程序

运行 trickquestions.py,输出如下所示:

```
Trick Questions, by Al Sweigart al@inventwithpython.com

Can you figure out the answers to these trick questions?
(Enter QUIT to quit at any time.)

Press Enter to begin...
--snip--
Question: 1
Score: 0 / 54
QUESTION: A 39 year old person was born on the 22nd of February. What year is
their birthday?
ANSWER: 1981
Incorrect! The answer is: Their birthday is on February 22nd of every year.
Press Enter for the next question...
--snip--
Question: 2
Score: 0 / 54
QUESTION: If there are ten apples and you take away two, how many do you have?
ANSWER: Eight
Incorrect! The answer is: Two.
Press Enter for the next question...
--snip--
```

工作原理

QUESTIONS 变量包含一个字典列表。每个字典代表一个脑筋急转弯问题及其答案,并有键

'question' 'answer'和'accept'。'question'和'answer'的值分别为程序向玩家提出问题和显示答案的字符串,键'accept'的值是一个字符串列表。如果玩家输入的答案中包含这些字符串中的任何一个,就认为答案是正确的。这将允许玩家输入自由格式的文本作为答案,当玩家输入正确答案时,程序可以相当准确地检测到。

```python
1.  """脑筋急转弯,作者: Al Sweigart al@inventwithpython.com
2.  一些脑筋急转弯问题
3.  标签: 大, 诙谐"""
4.
5.  import random, sys
6.
7.  # QUESTIONS 是一个字典列表,每个字典代表一个
8.  # 脑筋急转弯问题及其答案。 字典有键 "question"
9.  # (保存问题的文本), 'answer' (保存文本
10. # 的答案) 和 "accept" (其中包含一个字符串列表,如果
11. # 玩家的答案包含其中任何一项,则他们回答正确)
12. # (!) 试着想出你自己的脑筋急转弯问题并添加到这里
13. QUESTIONS = [
14.     {'question': "How many times can you take 2 apples from a pile of 10 apples?",
15.      'answer': "Once. Then you have a pile of 8 apples.",
16.      'accept': ['once', 'one', '1']},
17.     {'question': 'What begins with "e" and ends with "e" but only has one letter in it?',
18.      'answer': "An envelope.",
19.      'accept': ['envelope']},
20.     {'question': "Is it possible to draw a square with three sides?",
21.      'answer': "Yes. All squares have three sides. They also have a fourth side.",
22.      'accept': ['yes']},
23.     {'question': "How many times can a piece of paper be folded in half by hand without unfolding?",
24.      'answer': "Once. Then you are folding it in quarters.",
25.      'accept': ['one', '1', 'once']},
26.     {'question': "What does a towel get as it dries?",
27.      'answer': "Wet.",
28.      'accept': ['wet']},
29.     {'question': "What does a towel get as it dries?",
30.      'answer': "Drier.",
31.      'accept': ['drier', 'dry']},
32.     {'question': "Imagine you are in a haunted house full of evil ghosts. What do you have
        to do to stay safe?",
33.      'answer': "Nothing. You're only imagining it.",
34.      'accept': ['nothing', 'stop']},
35.     {
36.      'question': "A taxi driver is going the wrong way down a one-way street. She passes ten cops
37.      but doesn't get a ticket. Why not?",
38.      'answer': "She was walking.",
39.      'accept': ['walk']},
40.     {'question': "What does a yellow stone thrown into a blue pond become?",
41.      'answer': "Wet.",
42.      'accept': ['wet']},
43.     {'question': "How many miles does must a cyclist bike to get to training?",
44.      'answer': "None. They're training as soon as they get on the bike.",
45.      'accept': ['none', 'zero', '0']},
46.     {'question': "What building do people want to leave as soon as they enter?",
47.      'answer': "An airport.",
48.      'accept': ['airport', 'bus', 'port', 'train', 'station', 'stop']},
49.     {
50.      'question': "If you're in the middle of a square house facing the west side with the south
51.      side to your left and the north side to your right, which side of the house are you next to?",
52.      'answer': "None. You're in the middle.",
53.      'accept': ['none', 'middle', 'not', 'any']},
54.     {'question': "How much dirt is in a hole 3 meters wide, 3 meters long, and 3 meters deep?",
```

```
 55.        'answer': "There is no dirt in a hole.",
 56.        'accept': ['no', 'none', 'zero']},
 57.    {'question': "A girl mails a letter from America to Japan. How many miles did the stamp move?",
 58.        'answer': "Zero. The stamp was in the same place on the envelope the whole time.",
 59.        'accept': ['zero', '0', 'none', 'no']},
 60.    {'question': "What was the highest mountain on Earth the day before Mount Everest was discovered?",
 61.        'answer': "Mount Everest was still the highest mountain of Earth the day before it was discovered.",
 62.        'accept': ['everest']},
 63.    {'question': "How many fingers do most people have on their two hands?",
 64.        'answer': "Eight. They also have two thumbs.",
 65.        'accept': ['eight', '8']},
 66.    {'question': "The 4th of July is a holiday in America. Do they have a 4th of July in England?",
 67.        'answer': "Yes. All countries have a 4th of July on their calendar.",
 68.        'accept': ['yes']},
 69.    {'question': "Which letter of the alphabet makes honey?",
 70.        'answer': "None. A bee is an insect, not a letter.",
 71.        'accept': ['no', 'none', 'not']},
 72.    {'question': "How can a doctor go 30 days without sleep?",
 73.        'answer': "By sleeping at night.",
 74.        'accept': ['night', 'evening']},
 75.    {'question': "How many months have 28 days?",
 76.        'answer': "12. All months have 28 days. Some have more days as well.",
 77.        'accept': ['12', 'twelve', 'all']},
 78.    {'question': "How many two cent stamps are in a dozen?",
 79.        'answer': "A dozen.",
 80.        'accept': ['12', 'twelve', 'dozen']},
 81.    {'question': "Why is it illegal for a person living in North Dakota to be buried in South Dakota?",
 82.        'answer': "Because it is illegal to bury someone alive.",
 83.        'accept': ['alive', 'living', 'live']},
 84.    {'question': "How many heads does a two-headed coin have?",
 85.        'answer': "Zero. Coins are just circular pieces of metal. They don't have heads.",
 86.        'accept': ['zero', 'none', 'no', '0']},
 87.    {'question': "What kind of vehicle has four wheels and flies?",
 88.        'answer': "A garbage truck.",
 89.        'accept': ['garbage', 'dump', 'trash']},
 90.    {'question': "What kind of vehicle has four wheels and flies?",
 91.        'answer': "An airplane.",
 92.        'accept': ['airplane', 'plane']},
 93.    {'question': "What five-letter word becomes shorter by adding two letters?",
 94.        'answer': "Short.",
 95.        'accept': ['short']},
 96.    {
 97.        'question': "Gwen's mother has five daughters. Four are named Haha, Hehe, Hihi, and Hoho.
 98.        What's the fifth daughter's name?",
 99.        'answer': "Gwen.",
100.        'accept': ['gwen']},
101.    {'question': "How long is a fence if there are three fence posts each one meter apart?",
102.        'answer': "Two meters long.",
103.        'accept': ['2', 'two']},
104.    {'question': "How many legs does a dog have if you count its tail as a leg?",
105.        'answer': "Four. Calling a tail a leg doesn't make it one.",
106.        'accept': ['four', '4']},
107.    {'question': "How much more are 1976 pennies worth compared to 1975 pennies?",
108.        'answer': "One cent.",
109.        'accept': ['1', 'one']},
110.    {'question': "What two things can you never eat for breakfast?",
111.        'answer': "Lunch and dinner.",
112.        'accept': ['lunch', 'dinner', 'supper']},
113.    {'question': "How many birthdays does the average person have?",
114.        'answer': "One. You're only born once.",
115.        'accept': ['one', '1', 'once', 'born']},
116.    {'question': "Where was the United States Declaration of Independence signed?",
```

```
117.          'answer': "It was signed at the bottom.",
118.          'accept': ['bottom']},
119.         {
120.          'question': "A person puts two walnuts in their pocket but only has one thing in their
121.          pocket five minutes later. What is it?",
122.          'answer': "A hole.",
123.          'accept': ['hole']},
124.         {'question': "What did the sculptor make that no one could see?",
125.          'answer': "Noise.",
126.          'accept': ['noise']},
127.         {'question': "If you drop a raw egg on a concrete floor, will it crack?",
128.          'answer': "No. Concrete is very hard to crack.",
129.          'accept': ['no']},
130.         {
131.          'question': "If it takes ten people ten hours to build a fence, how many hours does it take
132.          five people to build it?",
133.          'answer': "Zero. It's already built.",
134.          'accept': ['zero', 'no', '0', 'already', 'built']},
135.         {'question': "Which is heavier, 100 pounds of rocks or 100 pounds of feathers?",
136.          'answer': "Neither. They weigh the same.",
137.          'accept': ['neither', 'none', 'no', 'same', 'even', 'balance']},
138.         {'question': "What do you have to do to survive being bitten by a poisonous snake?",
139.          'answer': "Nothing. Only venomous snakes are deadly.",
140.          'accept': ['nothing', 'anything']},
141.         {'question': "What three consecutive days don't include Sunday, Wednesday, or Friday?",
142.          'answer': "Yesterday, today, and tomorrow.",
143.          'accept': ['yesterday', 'today', 'tomorrow']},
144.         {'question': "If there are ten apples and you take away two, how many do you have?",
145.          'answer': "Two.",
146.          'accept': ['2', 'two']},
147.         {'question': "A 39 year old person was born on the 22nd of February. What year is their birthday?",
148.          'answer': "Their birthday is on February 22nd of every year.",
149.          'accept': ['every', 'each']},
150.         {'question': "How far can you walk in the woods?",
151.          'answer': "Halfway. Then you are walking out of the woods.",
152.          'accept': ['half', '1/2']},
153.         {'question': "Can a man marry his widow's sister?",
154.          'answer': "No, because he's dead.",
155.          'accept': ['no']},
156.         {'question': "What do you get if you divide one hundred by half?",
157.          'answer': "One hundred divided by half is two hundred. One hundred divided by two is fifty.",
158.          'accept': ['two', '200']},
159.         {'question': "What do you call someone who always knows where their spouse is?",
160.          'answer': "A widow or widower.",
161.          'accept': ['widow', 'widower']},
162.         {'question': "How can someone take a photo but not be a photographer?",
163.          'answer': "They can be a thief.",
164.          'accept': ['thief', 'steal', 'take', 'literal']},
165.         {
166.          'question': "An electric train leaves the windy city of Chicago at 4pm on a Monday heading south
167.          at 100 kilometers per hour. Which way does the smoke blow from the smokestack?",
168.          'answer': "Electric trains don't have smokestacks.",
169.          'accept': ["don't", "doesn't", 'not', 'no', 'none']},
170.         {'question': 'What is the only word that rhymes with "orange"?',
171.          'answer': "Orange.",
172.          'accept': ['orange']},
173.         {'question': "Who is the U.S. President if the U.S. Vice President dies?",
174.          'answer': "The current U.S. President.",
175.          'accept': ['president', 'current', 'already']},
176.         {
177.          'question': "A doctor gives you three pills with instructions to take one every half-hour.
178.          How long will the pills last?",
```

```
179.        'answer': "One hour.",
180.        'accept': ['1', 'one']},
181.     {'question': "Where is there an ocean with no water?",
182.        'answer': "On a map.",
183.        'accept': ['map']},
184.     {'question': "What is the size of a rhino but weighs nothing?",
185.        'answer': "A rhino's shadow.",
186.        'accept': ['shadow']},
187.     {'question': "The clerk at a butcher shop is exactly 177 centimeters tall. What do they weigh?",
188.        'answer': "The clerk weighs meat.",
189.        'accept': ['meat']}]
190.
191. CORRECT_TEXT = ['Correct!', 'That is right.', "You're right.",
192.                 'You got it.', 'Righto!']
193. INCORRECT_TEXT = ['Incorrect!', "Nope, that isn't it.", 'Nope.',
194.                   'Not quite.', 'You missed it.']
195.
196. print('''Trick Questions, by Al Sweigart al@inventwithpython.com
197.
198. Can you figure out the answers to these trick questions?
199. (Enter QUIT to quit at any time.)
200. ''')
201.
202. input('Press Enter to begin...')
203.
204. random.shuffle(QUESTIONS)
205. score = 0
206.
207. for questionNumber, qa in enumerate(QUESTIONS):    # 主循环
208.     print('\n' * 40)  # 清空屏幕
209.     print('Question:', questionNumber + 1)
210.     print('Score:', score, '/', len(QUESTIONS))
211.     print('QUESTION:', qa['question'])
212.     response = input('   ANSWER: ').lower()
213.
214.     if response == 'quit':
215.         print('Thanks for playing!')
216.         sys.exit()
217.
218.     correct = False
219.     for acceptanceWord in qa['accept']:
220.         if acceptanceWord in response:
221.             correct = True
222.
223.     if correct:
224.         text = random.choice(CORRECT_TEXT)
225.         print(text, qa['answer'])
226.         score += 1
227.     else:
228.         text = random.choice(INCORRECT_TEXT)
229.         print(text, 'The answer is:', qa['answer'])
230.     response = input('Press Enter for the next question...').lower()
231.
232.     if response == 'quit':
233.         print('Thanks for playing!')
234.         sys.exit()
235.
236. print("That's all the questions. Thanks for playing!")
```

输入源代码并运行几次后，请尝试对其进行更改。你可以参考标有!的注释对程序进行修改。

探索程序

这是一个比较基础的程序，没有太多可自定义的选项，不过你可以尝试思考问答程序格式的其他用途。

项目 79

2048

Gabriele Cirulli 是一位网络开发者,他用一个周末的时间发明了 2048 游戏。这款游戏的灵感来自 Veewo Studios 的 1024 游戏,后者的开发则受到了开发团队 Sirvo 的游戏 *Threes!* 的启发。在 2048 游戏中,你必须在一个 4×4 的方格板上合并数字,以将它们从屏幕中清除。两个 2 合并为一个 4,两个 4 合并为一个 8 等。每次合并时,游戏会在方格板上新增一个 2。游戏的目标是在整个方格板被填满之前在任一个方格中出现 2048。

运行程序

运行 twentyfortyeight.py,输出如下所示:

```
Twenty Forty-Eight, by Al Sweigart al@inventwithpython.com
--snip--
+-----+-----+-----+-----+
|     |     |     |     |
|     |     |  2  | 16  |
|     |     |     |     |
+-----+-----+-----+-----+
|     |     |     |     |
|     | 16  |  4  |  2  |
|     |     |     |     |
+-----+-----+-----+-----+
|     |     |     |     |
|  2  |     |  4  | 32  |
|     |     |     |     |
+-----+-----+-----+-----+
|     |     |     |     |
|     |     |     |  2  |
|     |     |     |     |
+-----+-----+-----+-----+
Score: 80
Enter move: (WASD or Q to quit)
--snip--
```

工作原理

本项目中的程序使用"列"数据结构实现滑动效果,其由包含 4 个字符串的列表表示:BLANK(单个空格字符串)、'2' '4'和'8'。列表中的第一个值表示列的最底端,而最后一个值则表示列的最顶端。组合在一列中的数字总是会向下滑动，无论玩家是向上还是向下、向左还是向右滑动滑块。我们可以根据重力的含义来理解这些方向拉动滑块的过程。例如，对于一块滑块向右滑动的方格板，我们会创建 4 个列表以表示列：

- ['2', '4', '8', ' ']；
- [' ', ' ', ' ', '4']；
- [' ', ' ', ' ', '2']；
- [' ', ' ', ' ', ' ']。

combineTilesInColumn() 函数接收一个 column 列表并返回另一个列表，将匹配的数字加以合并且移向底部。调用 combineTilesInColumn() 的代码负责创建有着正确方位的 column 列表，并使用返回的列表更新游戏面板。

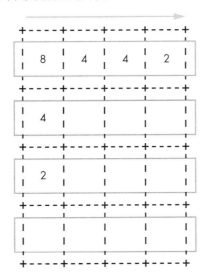

```
 1. """2048，Al Sweigart al@inventwithpython.com
 2. 一个滑动图块游戏
 3. 灵感来自 Gabriele Cirulli 的 2048，它的灵感来自 Veewo Studios 的 1024 游戏
 4. 1024 的灵感来自 Threes! 游戏
 5. 标签：大，游戏，谜题"""
 6.
 7. import random, sys
 8.
 9. # 创建常量
10. BLANK = ''   # 表示方格板上空格的值
11.
12.
13. def main():
14.     print('''Twenty Forty-Eight, by Al Sweigart al@inventwithpython.com
15.
16. Slide all the tiles on the board in one of four directions. Tiles with
17. like numbers will combine into larger-numbered tiles. A new 2 tile is
18. added to the board on each move. You win if you can create a 2048 tile.
19. You lose if the board fills up the tiles before then.''')
20.     input('Press Enter to begin...')
21.
22.     gameBoard = getNewBoard()
```

```
23.
24.     while True:    # 主循环
25.         drawBoard(gameBoard)
26.         print('Score:', getScore(gameBoard))
27.         playerMove = askForPlayerMove()
28.         gameBoard = makeMove(gameBoard, playerMove)
29.         addTwoToBoard(gameBoard)
30.
31.         if isFull(gameBoard):
32.             drawBoard(gameBoard)
33.             print('Game Over - Thanks for playing!')
34.             sys.exit()
35.
36.
37. def getNewBoard():
38.     """返回表示新方格板的数据结构
39.
40.     这是一个字典,包含 (x,y) 元组的键和方格的值
41.     方格的值是一个 2 的整数幂或 BLANK
42.     坐标布置为:
43.       X0 1 2 3
44.       Y+-+-+-+-+
45.       0| | | | |
46.       +-+-+-+-+
47.       1| | | | |
48.       +-+-+-+-+
49.       2| | | | |
50.       +-+-+-+-+
51.       3| | | | |
52.       +-+-+-+-+"""
53.
54.     newBoard = {}    # 包含要返回的方格板的数据结构
55.     # 循环遍历所有可能的方格并将所有方格设置为空
56.     for x in range(4):
57.         for y in range(4):
58.             newBoard[(x, y)] = BLANK
59.
60.     # 为两个初始值为 2 的滑块随机选择方格
61.     startingTwosPlaced = 0    # 选择的起始空格数
62.     while startingTwosPlaced < 2:    # 重复此操作,以获得同样的空格
63.         randomSpace = (random.randint(0, 3), random.randint(0, 3))
64.         # 确保随机选择的方格尚未被占用
65.         if newBoard[randomSpace] == BLANK:
66.             newBoard[randomSpace] = 2
67.             startingTwosPlaced = startingTwosPlaced + 1
68.
69.     return newBoard
70.
71.
72. def drawBoard(board):
73.     """在屏幕上绘制方格板"""
74.
75.     # 从左到右、从上到下,遍历每个可能的方格
76.     # 创建每个方格标签的列表
77.     labels = []    # 该方格的数字/空值的字符串列表
78.     for y in range(4):
79.         for x in range(4):
80.             tile = board[(x, y)]    # 获取当前方格
81.             # 确保标签长度为 5 个空格
82.             labelForThisTile = str(tile).center(5)
83.             labels.append(labelForThisTile)
84.
```

```
85.     # 将{}替换为方格的标签
86.     print("""
87. +-----+-----+-----+-----+
88. |     |     |     |     |
89. |{}|{}|{}|{}|
90. |     |     |     |     |
91. +-----+-----+-----+-----+
92. |     |     |     |     |
93. |{}|{}|{}|{}|
94. |     |     |     |     |
95. +-----+-----+-----+-----+
96. |     |     |     |     |
97. |{}|{}|{}|{}|
98. |     |     |     |     |
99. +-----+-----+-----+-----+
100. |     |     |     |     |
101. |{}|{}|{}|{}|
102. |     |     |     |     |
103. +-----+-----+-----+-----+
104. """.format(*labels))
105.
106.
107. def getScore(board):
108.     """返回方格板数据结构上所有方格的总和"""
109.     score = 0
110.     # 循环遍历每个方格并将方格添加到 score 中
111.     for x in range(4):
112.         for y in range(4):
113.             # 只在 score 中添加非空值得方格
114.             if board[(x, y)] != BLANK:
115.                 score = score + board[(x, y)]
116.     return score
117.
118.
119. def combineTilesInColumn(column):
120.     """该列是 4 个方格的列表。 索引 0 是列的 "底部"
121.     如果方格相同，则将其向下滑动并合并
122.     例如，combineTilesInColumn([2, BLANK, 2, BLANK])
123.     返回 [4, BLANK, BLANK, BLANK]"""
124.
125.     # 仅将列中的数字（而不是空值）复制到 combinedTiles 中
126.     combinedTiles = []    # 列中包含非空值的方格的列表
127.     for i in range(4):
128.         if column[i] != BLANK:
129.             combinedTiles.append(column[i])
130.
131.     # 继续添加空值，直到有 4 个方格
132.     while len(combinedTiles) < 4:
133.         combinedTiles.append(BLANK)
134.
135.     # 如果 "上面" 的数字相同，则将其合并并加倍
136.     for i in range(3):    # 跳过索引 3：它表示最顶层的方格
137.         if combinedTiles[i] == combinedTiles[i + 1]:
138.             combinedTiles[i] *= 2    # 将方格中的数字加倍
139.             # 将其上方的方格向下移动一格
140.             for aboveIndex in range(i + 1, 3):
141.                 combinedTiles[aboveIndex] = combinedTiles[aboveIndex + 1]
142.             combinedTiles[3] = BLANK    # 最上面的方格总为空
143.     return combinedTiles
144.
145.
146. def makeMove(board, move):
```

```
147.     """在方格上进行移动
148.
149.     移动参数是'W'、'A'、'S'或'D'
150.     函数返回更新后的方格板的数据结构"""
151.
152.     # 棋盘分为4列,各不相同,其
153.     # 取决于移动的方向
154.     if move == 'W':
155.         allColumnsSpaces = [[(0, 0), (0, 1), (0, 2), (0, 3)],
156.                             [(1, 0), (1, 1), (1, 2), (1, 3)],
157.                             [(2, 0), (2, 1), (2, 2), (2, 3)],
158.                             [(3, 0), (3, 1), (3, 2), (3, 3)]]
159.     elif move == 'A':
160.         allColumnsSpaces = [[(0, 0), (1, 0), (2, 0), (3, 0)],
161.                             [(0, 1), (1, 1), (2, 1), (3, 1)],
162.                             [(0, 2), (1, 2), (2, 2), (3, 2)],
163.                             [(0, 3), (1, 3), (2, 3), (3, 3)]]
164.     elif move == 'S':
165.         allColumnsSpaces = [[(0, 3), (0, 2), (0, 1), (0, 0)],
166.                             [(1, 3), (1, 2), (1, 1), (1, 0)],
167.                             [(2, 3), (2, 2), (2, 1), (2, 0)],
168.                             [(3, 3), (3, 2), (3, 1), (3, 0)]]
169.     elif move == 'D':
170.         allColumnsSpaces = [[(3, 0), (2, 0), (1, 0), (0, 0)],
171.                             [(3, 1), (2, 1), (1, 1), (0, 1)],
172.                             [(3, 2), (2, 2), (1, 2), (0, 2)],
173.                             [(3, 3), (2, 3), (1, 3), (0, 3)]]
174.
175.     # 移动后的方格板数据结构
176.     boardAfterMove = {}
177.     for columnSpaces in allColumnsSpaces:    # 循环遍历所有列
178.         # 获取该列的方格(第一次是列的"底部")
179.         firstTileSpace = columnSpaces[0]
180.         secondTileSpace = columnSpaces[1]
181.         thirdTileSpace = columnSpaces[2]
182.         fourthTileSpace = columnSpaces[3]
183.
184.         firstTile = board[firstTileSpace]
185.         secondTile = board[secondTileSpace]
186.         thirdTile = board[thirdTileSpace]
187.         fourthTile = board[fourthTileSpace]
188.
189.         # 形成列并组合其中的方格图块
190.         column = [firstTile, secondTile, thirdTile, fourthTile]
191.         combinedTilesColumn = combineTilesInColumn(column)
192.
193.         # 使用合并好的方格生成新的方格板数据结构
194.         boardAfterMove[firstTileSpace] = combinedTilesColumn[0]
195.         boardAfterMove[secondTileSpace] = combinedTilesColumn[1]
196.         boardAfterMove[thirdTileSpace] = combinedTilesColumn[2]
197.         boardAfterMove[fourthTileSpace] = combinedTilesColumn[3]
198.
199.     return boardAfterMove
200.
201.
202. def askForPlayerMove():
203.     """询问玩家下一步移动的方向(或退出)
204.
205.     确保玩家输入有效的移动:'W'、'A'、'S'或'D'"""
206.     print('Enter move: (WASD or Q to quit)')
207.     while True:    # 继续循环,直到玩家输入有效的移动
208.         move = input('> ').upper()
```

```
209.        if move == 'Q':
210.            # 程序结束
211.            print('Thanks for playing!')
212.            sys.exit()
213.
214.        # 要么返回有效的移动，要么并再次询问
215.        if move in ('W', 'A', 'S', 'D'):
216.            return move
217.        else:
218.            print('Enter one of "W", "A", "S", "D", or "Q".')
219.
220.
221. def addTwoToBoard(board):
222.     """随机将两个方格添加到方格板上"""
223.     while True:
224.         randomSpace = (random.randint(0, 3), random.randint(0, 3))
225.         if board[randomSpace] == BLANK:
226.             board[randomSpace] = 2
227.             return  # 找到一个非空值的方格并返回
228.
229.
230. def isFull(board):
231.     """如果方格板数据结构没有空值，则返回 True"""
232.     # 遍历方格板上的每个方格
233.     for x in range(4):
234.         for y in range(4):
235.             # 如果方格为空，则返回 False
236.             if board[(x, y)] == BLANK:
237.                 return False
238.     return True  # 如果没有方格，所以返回 True
239.
240.
241. # 程序运行入口（如果不是作为模块导入的话）
242. if __name__ == '__main__':
243.     try:
244.         main()
245.     except KeyboardInterrupt:
246.         sys.exit()  # 按下 Ctrl-C，程序结束
```

探索程序

请尝试找出以下问题的答案。你需要尝试对代码进行一些修改，再次运行程序，查看修改后的效果。

1. 如果将第 116 行的 `return score` 改为 `return 9999`，则会发生什么？
2. 如果将第 226 行的 `board[randomSpace] = 2` 改为 `board[randomSpace] = 256`，则会发生什么？

| 项目 80

弗吉尼亚密码

弗吉尼亚密码曾被误认为是由 19 世纪密码学家布莱斯·德·弗吉尼亚发明的（其实是前人在更早的时候独立发明的），数百年来都没能被破解。它本质上是恺撒密码，只不过用了大部分密钥。所谓的"弗吉尼亚密码"，是一个单词，甚至是随机的一系列字母。每个字母代表一个数字，用于表示移动消息中的字母：A 表示将消息中的字母移动 0 个位置，B 表示移动 1 个位置，C 表示移动 2 个位置，以此类推。

例如，如果弗吉尼亚密钥是单词"CAT"，则 C 表示移动 2 个位置，A 表示移动 0 个位置，T 表示移动 19 个位置。消息的第 1 个字母移动 2 个位置，第 2 个字母移动 0 个位置，第 3 个字母移动 19 个位置。对于第 4 个字母，重复移动 2 个位置。

使用多个恺撒密码密钥正是弗吉尼亚密码的强大之处。其可能的组合数量太大而无法使用蛮力破解。此外，弗吉尼亚密码没有"可以破解简单替换密码"的频数分析弱点。

你会注意到弗吉尼亚和恺撒密码程序的代码有许多相似之处。如果你希望了解有关密码和密码破解的更多信息，请阅读拙作《Python 密码学编程》（人民邮电出版社）。

运行程序

运行 vigenere.py，输出如下所示：

```
Vigenère Cipher, by Al Sweigart al@inventwithpython.com
The Vigenère cipher is a polyalphabetic substitution cipher that was
powerful enough to remain unbroken for centuries.
Do you want to (e)ncrypt or (d)ecrypt?
> e
Please specify the key to use.
It can be a word or any combination of letters:
> PIZZA
Enter the message to encrypt.
> Meet me by the rose bushes tonight.
Encrypted message:
Bmds mt jx sht znre qcrgeh bnmivps.
Full encrypted text copied to clipboard.
```

工作原理

由于加密和解密过程非常相似，因此在本项目的程序中，我们使用 translateMessage()

函数来处理加密和解密操作。encryptMessage()和decryptMessage()函数只是translateMessage()的包装函数。换句话说，它们只是调整了参数，并将参数转发给另一个函数，然后返回该函数的返回值。本项目中的程序使用这些包装函数，以便按照类似项目66简单替换密码中调用encryptMessage()和decryptMessage()的方式来调用它们。你可以将本项目的程序作为模块导入其他程序，以便使用它们的加密代码，从而无须直接将代码复制并粘贴到新程序中。

```python
1. """弗吉尼亚密码, 作者: Al Sweigart al@inventwithpython.com
2. 弗吉尼亚密码是一种多字母替换密码
3. 几百年来, 弗吉尼亚密码一直未得到破解
4. 标签: 简短, 密码学, 数学"""
5.
6. try:
7.     import pyperclip  # pyperclip 模块可以将文本复制到剪贴板
8. except ImportError:
9.     pass  # 如果未安装 pyperclip, 也没什么影响
10.
11. # 可以加密/解密的所有符号
12. LETTERS = 'ABCDEFGHIJKLMNOPQRSTUVWXYZ'
13.
14.
15. def main():
16.     print('''Vigenère Cipher, by Al Sweigart al@inventwithpython.com
17. The Viegenère cipher is a polyalphabetic substitution cipher that was
18. powerful enough to remain unbroken for centuries.''')
19.
20.     # 让玩家指定是加密还是解密
21.     while True:  # 继续询问, 直到用户输入 e 或 d
22.         print('Do you want to (e)ncrypt or (d)ecrypt?')
23.         response = input('> ').lower()
24.         if response.startswith('e'):
25.             myMode = 'encrypt'
26.             break
27.         elif response.startswith('d'):
28.             myMode = 'decrypt'
29.             break
30.         print('Please enter the letter e or d.')
31.
32.     # 让玩家指定要使用的密钥
33.     while True:  # 继续询问, 直到玩家输入有效的密钥
34.         print('Please specify the key to use.')
35.         print('It can be a word or any combination of letters:')
36.         response = input('> ').upper()
37.         if response.isalpha():
38.             myKey = response
39.             break
40.
41.     # 让玩家指定要加密/解密的消息
42.     print('Enter the message to {}.'.format(myMode))
43.     myMessage = input('> ')
44.
45.     # 执行加密/解密
46.     if myMode == 'encrypt':
47.         translated = encryptMessage(myMessage, myKey)
48.     elif myMode == 'decrypt':
49.         translated = decryptMessage(myMessage, myKey)
50.
```

```
51.     print('%sed message:' % (myMode.title()))
52.     print(translated)
53.
54.     try:
55.         pyperclip.copy(translated)
56.         print('Full %sed text copied to clipboard.' % (myMode))
57.     except:
58.         pass  # 如果未安装 pyperclip，则不用执行任何操作
59.
60.
61. def encryptMessage(message, key):
62.     """使用密钥加密消息"""
63.     return translateMessage(message, key, 'encrypt')
64.
65.
66. def decryptMessage(message, key):
67.     """使用密钥解密消息"""
68.     return translateMessage(message, key, 'decrypt')
69.
70.
71. def translateMessage(message, key, mode):
72.     """使用密钥加密或解密消息"""
73.     translated = []  # 存储加密/解密消息的字符串
74.
75.     keyIndex = 0
76.     key = key.upper()
77.
78.     for symbol in message:  # 循环遍历消息中的每个字符
79.         num = LETTERS.find(symbol.upper())
80.         if num != -1:  # -1 表示 symbol.upper()不在 LETTERS 中
81.             if mode == 'encrypt':
82.                 # 如果加密则索引加 1
83.                 num += LETTERS.find(key[keyIndex])
84.             elif mode == 'decrypt':
85.                 # 如果解密则索引减 1
86.                 num -= LETTERS.find(key[keyIndex])
87.
88.             num %= len(LETTERS)  # 处理潜在的环绕问题
89.
90.             # 将加密/解密符号添加到 translated 中
91.             if symbol.isupper():
92.                 translated.append(LETTERS[num])
93.             elif symbol.islower():
94.                 translated.append(LETTERS[num].lower())
95.
96.             keyIndex += 1  # 移至密钥中的下一个字母
97.             if keyIndex == len(key):
98.                 keyIndex = 0
99.         else:
100.            # 只需添加符号而不加密/解密
101.            translated.append(symbol)
102.
103.    return ''.join(translated)
104.
105.
106. # 程序运行入口（如果不是作为模块导入的话）
107. if __name__ == '__main__':
108.     main()
```

探索程序

请尝试找出以下问题的答案。你需要尝试对代码进行一些修改，再次运行程序，查看修改后的效果。

1. 如果使用密钥"A"加密，则会发生什么？
2. 如果删除或注释掉第 38 行的 myKey = response，则会导致什么错误？

项目 81

水桶谜题

在本项目中，我们会实现一个单人益智游戏。在这个游戏中，你必须用 3 个水桶（容量分别为 3L、5L 和 8L 的水桶）并在其中 1 个水桶中装恰好 4L 水。桶只能清空、完全装满或将水倒入另一个桶。例如，你可以将 5L 桶装满水，然后将水倒入 3L 桶，这样就可以得到装满水的 3L 桶和装有 2L 水的 5L 桶。

通过努力你一定能够解决这个难题。你能想出如何用最少的步骤来解决它吗？

运行程序

运行 waterbucket.py，输出如下所示：

```
Water Bucket Puzzle, by Al Sweigart al@inventwithpython.com

Try to get 4L of water into one of these
buckets:

8|      |
7|      |
6|      |
5|      | 5|      |
4|      | 4|      |
3|      | 3|      | 3|      |
2|      | 2|      | 2|      |
1|      | 1|      | 1|      |
+------+ +------+ +------+
    8L       5L       3L

You can:
  (F)ill the bucket
  (E)mpty the bucket
  (P)our one bucket into another
  (Q)uit
> f
Select a bucket 8, 5, 3, or QUIT:
> 5

Try to get 4L of water into one of these
buckets:
```

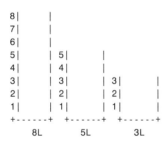

```
8|      |       |
7|      |       |
6|      |       |
5|      |  5|WWWWWW|
4|      |  4|WWWWWW|
3|      |  3|WWWWWW|   3|      |
2|      |  2|WWWWWW|   2|      |
1|      |  1|WWWWWW|   1|      |
+------+  +------+     +------+
   8L        5L           3L
--snip--
```

工作原理

waterInBucket 变量用于存储一个表示水桶状态的字典。该字典的键是字符串 '8'、'5'和 '3'（表示桶），它们的值是整数（表示该桶中水的升数）。

第 45~56 行使用这个字典在屏幕上渲染桶和水。waterDisplay 列表包含"WWWWWW"（代表水）或"　　　　"（代表空气）并传递给 format() 字符串方法。waterDisplay 列表中的前 8 个字符串填充 8L 桶，接下来的 5 个字符串填充 5L 桶，最后 3 个字符串填充 3L 桶。

```
 1. """水桶谜题，作者: Al Sweigart al@inventwithpython.com
 2. 一个水桶谜题游戏
 3. 标签：简短，游戏，数学，谜题"""
 4.
 5. import sys
 6.
 7. print('Water Bucket Puzzle, by Al Sweigart al@inventwithpython.com')
 8.
 9. GOAL = 4   # 桶中装恰好 4L 的水才能获胜
10. steps = 0  # 记录玩家为解决这个问题所用的步数
11.
12. # 每个桶的初始水量
13. waterInBucket = {'8': 0, '5': 0, '3': 0}
14.
15. while True:  # 主循环
16.     # 显示桶内当前水量
17.     print()
18.     print('Try to get ' + str(GOAL) + 'L of water into one of these')
19.     print('buckets:')
20.
21.     waterDisplay = []  # 用于存储表示水或剩余空间的字符串
22.
23.     # 获得 8L 水
24.     for i in range(1, 9):
25.         if waterInBucket['8'] < i:
26.             waterDisplay.append('      ')  # 添加空白空间
27.         else:
28.             waterDisplay.append('WWWWWW')  # 加水
29.
30.     # 获得 5L 水
31.     for i in range(1, 6):
32.         if waterInBucket['5'] < i:
33.             waterDisplay.append('      ')  # 添加空白空间
34.         else:
35.             waterDisplay.append('WWWWWW')  # 加水
36.
37.     # 获得 3L 水
```

```
38.     for i in range(1, 4):
39.         if waterInBucket['3'] < i:
40.             waterDisplay.append('      ')  # 添加空白空间
41.         else:
42.             waterDisplay.append('WWWWWW')  # 加水
43.
44.     # 显示每个桶的水量
45.     print('''
46. 8|{7}|
47. 7|{6}|
48. 6|{5}|
49. 5|{4}|   5|{12}|
50. 4|{3}|   4|{11}|
51. 3|{2}|   3|{10}|   3|{15}|
52. 2|{1}|   2|{9}|    2|{14}|
53. 1|{0}|   1|{8}|    1|{13}|
54. +------+  +------+  +------+
55.    8L       5L        3L
56. '''.format(*waterDisplay))
57.
58.     # 判断是否有满足目标水量的桶
59.     for waterAmount in waterInBucket.values():
60.         if waterAmount == GOAL:
61.             print('Good job! You solved it in', steps, 'steps!')
62.             sys.exit()
63.
64.     # 让玩家选择一个动作来处理一个桶
65.     print('You can:')
66.     print('  (F)ill the bucket')
67.     print('  (E)mpty the bucket')
68.     print('  (P)our one bucket into another')
69.     print('  (Q)uit')
70.
71.     while True:  # 不断询问，直到玩家输入有效的动作
72.         move = input('> ').upper()
73.         if move == 'QUIT' or move == 'Q':
74.             print('Thanks for playing!')
75.             sys.exit()
76.
77.         if move in ('F', 'E', 'P'):
78.             break  # 玩家选择了一个有效的动作
79.         print('Enter F, E, P, or Q')
80.
81.     # 让玩家选择一个桶
82.     while True:  # 继续询问，直到玩家选择有效的桶
83.         print('Select a bucket 8, 5, 3, or QUIT:')
84.         srcBucket = input('> ').upper()
85.
86.         if srcBucket == 'QUIT':
87.             print('Thanks for playing!')
88.             sys.exit()
89.
90.         if srcBucket in ('8', '5', '3'):
91.             break  # 玩家选择了一个有效的桶
92.
93.     # 执行选择的操作
94.     if move == 'F':
95.         # 将水量设置为水桶最大容量
96.         srcBucketSize = int(srcBucket)
97.         waterInBucket[srcBucket] = srcBucketSize
98.         steps += 1
99.
```

```
100.     elif move == 'E':
101.         waterInBucket[srcBucket] = 0  # 将水量设置为 0
102.         steps += 1
103.
104.     elif move == 'P':
105.         # 让玩家选择一个桶并将水倒出
106.         while True:  # 继续询问，直到玩家选择有效的桶
107.             print('Select a bucket to pour into: 8, 5, or 3')
108.             dstBucket = input('> ').upper()
109.             if dstBucket in ('8', '5', '3'):
110.                 break  # 玩家选择了一个有效的桶
111.
112.         # 计算要倒出的水量
113.         dstBucketSize = int(dstBucket)
114.         emptySpaceInDstBucket = dstBucketSize - waterInBucket[dstBucket]
115.         waterInSrcBucket = waterInBucket[srcBucket]
116.         amountToPour = min(emptySpaceInDstBucket, waterInSrcBucket)
117.
118.         # 从这个桶中倒出水
119.         waterInBucket[srcBucket] -= amountToPour
120.
121.         # 将倒出的水倒入另一个桶
122.         waterInBucket[dstBucket] += amountToPour
123.         steps += 1
124.
125.     elif move == 'C':
126.         pass  # 如果玩家选择取消，则什么都不做
```

输入源代码并运行几次后，请试着对其进行更改。你也可以尝试以下操作。

❏ 通过使游戏满足可配置来增加多样性，以便你为 3 个桶指定任意容量，为目标水量指定任意值。

❏ 添加"提示"功能，检查每个桶中的水量并提供下一步要采取的措施。如果程序无法确定下一步要执行的操作，它可以简单地显示"我不知道你接下来应该做什么，也许重新开始？"

探索程序

请尝试找出以下问题的答案。你需要尝试对代码进行一些修改，再次运行程序，查看修改后的效果。

1. 如果将第 101 行的 waterInBucket[srcBucket] = 0 改为 waterInBucket[srcBucket] = 1，则会发生什么？
2. 如果将第 13 行的{'8': 0, '5': 0, '3': 0}改为{'8': 0, '5': 4, '3': 0}，则会发生什么？
3. 如果将第 13 行的{'8': 0, '5': 0, '3': 0}改为{'8': 9, '5': 0, '3': 0}，则会发生什么？

附录 A 标签索引

本书中的项目标有标签，用于描述它们的程序类型。第一个标签表示程序的规模，包括小型（包含1~63行代码）、简短型（包含64~127行代码）、大型（包含128~255行代码）和特大型（包含256行及其以上的代码）。

- 小型：项目3、项目7、项目12、项目14、项目15、项目16、项目19、项目20、项目24、项目25、项目31、项目32、项目35、项目40、项目42、项目46、项目49、项目50、项目52、项目56、项目57、项目58、项目60、项目61、项目65、项目67、项目72、项目74。
- 简短型：项目1、项目2、项目5、项目6、项目8、项目10、项目13、项目18、项目21、项目26、项目29、项目51、项目53、项目54、项目55、项目59、项目64、项目66、项目69、项目71、项目76、项目77、项目80。
- 大型：项目4、项目9、项目11、项目17、项目22、项目23、项目28 项目30、项目33、项目34、项目36、项目37、项目39、项目41、项目43、项目44、项目47、项目48、项目62、项目63、项目68、项目70、项目73、项目75、项目78、项目79、项目81。
- 特大型：项目27、项目38、项目45。

其余标签表示程序的特点，如下所示。

- 艺术类：项目3、项目5、项目13、项目14、项目15、项目16、项目17、项目19、项目20、项目21、项目22、项目23、项目27、项目33、项目35、项目36、项目39、项目45、项目47、项目58、项目62、项目65、项目67、项目69、项目70。
- 初学者类：项目3、项目6、项目7、项目9、项目10、项目11、项目12、项目15、项目16、项目20、项目24、项目25、项目31、项目32、项目35、项目40、项目42、项目46、项目49、项目50、项目58、项目65、项目69、项目71、项目72、项目74。
- bext 模块类：项目5、项目27、项目28、项目29、项目36、项目39、项目47、项目58。
- 棋盘类：项目30、项目43、项目63、项目76。
- 卡牌类：项目4、项目75。
- 密码学类：项目6、项目7、项目61、项目66、项目80。

- 游戏类：项目 1、项目 4、项目 9、项目 10、项目 17、项目 25、项目 28、项目 30、项目 31、项目 33、项目 34、项目 37、项目 38、项目 41、项目 43、项目 44、项目 45、项目 48、项目 59、项目 60、项目 63、项目 68、项目 69、项目 71、项目 73、项目 75、项目 76、项目 77、项目 79、项目 81。
- 诙谐类：项目 11、项目 32、项目 38、项目 42、项目 55、项目 60、项目 78。
- 数学类：项目 2、项目 6、项目 7、项目 12、项目 17、项目 24、项目 26、项目 46、项目 48、项目 49、项目 52、项目 56、项目 62、项目 66、项目 70、项目 80、项目 81。
- 迷宫类：项目 44、项目 45。
- 模块类：项目 57、项目 64。
- 多人游戏类：项目 41、项目 69。
- 面向对象类：项目 22、项目 73。
- 谜题类：项目 1、项目 33、项目 34、项目 38、项目 68、项目 73、项目 77、项目 79、项目 81。
- 科学类：项目 21、项目 53。
- 动画类：项目 15、项目 20、项目 21、项目 22、项目 50、项目 51、项目 56、项目 58。
- 模拟类：项目 2、项目 13、项目 18、项目 29、项目 36、项目 39、项目 46、项目 48、项目 55、项目 70。
- 双人游戏类：项目 9、项目 30、项目 43、项目 63、项目 76。
- 文字类：项目 11、项目 34、项目 40、项目 51、项目 54、项目 72。

附录 B 字符映射表

print()函数可以轻松地将我们通过键盘输入的任何字符显示在计算机屏幕上。不过，你可能还想展示许多其他字符，包括红心、方块、梅花和黑桃，以及线条、阴影框、箭头、音符等。你可以通过将它们的数字代码（称为 Unicode 代码点）传递给 chr()函数来获取这些字符的字符串值。文本作为一系列数字存储在计算机上，每个字符用不同的数字表示。本附录包含此类代码点的列表。

使用 chr()和 ord()函数

Python 内置的 chr()函数接收一个整数参数并返回该整数对应的字符。ord()函数作用与 chr()的不同，它接收单个字符参数并返回该字符对应的数值，后者是该字符在 Unicode 标准中的代码点。

例如，在交互式 Shell 中输入以下内容：

```
>>> chr(65)
'A'
>>> ord('A')
65
>>> chr(66)
'B'
>>> chr(9829)
'♥'
```

并非所有数字都是可输出字符的有效代码点。显示程序文本输出的终端窗口的可显示字符可能有局限。终端窗口使用的字体也必须能支持程序输出的字符。终端窗口会为无法输出的任何字符输出 Unicode 替换字符❷。

Windows 终端窗口可以显示的字符范围更有限。该集合称为 Windows Glyph List 4，参见本附录末尾的介绍以及维基百科的相应词条。

字符的代码点通常以十六进制数的方式列出，而不是我们通常使用的十进制数。不同于十进制数位的 0 到 9，十六进制数位包含 0 到 9 以及字母 A 到 F。十六进制数通常标有 0x 前缀，表示后面的数字为十六进制的。

我们可以使用 hex()函数将十进制整数转换为十六进制字符串。还可以使用 int()函数将十六进制字符串转换为十进制整数，将 16 作为第二个参数传入即可。例如，在交互式 Shell 中输入以下内容：

```
>>> hex(9)
'0x9'
>>> hex(10)
'0xa'
>>> hex(15)
'0xf'
>>> hex(16)
'0x10'
>>> hex(17)
'0x11'
>>> int('0x11', 16)
17
>>> int('11', 16)
17
```

调用 chr() 函数时，必须传入一个十进制整数作为参数，而不能传入十六进制字符串。

代码点表

以下是 Windows Glyph List 4 集合中的所有 Unicode 代码点，它们是 Windows 终端程序命令提示符窗口支持的字符。macOS 和 Linux 可以显示比该表中更多的字符，但为了保持 Python 程序兼容，建议总是使用该表中的字符。

32	<space>	66	B	100	d	168	¨
33	!	67	C	101	e	169	©
34	"	68	D	102	f	170	ª
35	#	69	E	103	g	171	«
36	$	70	F	104	h	172	¬
37	%	71	G	105	i	173	-
38	&	72	H	106	j	174	®
39	'	73	I	107	k	175	¯
40	(	74	J	108	l	176	°
41	)	75	K	109	m	177	±
42	*	76	L	110	n	178	²
43	+	77	M	111	o	179	³
44	,	78	N	112	p	180	´
45	-	79	O	113	q	181	µ
46	.	80	P	114	r	182	¶
47	/	81	Q	115	s	183	·
48	0	82	R	116	t	184	¸
49	1	83	S	117	u	185	¹
50	2	84	T	118	v	186	º
51	3	85	U	119	w	187	»
52	4	86	V	120	x	188	¼
53	5	87	W	121	y	189	½
54	6	88	X	122	z	190	¾
55	7	89	Y	123	{	191	¿
56	8	90	Z	124	|	192	À
57	9	91	[	125	}	193	Á
58	:	92	\	126	~	194	Â
59	;	93	]	161	¡	195	Ã
60	<	94	^	162	¢	196	Ä
61	=	95	_	163	£	197	Å
62	>	96	`	164	¤	198	Æ
63	?	97	a	165	¥	199	Ç
64	@	98	b	166	¦	200	È
65	A	99	c	167	§	201	É

202	Ê	272	Đ	368	Ű	947	γ
203	Ë	273	đ	369	ű	948	δ
204	Ì	274	Ě	370	Ų	949	ε
205	Í	275	ě	371	ų	950	ζ
206	Î	278	Ė	376	Ÿ	951	η
207	Ï	279	ė	377	Ź	952	θ
209	Ñ	280	Ę	378	ź	953	ι
210	Ò	281	ę	379	Ż	954	κ
211	Ó	282	Ĕ	380	ż	955	λ
212	Ô	283	ĕ	381	Ž	956	μ
213	Õ	286	Ğ	382	ž	957	ν
214	Ö	287	ğ	402	ƒ	958	ξ
215	×	290	Ģ	710	ˆ	959	ο
216	Ø	291	ģ	711	ˇ	960	π
217	Ù	298	Ĭ	728	˘	961	ρ
218	Ú	299	ī	729	˙	962	ς
219	Û	302	Į	731	˛	963	σ
220	Ü	303	į	732	˜	964	τ
221	Ý	304	İ	733	˝	965	υ
223	ß	305	ı	900	΄	966	φ
224	à	310	Ķ	901	΅	967	χ
225	á	311	ķ	902	Ά	968	ψ
226	â	313	Ĺ	904	Έ	969	ω
227	ã	314	ĺ	905	Ή	970	ϊ
228	ä	315	Ļ	906	Ί	971	ϋ
229	å	316	ļ	908	Ό	972	ό
230	æ	317	Ľ	910	Ύ	973	ύ
231	ç	318	ľ	911	Ώ	974	ώ
232	è	321	Ł	912	ΐ	1025	Ё
233	é	322	ł	913	Α	1026	Ђ
234	ê	323	Ń	914	Β	1027	Ѓ
235	ë	324	ń	915	Γ	1028	Є
236	ì	325	Ņ	916	Δ	1029	Ѕ
237	í	326	ņ	917	Ε	1030	І
238	î	327	Ň	918	Ζ	1031	Ї
239	ï	328	ň	919	Η	1032	Ј
241	ñ	332	Ō	920	Θ	1033	Љ
242	ò	333	ō	921	Ι	1034	Њ
243	ó	336	Ő	922	Κ	1035	Ћ
244	ô	337	ő	923	Λ	1036	Ќ
245	õ	338	OE	924	Μ	1038	Ў
246	ö	339	oe	925	Ν	1039	Џ
247	÷	340	Ŕ	926	Ξ	1040	А
248	ø	341	ŕ	927	Ο	1041	Б
249	ù	342	Ŗ	928	Π	1042	В
250	ú	343	ŗ	929	Ρ	1043	Г
251	û	344	Ř	931	Σ	1044	Д
252	ü	345	ř	932	Τ	1045	Е
253	ý	346	Ś	933	Υ	1046	Ж
255	ÿ	347	ś	934	Φ	1047	З
256	Ā	350	Ş	935	Χ	1048	И
257	ā	351	ş	936	Ψ	1049	Й
258	Ă	352	Š	937	Ω	1050	К
259	ă	353	š	938	Ϊ	1051	Л
260	Ą	354	Ţ	939	Ϋ	1052	М
261	ą	355	ţ	940	ά	1053	Н
262	Ć	356	Ť	941	έ	1054	О
263	ć	357	ť	942	ή	1055	П
268	Č	362	Ū	943	ί	1056	Р
269	č	363	ū	944	ΰ	1057	С
270	Ď	366	Ů	945	α	1058	Т
271	ď	367	ů	946	β	1059	У

1060 ф	1102 ю	8745 ∩	9575 ┴
1061 х	1103 я	8776 ≈	9576 ┴
1062 ц	1105 ё	8801 ≡	9577 ┼
1063 ч	1106 ђ	8804 ≤	9578 ┼
1064 ш	1107 ѓ	8805 ≥	9579 ┼
1065 щ	1108 є	8976 ⌐	9580 ┼
1066 ъ	1109 ѕ	8992 ⌠	9600 ▀
1067 ы	1110 і	8993 ⌡	9604 ▄
1068 ь	1111 ї	9472 ─	9608 █
1069 э	1112 ј	9474 │	9612 ▌
1070 ю	1113 љ	9484 ┌	9616 ▐
1071 я	1114 њ	9488 ┐	9617 ░
1072 а	1115 ћ	9492 └	9618 ▒
1073 б	1116 ќ	9496 ┘	9619 ▓
1074 в	1118 ў	9500 ├	9632 ■
1075 г	1119 џ	9508 ┤	9633 □
1076 д	1168 Ґ	9516 ┬	9642 ▪
1077 е	1169 ґ	9524 ┴	9643 ▫
1078 ж	8211 –	9532 ┼	9644 ▬
1079 з	8212 —	9552 ═	9650 ▲
1080 и	8213 ―	9553 ║	9658 ►
1081 й	8216 '	9554 ╒	9660 ▼
1082 к	8217 '	9555 ╓	9668 ◄
1083 л	8218 ‚	9556 ╔	9674 ◊
1084 м	8220 "	9557 ╗	9675 ○
1085 н	8221 "	9558 ╕	9679 ●
1086 о	8222 „	9559 ╖	9688 ◘
1087 п	8224 †	9560 ╚	9689 ◙
1088 р	8225 ‡	9561 ╘	9702 ◦
1089 с	8226 •	9562 ╙	9786 ☺
1090 т	8230 …	9563 ╝	9787 ☻
1091 у	8240 ‰	9564 ╛	9788 ☼
1092 ф	8249 ‹	9565 ╜	9792 ♀
1093 х	8250 ›	9566 ╞	9794 ♂
1094 ц	8319 ⁿ	9567 ╟	9824 ♠
1095 ч	8359 ₧	9568 ╠	9827 ♣
1096 ш	8364 €	9569 ╡	9829 ♥
1097 щ	8470 №	9570 ╢	9830 ♦
1098 ъ	8482 ™	9571 ╣	9834 ♪
1099 ы	8729 ∙	9572 ╤	9835 ♫
1100 ь	8730 √	9573 ╥	
1101 э	8734 ∞	9574 ╦	